Deformation of Artinian Algebras and Jordan Type

CONTEMPORARY MATHEMATICS

805

Deformation of Artinian Algebras and Jordan Type

AMS-EMS-SMF Special Session
Deformation of Artinian Algebras and Jordan Type
July 18–22, 2022
Université Grenoble Alpes, Grenoble, France

Anthony Iarrobino
Pedro Macias Marques
Maria Evelina Rossi
Jean Vallès
Editors

AMS AMERICAN MATHEMATICAL SOCIETY

2020 *Mathematics Subject Classification.* Primary 13-06, 13E10, 14B12, 14C05.

Library of Congress Cataloging-in-Publication Data

Names: AMS-EMS-SMF Special Session on Deformation of Artinian Algebras and Jordan Type (2022 : Grenoble, France) | Iarrobino, Anthony A. (Anthony Ayers), 1943- editor. | Macias Marques, Pedro, 1975- editor. | Rossi, Maria Evelina, editor. | Vallès, Jean, 1967- editor.

Title: Deformation of Artinian algebras and Jordan type : AMS-EMS-SMF Special Session on Deformation of Artinian Algebras and Jordan Type, July 18-22, 2022, University of Grenoble, Grenoble, France / Anthony Iarrobino, Pedro Macias Marques, Maria Evelina Rossi, Jean Vallès, editors.

Description: Providence, Rhode Island : American Mathematical Society, [2024] | Series: Contemporary mathematics, 0271-4132 ; volume 805 | Includes bibliographical references.

Identifiers: LCCN 2023053397 | ISBN 9781470473563 (paperback) | ISBN 9781470476656 (ebook)

Subjects: LCSH: Commutative algebra–Congresses. | AMS: Commutative algebra – Proceedings, conferences, collections, etc. | Commutative algebra – Chain conditions, finiteness conditions – Artinian rings and modules, finite-dimensional algebras. | Algebraic geometry – Local theory – Local deformation theory, Artin approximation, etc. | Algebraic geometry – Cycles and subschemes – Parametrization (Chow and Hilbert schemes).

Classification: LCC QA251.3 .D43 2024 | DDC 512/.44–dc23/eng/20240613

LC record available at https://lccn.loc.gov/2023053397

Contemporary Mathematics ISSN: 0271-4132 (print); ISSN: 1098-3627 (online)

DOI: https://doi.org/10.1090/conm/805

---

10 9 8 7 6 5 4 3 2 1   29 28 27 26 25 24

Contents

Introduction

There are lots of papers devoted to the study of Artinian algebras or zero-dimensional singularities. The motivations come from different areas of Mathematics as Commutative Algebra, Algebraic Geometry, Singularity theory and Combinatorics. In the last years the interest is even increasing due to several open problems on the topic. Hence in July 18-22, 2022, a group of mathematicians from all over the world gathered for a special session on Deformation of Artinian Algebras and Jordan Type at the AMS-EMS-SMF joint international meeting in Grenoble, France. The session was organized by Anthony Iarrobino, Pedro Macias Marques, Maria Evelina Rossi and Jean Vallès.

Photo courtesy of Nancy Abdallah

FIGURE 1. From left to right: Michel Brion, Roser Homs Pons, Oana Veliche, Liena Colarte Gómez, Gianfranco Casnati, Martí Salat Moltó, Liana Şega, Nasrin Altafi, Maria Evelina Rossi, Emanuela De Negri, Pedro Macias Marques, Mats Boij, Jean Vallès, Alessandro Oneto, Chris McDaniel, Bernard Mourrain, Emanuele Ventura, Klemen Šivic, Filippo Favale, Alessandra Bernardi, and Nancy Abdallah. Also organized or participated in the session, but are not in the photo: Tony Iarrobino, Rosa Maria Miró-Roig, and Matteo Varbaro.

Despite the record-breaking heat wave that struck the region at the time, the participants enjoyed many interesting and fascinating talks followed by fruitful discussions.

This volume contains contributions from participants and invitees.

Acknowledgment. We thank our outside readers for their help with the volume: Alessandra Bernardi, Christina Bertone, Enrico Carlini, Federico Galetto, Steve Kleiman, Francesco Russo, Hal Schenck, Alexandra Seceleanu, Klemen Šivic, Daniele Taufer, Zach Teitler, Junzo Watanabe, Joachim Yaméogo. We also thank Christine Thivierge and Sergei Gelfand of American Mathematical Society Publications for their assistance, and the CONM Board.

Joint International Meeting of the American Mathematical Society (AMS), European Mathematical Society (EMS), and Société Mathématique de France (SMF), Université Grenoble Alpes, July 18-22, 2022, special session *Deformation of Artinian algebras and Jordan type.*

1. Introduction to the articles

Part I is comprised of 3 surveys.

Open problems in deformations of Artinian algebras, Hilbert schemes and around, by Joachim Jelisiejew.

This ambitious survey connects the work on deforming Artin algebras with the contexts of punctual Hilbert schemes and deformations of tensor products, emphasizing applications, in particular the problem of determining new bounds on the asymptotic complexity of matrix multiplication. Joachim with coauthors has been a leader in introducing new methods to the study of Hilbert schemes, in particular intensifying the application of the Białynicki-Birula decomposition, earlier used by G. Ellingsrud and G. Strömme. The 38 problems he proposes range from study of the intersections of irreducible components of the Hilbert scheme, and determining smoothable Artin algebras, to understanding the moduli space for bilinear maps. In particular he considers a "dividing wall" between moduli with a single irreducible component, and those with more in various contexts (Table 1); and he proposes that the tangent space to the moduli scheme is especially important in the dividing wall cases. He includes an extensive list of 135 references: each is specifically cited and given context in the text.

Commuting Jordan types: A survey, by Leila Khatami.

The study of which pairs of Jordan block decompositions are possible for two commuting $n \times n$ matrices reduces to the case where the matrices are nilpotent. The nilpotent commuting variety was known to be irreducible, but it was only in 2008 that Polona Oblak conjectured what the generic Jordan type $Q(P)$ in the commuting variety would be for a matrix of Jordan type P, a conjecture only recently established in full by Roberta Basili. In between there were results by several groups on the possible pairs of commuting matrices, including work of Leila Khatami and coauthors, Roberta Basili and coauthors, Tomaz Košir and Polona Oblak, and V. Panyushev. There is a feeling of workers in the area - including an editor - that there is, hidden, some deeper theory waiting to be found that explains the somewhat striking, but limited results: for example the table of pairs $(P, Q(P))$ for $Q(P)$ fixed was known only for $Q(P)$ having two parts, but very recently has been resolved in general. Here Leila Khatami surveys the current results and concludes with open problems.

A brief survey of Specht ideals and their perfection in the two-rowed case, by Chris McDaniel.

Specht polynomials corresponding to a partition λ of n are polynomials attached to the tableaux of shape λ: they were introduced by Wilhelm Specht in 1935 as basis for the irreducible representations of the symmetric group (in characteristic other than two). There has been substantial study of their relation to other bases of representations, and applications since. A Specht module corresponding to λ is the linear span of the Specht polynomials and a Specht ideal is the ideal they generate. Although touched on as early as 1981 in a paper of Shuo Li and Wen Li on the independence number of graphs, Specht ideals were systematically studied in 2021 by Kohji Yanagawa in "When is a Specht ideal Cohen-Macaulay?" (reference [26] of the survey), and by Philippe Moustrou, Cordian Riener, and Hugues Verdure [16] relating a symmetric ideal I to the Specht ideals that I contains. Chris McDaniel introduces the Specht ideals, collecting general results on their prime decompositions, radicals, and perfection; he then focuses on the perfection of those determined by two rowed partitions, in particular his work with Junzo Watanabe determining the perfection of these ideals in the two-rowed case, which depends on the characteristic of the field [13]. Their work is based on "principal perfection systems", related to the principal radical systems introduced by Mel Hochster and John Eagon in their influential 1971 paper [8]. He ends with comments on a conjecture of K. Yanagawa that the two rowed Specht ideal is not perfect in (specific) low enough characteristic.

Part II is comprised of 12 articles. We introduce them in alphabetic order by first author.

A note on Artin Gorenstein algebras of Hilbert function $(1, 4, k, k, 4, 1)$, by Nancy Abdallah.

Nancy Abdallah studies Lefschetz Properties of Artinian Gorenstein (AG) algebras having a fixed Hilbert function $H = (1, 4, k, k, 4, 1)$, using their minimal free resolutions. In a previous article, the author and Hal Schenck studied the minimal free resolution of certain AG algebras of codimension 4 as well as the connection to the Lefschetz properties and Jordan types. Here, the author proves that for such algebras of socle degree 5, the weak and the strong Lefschetz Properties are equivalent. For algebras A having the Hilbert function $(1, 4, k, k, 4, 1)$ the authors of the previous article had conjectured that there are 36 possible Betti tables. In the current paper, it is shown that there is only one Betti table for $k = 4$ and verifies that there is only one when A is a complete intersection algebra (here $k = 7$). The author uses the theory of Macaulay inverse system for AG algebras; for determining the Lefschetz properties she uses the Hessian criteria introduced by Toshiaki Maeno and Junzo Watanabe. In determining the possible Betti tables for AG algebras the author uses theorems of Macaulay and Gotzmann.

A note on Jordan types and Jordan degree types, by Nasrin Altafi.

The Jordan type (JT) of an Artinian algebra A and a linear form ℓ encodes the ranks of multiplication maps by ℓ^i for every i; the Jordan degree type (JDT) determines the ranks of ℓ^i in various degrees of A. Nasrin Altafi studies the comparison between JT and the JDT of graded Artinian Gorenstein (AG) algebras. It was shown by Anthony Iarrobino, Pedro Macias Marques, and Chris McDaniel that JT and JDT are equivalent for Artinian algebras of codimension two. The author recovers this result for AG algebras of codimension two; the new proof uses the method of rank matrices for AG algebras the author introduced in a previous work. The main new result here is that JDT and JT are not equivalent for AG algebras of codimension higher than two. Previously, the author had provided necessary conditions on the rank matrices and equivalently on the JDTs of AG algebras. Nasrin describes, given a JDT $\mathcal{S}$, satisfying the necessary conditions, how to determine an AG algebra and a linear form for which the corresponding JDT is equal to $\mathcal{S}$. Her construction yields AG algebras A and B of socle degree 9 and linear forms $\ell_1 \in A$ and $\ell_2 \in B$, so that A and B have the same JT but different JDT. Using algebra software Macaulay 2, she proves that this is the smallest possible socle degree with this property in codimension three; however in higher codimensions, there are such examples having smaller socle degree.

Waring decompositions of special ternary forms with different Hilbert functions, by Elena Angelini, Luca Chiantini, and Alessandro Oneto.

A Waring decomposition of a homogenous polynomial f of degree d is an expression as a sum of d-th powers of linear forms. By Macaulay's result there is a one-to-one correspondence between homogeneous forms of degree d and Artinian Gorenstein algebras of socle degree d. This correspondence, and the Apolarity Lemma affirm that Waring decompositions of a homogeneous polynomial f correspond to sets of points whose ideal is contained in the Gorenstein ideal associated to f. For forms in polynomial rings with two variables all sets of points computing the Waring rank of f share the same regularity (in the sense of degree where the Sperner number is achieved) and, even the same Hilbert function. The last author and Bernard Mourrain asked if this might be generalized to forms in polynomial rings with more variables. The authors provide a negative answer, by constructing examples of certain polynomials in three variables and having sufficiently large degree. The authors employ the construction obtained from liaison theory and use the Cayley-Bacharach properties for sets of points on projective space; they obtain properties of Hilbert functions of sets of points $A \cup B$ where A and B are apolar to the same homogeneous form.

Some remarks on varieties whose twisted normal bundle is an instanton, by Vincenzo Antonelli and Gianfranco Casnati.

The authors characterise those varieties whose normal bundle admits a twist that is instanton. They show that a result of A. F. Lopez on Ulrich bundles can be adapted to the case of instanton bundles. They conclude that, apart from linear subspaces, the only projective varieties that admit an instanton twist of the normal bundle have codimension two, and they also describe further cohomology conditions for that to happen.

They go on to study codimension-two varieties, and discuss nice examples of surfaces and threefolds, relating these varieties to Steiner bundles. They also list several examples of varieties having a twist of the normal bundle that is instanton, and having strictly positive quantum number.

On the Koiran-Skomra's question about Hessians, by Edoardo Ballico and Emanuele Ventura.

The paper is concerned with the following question:

QUESTION 1. (Koiran-Skomra). Let $n \geq 2$ and $d \geq 3$. Let $f \in \mathbb{C}[x_1, \ldots, x_n]_d$ be such that $\det \operatorname{Hess}(f) = \alpha(x_1 \cdots x_n)^{d-2}$ for some $\alpha \in \mathbb{C}^*$ where $\operatorname{Hess}(f)$ is the Hessian matrix of f. Does it follow that $f = \alpha_1 x_1^d + \cdots + \alpha_n x_n^d$, for some constants $\alpha_j \in \mathbb{C}$?

Before its formulation Beniamino Segre found a family of counterexamples in four variables of degree $d = 2k$:

$$g(x_1, x_2, x_3, x_4) = x_1^{2k} + (1 - 2k)x_2^{2k} - (x_3x_4)^k.$$

The authors extend Segre's counterexamples to an arbitrary number of variables and composite degrees; more precisely, in Theorem 2.1, they show that the polynomials

$$(x_1 \cdots x_q)^k + x_{q+1}^d + \cdots + x_n^d,$$

under the numerical conditions $d = kq$ with $k \geq 1$ (or $k \geq 2$ if $q = 2$), give a negative answer to the Koiran-Skomra question.

In Theorem 2.8 and Theorem 2.9, they deal more particularly with the case of singular binary homogeneous polynomials, showing that for d odd greater than 3 there are no singular binary homogeneous polynomials satisfying the assumptions in Question 1; and there is only one, which is the monomial $x_1^{d/2}x_2^{d/2}$, for d even greater than 4. The authors conjecture (see Conjecture 2.16) that for any $n \geq 2$ the only smooth homogeneous polynomials satisfying the assumptions of Question 1 are the Fermat polynomials; this conjecture is supported by Theorem 2.8 and Theorem 2.9 and by the well known result on the Hesse pencil of ternary cubics (recalled in Proposition 2.6), that implies that the described counterexamples are the only ones.

Standard Artinian algebras and Lefschetz properties: A geometric approach, by Davide Bricalli and Filippo Francesco Favale.

The authors study the ranks of multiplication maps by forms of degree one and two on standard Artinian Gorenstein algebras. In particular, they focus on complete intersection algebras that are presented by quadrics. They use the machinery of their previous article with Gian Pietro Pirola, analyzing the non-Lefschetz loci of standard graded AG algebras, and using projective geometry. In the previous work they first showed that a complete intersection presented by quadrics can be lifted to one of a smaller codimension by a linear element in the non-Lefschetz locus. They also showed that if the smaller algebra satisfies the weak Lefschetz property in degree 2 (WLP_2) then the same holds for the original algebra. Here the authors generalize the first result: they consider the quotient algebra by certain ideals generated by two linear elements. They conjecture that the WLP_2 of the smaller algebra implies the same for the original algebra. They also describe the relation between the linear factors of the Hessian of f and nilpotent elements of the Jacobian ring associated to f, where f has degree 3.

Decompositions and Terracini loci of cubic forms of low rank, by Luca Chiantini and Fulvio Gesmundo.

Finding Waring decompositions for a homogeneous polynomial f is a long-standing problem. One key question is whether they are unique, in which case we say that f is identifiable, and, if not, what is the set of subsets of linear forms giving a Waring decomposition for f. The authors study the case of cubics in $n+1$ variables having rank at most $n+2$. The rank of a concise cubic in $n+1$ variables (i.e. a cubic that cannot be written in fewer than $n+1$ variables, even after a linear change of variables) is at least $n+1$. By a well-known result of J. Alexander and A. Hirschowitz, we know that the rank of a general cubic in $n+1$ variables is $\lceil \frac{1}{n+1}\binom{n+3}{n} \rceil$, except for $n=4$, when the rank is 8 instead of 7. So $n+2$ is the second lowest possible rank and it gets further apart from the typical rank as n grows.

The main result of the paper gives the possible configurations of decompositions for a concise cubic in $n+1$ variables, relating them to the Kruskal rank of the set A of linear forms in each decomposition, i.e. the highest number k such that any subset of k linear forms in A is linearly independent.

The authors go on to describe a subvariety of the $(n+2)$th Terracini locus of the third Veronese embedding of n-dimensional projective space. They show that the concise $(n+2)$th Terracini locus is the closure of the set of sets of cubes of linear forms that span the whole space and have four coplanar points.

Almost finitely generated inverse systems and reduced k-algebras, by Joan Elias and Maria Evelina Rossi.

Macaulay's inverse system is an effective tool for characterizing interesting classes of Artinian (local or graded) rings, such as Gorenstein or level k-algebras. The authors of this paper, and later S. Masuti, P. Schulze and L.Tozzo, extended this constructive method to any Cohen-Macaulay k-algebra, given the type - the dimension of the socle of an Artinian reduction.

Despite several advances on the topic, many basic problems remain open. For instance, in the Artinian case, it is not known how to characterize the inverse system of a complete intersection. In positive dimension, it would be interesting to describe Macaulay's inverse system for domains and, more generally, for reduced k-algebras. The purpose of this paper is to characterize one-dimensional local domains, or more in general reduced domains, in terms of its Macaulay's inverse system. This leads to a study of almost finitely generated modules in the divided power ring, a topic already introduced by W.D. Weakley. Interesting applications to the numerical semigroup rings and to reduced arithmetically Gorenstein 0-dimensional schemes are given.

Deformation of local Artin rings via Hilbert-Burch matrices, by Roser Homs and Anna-Lena Winz.

The authors study the local punctual Hilbert scheme $\mathrm{Hilb}^n k[[x,y]]$ parametrizing ideals of colength n in $k[[x,y]]$. Joel Briançon and Andrei Galligo (article reference [5]) had introduced the technique of vertical standard basis stratification of the Hilbert scheme, and, based on their study of the generators of a flat deformation of a given colength n monomial ideal, J. Yaméogo showed in [22] that each vertical stratum of $\mathrm{Hilb}^n k[[x,y]]$ is isomorphic to an affine space of dimension given by a simple formula. A description of a cellular decomposition of $\mathrm{Hilb}^n k[[x, y]]$ equivalent to the vertical stratification has been given by Giuseppe Valla and Aldo Conca [6] in terms of Hilbert-Burch matrices. Vertical strata or, equivalently, Gröbner cells of $\mathrm{Hilb}^n k[[x, y]]$ as described by Valla and Conca do not respect the Hilbert function: colength n ideals with different Hilbert functions may belong to the same Gröbner cell. The authors previously presented a cellular decomposition, up to a generic change of coordinates, of the zero-dimensional ideals of $k[[x, y]]$ having given Hilbert function [15] and initial monomial ideal. Here, they give their Betti strata explicitly, generalizing a result of Maria Evelina Rossi and Leila Sharifan [21]. They conjecture that these lead to a Betti stratification of $\mathrm{Hilb}^n k[[x, y]]$ consistent with Hilbert function.

The Hilbert function of local Artinian level algebras of codimension 3 and type 2, by Shreedevi Masuti.

S. Masuti studies the Hilbert functions and structure of local level Artinian algebras of codimension 3 and type 2 whose defining ideal has three order-two generators. Using Macaulay's inverse systems, the author shows that these Hilbert functions have jumps between two consecutive terms of at most 4. She applies this to the case of socle degree at most 6, obtaining the complete classification of such Hilbert functions. Level graded codimension three algebras had been previously studied in a 2007 monograph of A. Geramita, T. Harima, J. Migliore and Y. Shin. Previously, the author had with Joachim Jelisiejew and Maria Evelina Rossi answered analogous questions for codimension three local Gorenstein algebras, where there is an added tool of symmetric decomposition. S. Masuti's study here of local level algebras is rather more difficult than the graded case, as the associated graded algebra is rarely level; and in the type 2 local case there is no known analogue of symmetric decomposition. She here has created new tools.

Isolated singularities, inverse systems and the punctual Hilbert scheme, by Bernard Mourrain.

The paper addresses the problem of understanding the algebraic properties of isolated singular solutions to polynomial systems, by examining the inverse system of such solutions. The author recalls two distinct algorithms for computing it. The first one (Dialytic Method) recursively computes the inverse system as the kernel of a linear operator, degree by degree. The second method (Integration) computes the same object from its graded primal-dual pair. A comparison of the complexity of these two algorithms is provided. The primal-dual pair allows the author to define the punctual Hilbert scheme Hilb_δ as the δ-dimensional (dual) linear spaces that are stable under divided difference. The author then studies the affine charts Hilb_B of Hilb_δ, determined by a monomial basis B, which are shown to be characterized by integration and commutation relations. It is proved that the δ-linear spaces orthogonal to all the polynomials in the considered ideal intersect Hilb_B transversely at $\mathcal{L}$ when $\mathcal{L}^\perp$ is a primary isolated component of the ideal. The dimension of the irreducible components of Hilb_B is established in terms of the commutation rank of its regular points, and Hilb_B is shown to be a birational variety. The author ends with several explicit examples.

Dg module structures and minimal free resolutions modulo an exact zero-divisor, by Liana Şega and Deepak Sireeshan.

In this paper the authors are concerned with minimal resolutions of modules over quotient rings of the form $R = Q/(f)$ where $(Q, \mathfrak{n})$ is a local ring and $f \in \mathfrak{n} \backslash \mathfrak{n}^2$.

The minimal free resolution U of an R module M over Q is a differential graded algebra (for short DG-algebra) over K where

$$K = Q\langle y \,|\, \partial(y) = f\rangle.$$

By the celebrated work of Samash (see reference [10]) the minimal free resolution of M over R is $U \otimes_K R$ when f is regular. The main result presented here concerns the case when there exists $g \in Q$ so that $fg = 0$ (i.e. f is a zero divisor). For quotient rings the approach based on DG-algebras is new, in particular the DG-algebra K is replaced by the Tate resolution of R over Q

$$A = Q\langle y, t \,|\, \partial(y) = f,\, \partial(t) = gy\rangle.$$

The best possible generalizations of the regular case obtained in this paper are the following:

- If M is an R-module and U is a minimal free resolution of M over R then U is a semi-free DG A-module.
- If f and g form an exact pair of zero divisors then the minimal free resolution of M over R is $U \otimes_A R$.

Surveys

Contemporary Mathematics
Volume **805**, 2024
https://doi.org/10.1090/conm/805/16122

Open problems in deformations of Artinian algebras, Hilbert schemes and around

Joachim Jelisiejew

ABSTRACT. We review the open problems in the theory of deformations of zero-dimensional objects, such as algebras, modules or tensors. We list both the well-known ones and some new ones that emerge from applications. In view of many advances in recent years, we can hope that all of them are in the range of current methods.

On sait qu'on ne sait rien sur cette sorte d'algèbre.

André Weil [**133**, p.221]. Quoted thanks to [**123**].[1]

1. Introduction

The Hilbert scheme of points of a variety X (see, e.g., [**12, 58, 122, 129**]), occupies a special place among the moduli spaces investigated in algebraic geometry. On the one hand, for smooth curves or surfaces X it behaves very nicely and gives rise to beautiful global geometry [**6, 54, 55**]. On the other hand, for a smooth variety X with $\dim X \geq 3$ it presents a mixture of pathological and unknown behaviour which would almost surely doom it to oblivion if not for the applications.

1.1. Applications. Since the Hilbert scheme parameterizes relatively elementary objects: tuples of points, finite-dimensional algebras, etc., deep and important applications arise naturally even in mathematical areas far from algebra, such as motivic homotopy theory [**45**, §3.5], string theory [**51**] or classical topology [**21**].

From the personal perspective of the author, a less known but key source of applications, probably for many years to come, is the problem of finding new bounds on the asymptotic complexity of matrix multiplication. Finding this complexity is one of the most important problems in the whole field of computational complexity [**25**, §15], only a little less important than the famous "P = NP" problem. Roughly speaking, the complexity is n^{ω}, where ω is an unknown constant

2020 *Mathematics Subject Classification.* Primary 13D10, 14C05; Secondary 13H10.

Key words and phrases. Hilbert scheme of points, Quot scheme of points, zero-dimensional schemes, Artin algebras, deformations, tensors, algebraic geometry of complexity theory.

Partially supported by National Science Centre grant 2020/39/D/ST1/00132. This work is partially supported by the Thematic Research Programme "Tensors: geometry, complexity and quantum entanglement", University of Warsaw, Excellence Initiative – Research University and the Simons Foundation Award No. 663281 granted to the Institute of Mathematics of the Polish Academy of Sciences for the years 2021-2023.

[1] Even more engaged is Weil's polemic [**133**, footnote, p.221] with Weyl [**134**, p.711].

$2 \leq \omega < 2.373$ [**37, 93, 135**]. The leading algorithm for computing ω is called the laser method [**25**, §15.8]. The input of this algorithm is a minimal border rank tensor. The main source of such tensors are multiplication tensors of *smoothable* algebras and of modules in the principal component of Quot [**14, 82, 91**]. The most successful inputs so far are the *big Coppersmith-Winograd* tensors, which are indexed by q; to every q we associate the multiplication tensor of the apolar algebra of $x_1^2 + \ldots + x_q^2$. See [**60**] for examples of bounds from other tensors. Understanding smoothable algebras is thus understanding the possible inputs for the laser method.

1.2. Big picture. The geometry of the Hilbert scheme is so rich that it is easy to lose sight of the big picture. We do not plan to discuss it here too much. However, we mention two fundamental invariants. The first one is the degree: the intuition is that analysing the objects of degree d is roughly an exponential amount of work with respect to d. Consequently, the field divides into the "near cosmos" of small enough d and the vastness of general d. The second comes from the dimension n of the ambient space X, which we assume to be $\mathbb{A}^n$. This subdivides the spaces into the "nicely behaved", such as $\mathrm{Hilb}_d(\mathbb{A}^2)$, the "border" ones, such as $\mathrm{Hilb}_d(\mathbb{A}^3)$, and the "bad" ones, such as $\mathrm{Hilb}_d(\mathbb{A}^n)$, $n \geq 4$, see Table 1 below. Apart from the Hilbert scheme, there is the Quot scheme, with similar behavior, and other moduli spaces such as the nested Hilbert schemes. There is also the natural Bilinear scheme, which we introduce in §2.10, of which Hilbert and Quot scheme are special cases. Finally, the deformation theory is tightly tied to the theory of tensors. Regretfully, a review of connections of tensors and deformation theory is yet to come, see [**82, 92**] for some ideas on how it looks like. Below we keep the tensor language to a minimum, see §2.10 for some discussion.

1.3. Definitions and notation. This section contains the bare minimum of definitions necessary to avoid ambivalence in the formulations of the problems below. For readers willing to learn more about the moduli spaces, we recommend the sources [**12, 58, 122, 129**] mentioned already above.

We work over an algebraically closed field $\mathbb{k}$. We say that an algebra A or a module M is *finite of degree* d if $\dim_{\mathbb{k}} A = d$ or $\dim_{\mathbb{k}} M = d$, respectively. It is also common to use the name *length* or *multiplicity* for what we call degree. A finite algebra A (respectively, a module M) is a product of local algebras $A_{\mathfrak{m}}$ (respectively, a direct sum of localizations $M_{\mathfrak{m}}$).

Take a local algebra A with the maximal ideal $\mathfrak{m}$. The map $\mathbb{k} \to A/\mathfrak{m}$ is an isomorphism since $\mathbb{k}$ is algebraically closed. The *(local) Hilbert function* of A is given by $H_A(i) := \dim_{\mathbb{k}} \mathfrak{m}^i/\mathfrak{m}^{i+1}$. We have $H_A(i) = 0$ for i large enough and so we frequently write the Hilbert function of A as a vector of its nonzero values, for example for $A = \mathbb{k}[x, y]/(x^2, x - y^2)$ we have $H_A(0) = H_A(1) = H_A(2) = H_A(3) = 1$, $H_A(4) = 0$, so the Hilbert function is $(1, 1, 1, 1)$. The *associated graded algebra of* A is $\operatorname{gr} A := \bigoplus_{i \geq 0} \mathfrak{m}^i/\mathfrak{m}^{i+1}$. The Hilbert function H_A is arguably the most important numerical invariant of A. Below (for example after Problem VIII) we sometimes informally refer to components and loci by describing the Hilbert function of a general point, for example the "$(1, 7, 7)$ component" is a shorthand for "the (unique/known/referred to) component whose general point corresponds to a local algebra with Hilbert function $(1, 7, 7)$".

For a quasi-projective scheme X over $\mathbb{k}$, let $\mathrm{Hilb}_d(X)$ denote the Hilbert scheme of d points on X, for constructions see [**12, 47, 56**]. More generally, let $\mathrm{Quot}_d^r(X)$

denote the Quot scheme whose points are, up to equivalence, surjections $\mathcal{O}_X^{\oplus r} \twoheadrightarrow \mathcal{M}$ of $\mathcal{O}_X$-modules, where $\mathcal{M}$ is finite and has degree d. We will be mostly interested in the case when $X = \mathbb{A}^n$ and here the points of $\mathrm{Quot}_d^r(\mathbb{A}^n)$ are, up to equivalence, surjections $S^{\oplus r} \twoheadrightarrow M$ of S-modules, where $S = \Bbbk[x_1, \ldots, x_n]$ and M is finite of degree d. The "up to equivalence" above just means that a surjection $\pi\colon S^{\oplus r} \twoheadrightarrow M$ is identified with $S^{\oplus r} \twoheadrightarrow S^{\oplus r}/\ker\pi$. The Hilbert scheme is equal to the Quot scheme taken for $r = 1$ and frequently it is useful to think about it this way.

The *Gorenstein locus* $\mathrm{Hilb}_d^{\mathrm{Gor}}(\mathbb{A}^n)$ is an open [**29**, Proposition 2.5] locus parameterizing Gorenstein algebras. The Hilbert-Chow map [**12**, Theorem 2.16] is a projective morphism

$$\rho_{X,d}\colon \mathrm{Hilb}_d(X) \to \mathrm{Sym}^d(X)$$

that maps a subscheme Γ to its support counted with multiplicities. The *punctual Hilbert scheme*[2] $\mathrm{Hilb}_d(X, x)$ is by definition $\rho_{X,d}^{-1}(d[x])$. This is a projective scheme and its set of $\Bbbk$-points is the set of Γ supported only at x. The Hilbert-Chow morphism naturally generalizes to the "Quot-Chow" morphism and in particular yields a *punctual Quot scheme* $\mathrm{Quot}_d^r(\mathbb{A}^n, 0)$.

A variant of the Hilbert scheme is the *nested* or *flag Hilbert scheme* $\mathrm{Hilb}^{d_1,\ldots,d_k}(X)$. Its $\Bbbk$-points are tuples

$$\Gamma_1 \subseteq \Gamma_2 \subseteq \ldots \subseteq \Gamma_k \subseteq X$$

where each Γ_i is finite of degree d_i. The nested Hilbert scheme is a closed subscheme of the product of $\mathrm{Hilb}_{d_i}(X)$, see [**122**, §4.5], [**32**].

The Hilbert and Quot schemes exist also for schemes over $\mathrm{Spec}(\mathbb{Z})$, in particular $\mathrm{Hilb}_d(\mathbb{A}^n_{\mathbb{Z}})$ is a quasi-projective scheme over $\mathrm{Spec}(\mathbb{Z})$ and the fiber of $\mathrm{Hilb}_d(\mathbb{A}^n_{\mathbb{Z}}) \to \mathrm{Spec}(\mathbb{Z})$ over every prime $p \in \mathrm{Spec}(\mathbb{Z})$ is $\mathrm{Hilb}_d(\mathbb{A}^n_{\mathbb{Z}/p})$, while the fiber over the generic point $(0) \in \mathrm{Spec}(\mathbb{Z})$ is $\mathrm{Hilb}_d(\mathbb{A}^n_{\mathbb{Q}})$.

Below we sometimes use Macaulay's inverse systems (also known as apolarity). They are a useful tool for constructing finite algebras and modules. See for example [**42, 73, 76**] for an introduction.

An irreducible component of a finite type scheme is *generically nonreduced* if it is nonreduced everywhere, that is, all its local rings are (geometrically) nonreduced. Otherwise is it *generically reduced*. Being generically reduced is the same as having a smooth point. A finite type scheme is *generically reduced* if all its irreducible components are.

1.4. Acknowledgements. The sources [**4, 67, 73, 110**] were very useful for collecting the more classical problems. Equally important were the suggestions of Jarosław Buczyński, Ádám Gyenge, Mateusz Michałek, Ritvik Ramkumar, Maria Evelina Rossi, Alessio Sammartano, Klemen Šivic, and Balázs Szendrői for which I am very grateful. Finally, I thank the editors of the CONM volume: Anthony Iarrobino, Pedro Macias Marques, Maria Evelina Rossi and Jean Vallès for the possibility to contribute. I also thank two anonymous referees, especially one of them, for numerous most valuable remarks and suggestions which improved the presentation.

[2]Caution: some authors use the term "punctual Hilbert scheme" for what we call here "the Hilbert scheme of points".

2. Open problems

Below, we provide a list of open problems (coincidentally, one per each year since the conference that yielded the overview [**67**]!). We hope it does list most of the problems "known to experts", but it also includes a number of directions which we believe are very important for the future even though they do not appear much in the literature yet. We begin with two tantalizing classical problems, important also for applications, for example to the theory of counting on threefolds.

PROBLEM I. *Is* $\mathrm{Hilb}_d(\mathbb{A}^3)$ *nonreduced for some* d*?*

PROBLEM II. *Is the Quot scheme* $\mathrm{Quot}_d^r(\mathbb{A}^2)$ *reduced for all* r, d*?*

Note that the Quot scheme for a nontrivial sheaf can be nonreduced [**85**]. The preprint [**31**] claims to prove that the answer to Problem II is affirmative. As far as the author knows, it is not yet generally accepted but neither any particular issue is identified. The preprint [**127**] conjectures that not only $\mathrm{Quot}_d^r(\mathbb{A}^2)$ is reduced, but it has rational singularities. For $r = 1$ the answer is affirmative, as $\mathrm{Hilb}_d(\mathbb{A}^2)$ is smooth [**48**].

2.1. Irreducible components. Each of the moduli spaces defined above has a distinguished "nice" open locus whose closure is the "main component":

- for the Hilbert scheme $\mathrm{Hilb}_d(X)$ this is the locus of $\Gamma \subseteq X$ with Γ smooth. The closure is the *smoothable component* $\mathrm{Hilb}_d^{\mathrm{sm}}(X)$. A smooth $\Bbbk$-scheme Γ is isomorphic to d copies of $\mathrm{Spec}(\Bbbk)$, so it is a tuple of d points of X. The dimension of this component is $d \cdot \dim X$.
- for the nested Hilbert scheme $\mathrm{Hilb}^{d_1,\ldots,d_k}(X)$ this is the locus of $\Gamma_1 \subseteq \ldots \subseteq \Gamma_{d_k} \subseteq X$ with every Γ_i smooth (it is enough to assume that the largest one is smooth). Again, this locus parameterizes tuples of d_k points of X with additional data coming from selecting d_{k-1}, d_{k-2} etc. points. The dimension of this locus is $d_k \cdot \dim X$.
- for the Quot scheme $\mathrm{Quot}_d^r(\mathbb{A}^n)$ this is the locus of surjections $S^{\oplus r} \twoheadrightarrow M$ with $M \simeq \bigoplus_{i=1}^d S/\mathfrak{m}_i$ with $\mathfrak{m}_i$ maximal; these are the *semisimple* S-modules. The closure is called the *principal component* of $\mathrm{Quot}_d^r(\mathbb{A}^n)$. Its dimension is $d(n + r - 1)$, see for example [**84**, §3.4].

  Observe that an algebra S/I corresponds to a semisimple S-module exactly when $\mathrm{Spec}(S/I)$ is a tuple of d points of $\mathbb{A}^n$, so the above loci for Hilb and Quot schemes yield the same type of component.
- for the punctual Hilbert scheme $\mathrm{Hilb}_d(X, x)$ this is the locus of $\Gamma \subseteq X$ with $\Gamma \simeq \mathrm{Spec}\, \Bbbk[\varepsilon]/(\varepsilon^d)$ as abstract schemes, called the *curvilinear locus.* The closure is sometimes called the *curvilinear component.* The dimension is $(d - 1)(\dim X - 1)$ for X smooth.

In an ideal world, we would like to look at a main component only, however in the cases above we have little control over this component: we usually do not even understand the tangent space to the component at a given point! Therefore, the current gold standard is to understand when the main component is the whole space and, when this is not the case, find and describe the other components. In the case of Hilbert/Quot schemes, we do not need to directly consider all other components. By abstract nonsense, each component is obtained from elementary components [**66**, Lemma 1], [**84**, §4.4]. A component $\mathcal{Z}$ of the Hilbert (respectively,

Quot) scheme is *elementary* [**66**] if every point of $\mathcal{Z}$ corresponds to an algebra (respectively, a module) supported only at one point, see below for examples.

The current state of the art is described in Table 1 below. *None* means that the main component is the whole space, *exist* means that we know that the space is reducible (for some d) but we do not have a description of any component, while *known* means that we do have a description of general points of some components. As of today, in all cases (except Gorenstein in codimension 4) where one component is known, many more are known as well for perhaps larger d.

If no reference is given, this means that the result follows from results in other rows. A more thorough discussion is given below the table.

The table shows a curious "dividing wall" of unknown but existing components. It would be important to understand whether this is an artifact of our current state of knowledge or a genuinely existing "transitory phase" between the irreducible moduli for small n and the highly reducible ones for larger n.

	$n = 1$	$n = 2$	$n = 3$	$n \geq 4$
Gorenstein locus	none	none	none [**88, 89**]	exist [**18, 72**]
Hilbert scheme	none	none [**48**]	exist [**68, 70**]	known [**26**]
punctual Hilbert scheme	none	none [**17, 44**]	exist	known
Quot scheme	none	none [**102**]	exist	known [**84**]
punctual Quot scheme	none	none [**5, 43**]	exist [**106**, 7.10.5]	known
flag Hilbert scheme	none	exist [**119**]	exist	known

TABLE 1. Nontrivial components of moduli spaces. Exist/known refer to large enough d.

With respect to d, the known reducibility statements (in characteristic zero) are as follows. By [**28, 30**] the Gorenstein locus is irreducible for $d \leq 14$, $n \leq 5$ as well as for $d \leq 13$, any n. This locus is reducible for $d \geq 14$, $n \geq 6$ by [**72**]. It is reducible also for $n = 4$, $d \geq 140$, $n = 5$, $d \geq 42$ [**18**, §6] by a parameter count using compressed algebras [**70**]. Actually, for $n = 4$ a single component is known, with $d = 140$; see comment below Problem XVIII.

For $n \geq 4$, the Hilbert scheme is reducible precisely when $d \geq 8$, see [**26**]. For $n = 3$ it is irreducible for $d \leq 11$, see [**40**] and reducible for $d \geq 78$, see [**70**, Example 4.3], see also §2.3. The punctual Hilbert scheme for $n \geq 4$ is reducible when $d \geq 8$ and irreducible for $d \leq 7$, see [**100**] and also [**80**, Proposition 2.7 and proof of Theorem 1.2]. For $n = 3$ it is irreducible for $d \leq 11$ [**80**, Theorem 1.3], while reducible for $d = 18$, as the locus of compressed algebras with Hilbert function $(1, 3, 6, 6, 2)$ has dimension 34, by direct check or [**70**, Theorem II.C], and $34 = (18 - 1) \cdot (3 - 1)$ is equal to the dimension of the curvilinear locus.

PROBLEM III. *What is the smallest d for which the punctual Hilbert scheme* $\mathrm{Hilb}_d(\mathbb{A}^3, 0)$ *is reducible? We have* $12 \leq d \leq 18$.

PROBLEM IV. *Does there exist a component of* $\mathrm{Hilb}_d(\mathbb{A}^3)$ *of dimension smaller than* $3d$*? Does there exist a component of* $\mathrm{Hilb}_d(\mathbb{A}^3, 0)$ *of dimension smaller than* $2(d-1)$*?*

Low-dimensional components of $\mathrm{Hilb}_d(\mathbb{A}^3, 0)$ are of interest also for the study of $\mathrm{Hilb}_d(\mathbb{A}^3)$, as they can only occur for d such that $\mathrm{Hilb}_d(\mathbb{A}^3)$ is reducible, see [**49**, Theorem 3.5] or [**80**, Proposition 2.7]. It seems unknown whether existence of a small component of $\mathrm{Hilb}_d(\mathbb{A}^3, 0)$ implies the existence of such for $\mathrm{Hilb}_d(\mathbb{A}^3)$.

PROBLEM V. *What is the smallest d such that* $\mathrm{Hilb}_d(\mathbb{A}^3)$ *is reducible? We have* $12 \leq d \leq 78$.

While the Hilbert scheme of $\mathbb{A}^3$ was investigated for many years, it seems that the Gorenstein locus in $\mathbb{A}^4$ received less attention. It is however important for applications to secant varieties, see [**18, 20**].

PROBLEM VI. *What is the smallest d such that* $\mathrm{Hilb}_d^{\mathrm{Gor}}(\mathbb{A}^4)$ *is reducible? We have* $15 \leq d \leq 140$.

It is puzzling that despite proofs of reducibility, no components other than the smoothable one are known (see also §2.3) for $\mathrm{Hilb}_d(\mathbb{A}^3)$. We put forward the following problem.

PROBLEM VII. *Describe (an open subset of) a component of* $\mathrm{Hilb}_d(\mathbb{A}^3)$ *other than the smoothable one. Here d can be a large integer, even* $d \gg 78$.

Finding the components on $\mathbb{A}^3$ means working with large d. However, it would be really interesting to have a full description of what happens "near home". This is the motivation for the following problem.

PROBLEM VIII. *Classify all components of* $\mathrm{Hilb}_d(\mathbb{A}^n)$ *for d reasonably small, for example* $d = 10$ *or a little higher, n arbitrary.*

It is enough to classify elementary components: those which parameterize irreducible finite subschemes. The known elementary components for small d come from Shafarevich [**124**], which generalizes [**72**]. The components parameterize algebras with Hilbert function $(1, n, r)$ for certain r, see discussion below Problem XVI. The component $(1, 7, 7)$ was also found independently later in [**13**, Corollary 4.10]. The other known elementary components of degree $d \leq 15$, according to the author's knowledge, are the $(1, 5, 3, 4)$ obtained by Huibregtse [**63**], the $(1, 5, 6, 1)$ obtained by Kleiman-Kleppe [**87**], the $(1, 6, 6, 1)$ obtained earlier by Iarrobino-Emsalem by the method of [**72**], and the $(1, 4, 6, 4)$ obtained by Satriano-Staal in [**121**, Theorem 1.3]. It would be very interesting to understand whether these are all elementary components for small d. (For example, Huibregtse generalized his result in [**64**], but d is larger in the examples obtained there). The recent advances in Białynicki-Birula decompositions, see for example [**78**], make such a classification project ambitious but not insane.

We now pass to discussing the Quot scheme. We consider the case of r large enough, say $r \geq d$. We still assume characteristic zero. For $n \geq 4$ the scheme $\mathrm{Quot}_d^r(\mathbb{A}^n)$ is reducible for every $d \geq 4$ [**53**, p.72]. For $n = 3$, the scheme $\mathrm{Quot}_d^r(\mathbb{A}^3)$ is irreducible for $d \leq 10$ [**125**] and actually for $d \leq 11$ (personal communication from Klemen Šivic), while this scheme is reducible for $d \geq 29$, see [**59**] or [**104**, p.238]. The punctual Quot scheme $\mathrm{Quot}_d^r(\mathbb{A}^3, 0)$ behaves similarly. It is irreducible for $d \leq 6$ by [**104**] and reducible for $d \geq 13$ by [**106**, 7.10.5].

PROBLEM IX. *What is the smallest d such that* $\mathrm{Quot}_d^r(\mathbb{A}^3)$ *is reducible for some r? We have* $12 \leq d \leq 29$.

A similar question can be asked for the punctual Quot scheme.

In the case of the flag or nested Hilbert scheme, we have an additional parameter: the length of the flag. Applications mostly concern the case of a two-element flag, see for example [**61**] and references therein. Partially this may be because the two-element flag is used to construct Nakajima operators, see [**103**, §8.3].

PROBLEM X. *What are the singularities of* $\mathrm{Hilb}^{d_1,d_2}(\mathbb{A}^2)$*? Can it be nonreduced?*

For $d_1 \leq 2$ the singularities are rational, see [**114, 126**].

Finally, for the Gorenstein locus we have also an additional structure: Iarrobino's symmetric decomposition [**71**]. Call a local algebra *graded* if it is isomorphic to its associated graded $\operatorname{gr} A$. Iarrobino's decomposition yields, for a local Gorenstein algebra A, a canonical graded Gorenstein algebra $Q(0)$ with a surjection $\operatorname{gr} A \twoheadrightarrow Q(0)$. This surjection is an isomorphism precisely when A is isomorphic to its associated graded. The author is not aware of any examples of Gorenstein nonsmoothable algebras A with $\operatorname{gr} A$ smoothable. For $H_A(4) = 0$ they do not exist by [**27**]. However, they should exist in general.

PROBLEM XI. *Find a nonsmoothable local Gorenstein algebra A with $Q(0)$ smoothable, preferably of small degree.*

Many of the irreducibility results above use the characteristic zero assumption. However, remarkably, there are no counterexamples for these results in positive characteristic and, in general, the author is unaware of any interesting phenomena for small d that exist only in positive characteristic.

2.2. Singular surfaces. In general, the relation between the singularities of X and the topological properties of $\mathrm{Hilb}_d(X)$ for varying d is complicated. Much is known for singular curves, we point to [**3**, p.7], [**101, 105, 117**] as example references.

Recall that by Fogarty's result, the Hilbert scheme of a smooth surface is smooth. Below we consider normal singular surfaces. By a *singularity* we mean the complete local ring at a singular point. The nicest class of surface singularities are du Val singularities [**41, 116**]. In algebraic terms, over an algebraically closed field $\Bbbk$ of characteristic zero, a surface singularity is *du Val* if both of the following hold

(1) the tangent space is three-dimensional, so that the complete local ring is $\Bbbk[[x, y, z]]/(f)$ for some local coordinates x, y, z.
(2) If we write $f = \sum_{i,j,k} a_{ijk} x^i y^j z^k$ then all of the conditions below hold
    - there is some nonzero a_{ijk} with $i + j < 2$,
    - there is some nonzero a_{ijk} with $i + j + k < 3$,
    - there is some nonzero a_{ijk} with $2i + j + k < 4$,
    - there is some nonzero a_{ijk} with $3i + 2j + k < 6$.

Moreover, the characterization does not depend on the choice of local coordinates x, y, z above. See [**116**, Theorem 2.1(3)] for a sketch of proof.

For a normal surface S with all singularities du Val the Hilbert scheme $\mathrm{Hilb}_d(S)$ is irreducible by [**39, 136**], see also [**38**]. In general, a normal surface S has only finitely many singular points $p_1, \ldots, p_s$ and the irreducibility of $\mathrm{Hilb}_d(S)$ depends only on the singularities $\widehat{\mathcal{O}}_{S,p_i}$ for $i = 1, \ldots, s$. In particular, using quasi-projectivity one can reduce to studying affine S with a unique singular point at the origin.

PROBLEM XII. *Is* $\mathrm{Hilb}_d(S)$ *reducible for every normal surface* S *with a singularity which is not a du Val singularity?*

PROBLEM XIII. *In the setting of Problem* XII, *assume that the singularities of* S *are du Val. Is* $\mathrm{Hilb}_d(S)$ *reduced for every* d*?*

The proof of irreducibility [**39**] does not imply reducedness, however it gives some ideas about a possible proof. It is enough to make the proof for the cases $S \subseteq \mathbb{A}^3$ listed in [**116**, Table 1, p.5].

2.3. Very compressed algebras and modules. Compressed algebras, developed and employed successfully by Iarrobino [**68, 70, 87**] are the main known tool to construct large loci in the Hilbert scheme and in its Gorenstein locus. The idea is very natural and has numerous variants and adaptations.

For fixed n and d we can form the *locus of very compressed algebras* $\mathrm{VC}^{\circ}_{n,d}$. It consists of degree d algebras $A = \Bbbk[x_1, \ldots, x_n]/I$, where the radical of I is $(x_1, \ldots, x_n)$ and with Hilbert function maximal for "as long as possible" given the degree. For example $\mathrm{VC}^{\circ}_{4,21}$ parameterizes algebras with Hilbert function $(1, 4, 10, 6)$. More formally, if we pick an s so that $\binom{n+s-1}{n} < d \leq \binom{n+s}{n}$, then the Hilbert function of A satisfies

$$H_A(i) = \begin{cases} \binom{i+n-1}{n-1} & \text{for } i < s,\\ d - \binom{n+s-1}{n} & \text{for } i = s,\\ 0 & \text{for } i > s. \end{cases}$$

In this case $\mathrm{VC}^{\circ}_{n,d}$ is isomorphic to the Grassmannian $\mathrm{Gr}(d - \binom{n+s-1}{n}, \binom{n-1+s}{s})$. By moving the support, we obtain a yet bigger locus $\mathrm{VC}_{n,d} = \mathrm{VC}^{\circ}_{n,d} \times \mathbb{A}^n$. Already for $n = 3$ one can check [**68**] that the dimension of $\mathrm{VC}_{3,d}$ grows faster than $3d$ so the existence of such a large locus proves reducedness of $\mathrm{Hilb}_d(\mathbb{A}^3)$ for large d. It turns out that $d = 96$ is the first value when $\dim \mathrm{VC}_{3,d} \geq 3d$.

A generalization of very compressed algebras are *compressed* algebras [**70**]. We do not discuss them at length here, mostly because a natural treatment of them would require introducing apolarity. We only recall one specific important locus constructed by Iarrobino in [**70**, Example 4.3]: it is a locus $\mathcal{L}_{78} \subseteq \mathrm{Hilb}_{78}(\mathbb{A}^3)$ which has dimension $\dim \mathcal{L}_{78} = 235 > 3 \cdot 78$, so it proves that $\mathrm{Hilb}_{78}(\mathbb{A}^3)$ is reducible.

The compressed algebras allow us to formulate a more specific incarnation of Problem I, actually two slightly different ones.

PROBLEM XIV. *Consider the locus* $\mathcal{L}_{78}$. *Is this locus a component? (If so, the component is generically nonreduced.) Consider the locus* $\mathrm{VC}_{3,96}$. *Is this locus a component? (If so, the component is generically nonreduced.)*

The rationale behind this is that since [**70**] there has been no progress on finding large loci and it seems hard to imagine how to construct any locus containing $\mathcal{L}_{78}$ as a proper subset. For $\mathrm{VC}_{3,96}$ the rationale is similar, but the evidence is weaker as there may be loci of compressed algebras in $\mathrm{Hilb}_{96}(\mathbb{A}^3)$ which have dimension higher than $\dim \mathrm{VC}_{3,96}$. An even bolder version of Problem XIV is the following.

PROBLEM XV. *Is the following alternative true for all* (n, d)*:* $\mathrm{VC}_{n,d}$ lies in the smoothable component or it forms a component*?*

For $n = 3$, $d \leq 95$ the locus $\mathrm{VC}_{n,d}$ is contained in the smoothable component, at least in characteristic zero [**40**]. It can be checked as in [**84**, §6.5] that for $n = 4$,

$d = 21, 22, 23, 24$ we indeed get that $\mathrm{VC}_{4,d}$ is a (generically nonreduced) component of $\mathrm{Hilb}_d(\mathbb{A}^4)$; this solves [**4**, Problem 3.8].

A subproblem of the above is the case when $d \leq \binom{n+2}{2}$, so that $s \leq 2$.

PROBLEM XVI (Shafarevich's gap, algebras with Hilbert function $(1, n, r)$). *Let $r = d - (1 + n)$ and assume that $\frac{(n-1)(n-2)}{6} + 2 < r \leq \binom{n+1}{2}$. Decide for which (n, d) the locus $\mathrm{VC}_{n,d}$ is a component or lies in the smoothable component.*

For r outside the range above, the situation is as follows. Values $r = 1, 2$ give smoothable algebras, see [**26**, Proposition 4.10]. A choice $3 \leq r \leq \frac{(n-1)(n-2)}{6} + 2$ yields a generically reduced elementary component $\mathrm{VC}_{n,d}$ [**124**, p.192]. The choice $r > \frac{n^2-1}{3}$ cannot lead to a generically reduced component [**124**, p.184] (but perhaps a nonreduced component is still possible).

One can also consider compressed and very compressed modules. Here, care is necessary to define the apolarity action properly, see [**90**, §2], [**84**, §4.1], [**65**, §3]. We leave the precise analysis of compressed modules to future researchers. The definition of the very compressed locus is easier. Namely, we define a sublocus $(\mathrm{VC}^r_{n,d})^\circ \subseteq \mathrm{Quot}^r_d(\mathbb{A}^n, 0)$, which consists of modules M with Hilbert function

$$H_A(i) = \begin{cases} r \cdot \binom{i+n-1}{n-1} & \text{for } i < s, \\ d - r \cdot \binom{n+s-1}{n} & \text{for } i = s, \\ 0 & \text{for } i > s, \end{cases}$$

where s is such that $r\binom{n+s-1}{n} < d \leq r\binom{n+s}{n}$ and define $\mathrm{VC}^r_{n,d}$ as $(\mathrm{VC}^r_{n,d})^\circ \times \mathbb{A}^n \subseteq \mathrm{Quot}^r_d(\mathbb{A}^n)$, where the embedding sends $([M], v)$ to M translated by the vector v, so that the resulting module is supported at v.

PROBLEM XVII. *Is the following alternative true for most (n, d, r): $\mathrm{VC}^r_{n,d}$ lies in the principal component or forms a component?*

Here "most" means: "all except a short finite list of exceptions". By [**84**], the alternative holds for $d \leq 7$ except for $(n, d, r) = (4, 7, 3), (4, 7, 4)$; in many cases we do get a component.

2.4. Compressed Gorenstein algebras.

For a fixed n and j we can consider the apolar algebra of a general form $F \in \Bbbk[y_1, \ldots, y_n]_j$ where $y_\bullet$ are variables dual to variables $x_\bullet$ in the ring $S = \Bbbk[x_1, \ldots, x_n]$. Taking F to its apolar algebra yields a map $U \to \mathrm{Hilb}^{\mathrm{Gor}}_d(\mathbb{A}^n)$, where U is an open subset of the affine space $\Bbbk[y_1, \ldots, y_n]_j$ consisting of F whose apolar algebras have the largest degree, see [**46, 77, 87**]. See [**15**] for a description of the Betti tables in the image.

PROBLEM XVIII (General forms of odd degree [**73**, Conjecture 6.30]). *Fix an odd j. Let F be a general form of degree j in n variables, where $n \geq 6$ or $(n, j) = (5, \geq 5)$ or $(n, j) = (4, \geq 15)$ and moreover $(n, j) \neq (7, 3)$. Let $\Gamma = \mathrm{Spec}(S/\mathrm{Ann}(F)) \subseteq \mathbb{A}^n$. Is it true that $\dim(T_{[\Gamma]})_{<0} = n$?*

If the claim is true, then $[\Gamma]$ lies on a unique elementary component of $\mathrm{Hilb}_d(\mathbb{A}^n)$ and is a smooth point there. The case $j = 3$ is solved in [**132**]. It is likely that the method generalizes. By semicontinuity, it is enough to check that $\dim(T_{[\Gamma]})_{<0} = n$ for a single F, which can be done using a computer algebra system. The author verified the cases $(n, j) = (4, 15), (5, 5)$.

PROBLEM XIX (Odd behaviour of even degrees). *Fix an even $j > 2$. Let F be a general form of degree j in n variables. Let $\Gamma = \operatorname{Spec}(S/\operatorname{Ann}(F)) \subseteq \mathbb{A}^n$. Is it true that then $[\Gamma] \in \operatorname{Hilb}_d(\mathbb{A}^n)$ lies on a unique elementary component which is everywhere nonreduced for "most" pairs (j, n)? Can one at least find one such pair? (Here n should probably be taken large with respect to j.)*

The reason behind the different expectations in Problems XVIII–XIX is numerical. For $j = 2k$ and general F let $I := \operatorname{Ann}(F)$ and $\Gamma := \operatorname{Spec}(S/I)$. Then we have $I_{\leq k} = 0$ so that $(I^2)_{\leq 2k+1} = 0$. Moreover $I_{2k+1} = S_{2k+1}$, since $\deg(F) < 2k + 1$. It follows that $\dim_{\Bbbk}(I/I^2)_{2k+1} = \dim_{\Bbbk} S_{2k+1}$ is large. By duality for Gorenstein algebras, the space $(I/I^2)_{2k+1}$ has the same dimension as the degree -1 part of the tangent space at $[\Gamma] \in \operatorname{Hilb}_d(\mathbb{A}^n)$, see for example [**132**, Proposition 2.12]. This rules out the behaviour expected in the odd j case. Moreover, the feeling is that these degree -1 tangent directions have no geometric "meaning", so for most cases they should not extend to higher order. If this is the case, then an argument similar to [**131**] proves nonreducedness.

2.5. Special components. In this section we collect the odd problems concerning components. All of them are classical and of interest.

PROBLEM XX ([**4**, **96**]). *Does there exist a non-rational component of* $\operatorname{Hilb}_d(\mathbb{A}^n)$*?*

Recall that a component is rational if it contains an open subset isomorphic to an open subset of an affine space. The rationality is of interest also for more practical terms: some algorithms aiming at computing the local structure of the Hilbert scheme [**75**, **95**], when applied to a smooth point, will currently succeed in producing a universal family only if the component containing the point is rational.

PROBLEM XXI. *Does there exist a component of* $\operatorname{Hilb}_d(\mathbb{A}^n_{\mathbb{Z}})$ *that lives entirely in characteristic p, for small d?*

For d a few hundred, an example is given in [**79**, §5]. A smaller example would be useful for finding varieties over characteristic p nonliftable to characteristic zero.

PROBLEM XXII. *Fix n. What is the smallest dimension Δ of a component of* $\operatorname{Hilb}_d(\mathbb{A}^n, 0)$*, when $d \to \infty$? Or, a bit more precisely, what can be said asymptotically about $1/d \cdot (nd - \Delta)$?*

There seemed to be a consensus among experts that every component of $\operatorname{Hilb}_d(\mathbb{A}^n, 0)$ has dimension at least $(n-1)(d-1)$. This was recently disproved by Satriano-Staal [**121**] already for $n = 4$. The dimension of the *largest* component is known to grow asymptotically as $d^{2-2/n}$, see [**16**]. See also [**111**].

PROBLEM XXIII ([**67**, p.313]). *Does there exist an irreducible finite $\Bbbk$-scheme Γ with $T^1_\Gamma = 0$, other than $\Gamma = \operatorname{Spec}(\Bbbk)$?*

Recall that such a scheme is called *rigid* (to be safe, assume $\Bbbk$ has characteristic zero here). Here, T^1_Γ is the first Schlessinger's functor [**58**, §3]. If $\Gamma = \operatorname{Spec}(S/I)$, with $S = \Bbbk[x_1, \ldots, x_n]$ and $d = \dim_{\Bbbk} S/I$, then T^1_Γ is the quotient of the tangent space $T_{[\Gamma]} \operatorname{Hilb}_d(\mathbb{A}^n)$ by the S-submodule generated by (the images in $T_{[\Gamma]}$ of) the n-dimensional space of vector translations, see for example [**58**, Proposition 3.10]. It seems that for Gorenstein Γ the answer is negative by [**2**, Theorem 2.3 and Proposition 4.1].

2.6. Tangent spaces. The tangent space in general does not seem to carry much geometric information, except that a large tangent space makes explicit calculations very resource-consuming. However, on the "dividing wall", the tangent space seems to carry a lot more information, at witnessed by the following problems.

PROBLEM XXIV ([**16**]). *Which is the maximal tangent space dimension to a point of* $\mathrm{Hilb}_d(\mathbb{A}^3)$*?*

Briançon-Iarrobino [**16**, p.544] put forward a conjectural answer. The conjecture concerns every n, we discuss it here only for $n = 3$. The conjecture has two versions. The first one is formulated only for degrees of the form $d = \binom{k+2}{3}$ for some natural k; we call it the *triangular* version. This version claims that the maximal tangent space is attained at the point corresponding to $(x, y, z)^k$. The second, general conjecture was made for every d. The more general version was disproved and updated by Sturmfels [**130**]. Sturmfels' version was subsequently disproved by Ramkumar-Sammartano [**113**]. The triangular case of the original conjecture still stands and Ramkumar-Sammartano come close to proving it.

The following two problems have very different formulations, but come from the common motivation in enumerative geometry [**99**, **107**]. (The following problem has been solved [**50**] while the present paper was in review.)

(SOLVED) PROBLEM XXV. *Is it true that for every point* $[\Gamma] \in \mathrm{Hilb}_d(\mathbb{A}^3)$ *the parity of d and of the dimension of the tangent space at* $[\Gamma]$ *is the same?*

The answer is affirmative when $I = I(\Gamma)$ is monomial [**99**, Theorem 2] and more generally for ideals I homogeneous with respect to the standard $\mathbb{N}$-grading (and even some nonstandard gradings), see [**112**]. While this list was in review, a negative answer to the question in Problem XXV was given by Giovenzana-Giovenzana-Graffeo-Lella [**50**]; it is striking that their example has degree only $d = 12$ and is a (non-standard) graded ideal. The story of Problem XXV is not yet finished. A weaker question: "Do parities agree on an open dense subset?" is still open and well motivated by the enumerative geometry, see [**118**, Theorem 1.2].

The formulation of the second problem is very far from elementary, as it uses the Behrend function, a measure of the singularity, introduced by Behrend [**7**, Definition 1.4] and quite difficult to compute even for small examples. It is known that the function is equal to $(-1)^d$ for every point corresponding to a monomial ideal of degree d [**8**, Proposition 4.2]. See also [**108**, §4] for an example of computation.

(SOLVED) PROBLEM XXVI. *Is the Behrend function equal to* $(-1)^d$ *for every point of* $\mathrm{Hilb}_d(\mathbb{A}^3)$*?*

If true, this would imply that $\mathrm{Hilb}_d(\mathbb{A}^3)$ is generically reduced, see [**118**]. In general, the positive answer would restrict the possible singularities of this Hilbert scheme. A negative answer would be philosophically very important for the research into counting on threefolds, as the multiplicities of the components would need to be taken into account. (A negative answer was given in [**81**] while the present paper was being prepared for publication.)

The remaining problems in this section follow from Hu's striking idea [**62**] that the singularities of points with only a small surplus in the tangent space could be classified and understood. We follow Hu's decision to restrict to $\mathbb{A}^3$. Let $[\Gamma]$ be a singular point of the smoothable component of $\mathrm{Hilb}_d(\mathbb{A}^3)$. Then the tangent space $T_{[\Gamma]} \mathrm{Hilb}_d(\mathbb{A}^3)$ has dimension at least $3d + 1$. According to Problem XXV is could be at least $3d+2$. The smallest dimension known to be attained is $3d+6$, it occurs for example for $\Gamma = \operatorname{Spec}(\Bbbk[x_1, x_2, x_3]/(x_1, x_2, x_3)^2)$.

To formulate the problem below, we need the notion of smooth equivalence, see for example [**79**, §2]. Namely, two complete local rings A and B are *smoothly equivalent* if $A[[x_1, \dots, x_a]]$ and $B[[x_1', \dots, x_b']]$ are isomorphic for some a and b; the parentheses denote taking formal power series. For closed points $x \in X$, $y \in Y$ of finite type schemes X, Y we say that $x \in X$, $y \in Y$ are *smoothly equivalent* if their complete local rings are. This can be rephrased in geometric terms: closed points $x \in X$, $y \in Y$ are smoothly equivalent if and only if there is a finite type scheme Z with a closed point $z \in Z$ and *smooth* maps $Z \to X$, $Z \to Y$ which send z to x, y, respectively. Hence, smooth equivalence is a way to compare complete local rings of different dimensions. Properties of a singularity, such as being reduced, are preserved by smooth equivalence, although this needs to be carefully proven for each given property, see for example [**1**, Tag 033D].

PROBLEM XXVII (Mildly singular points of $\operatorname{Hilb}_d(\mathbb{A}^3)$, [**62**]). *Let* $[\Gamma] \in \operatorname{Hilb}_d(\mathbb{A}^3)$ *be a point of the smoothable component (see* §2.1*) with tangent space of dimension* $3d+6$. *Is it true that* $[\Gamma] \in \operatorname{Hilb}_d(\mathbb{A}^3)$ *is smoothly equivalent to the cone point of a cone over the Grassmannian* $\operatorname{Gr}(2,6)$*?*

For $d = 4$ this is true classically and is explained for example in [**86, 128**]. For $d \leq 7$ this is proven in [**62**, Proposition 4.30].

PROBLEM XXVIII (Mildly singular points of $\operatorname{Hilb}_d^{\mathrm{Gor}}(\mathbb{A}^4)$). *Let* $[\Gamma] \in \operatorname{Hilb}_d^{\mathrm{Gor}}(\mathbb{A}^4)$ *be a point of the smoothable component. Suppose that the surplus*

$$\dim T_{[\Gamma]} \operatorname{Hilb}_d(\mathbb{A}^4) - 4d$$

is small. Can we classify singularities of such points up to smooth equivalence?

For example, Ranestad-Schreyer [**115**, Cor 5.16] proved that for

$$\Gamma = \operatorname{Spec}\left(\frac{\Bbbk[x_1, \dots, x_4]}{(x_i x_j \mid i \neq j) + (x_i^2 - x_1^2 \mid i \neq 1)}\right)$$

the surplus is 5, the singularity is equivalent to a cone point of a cone over the 10-dimensional spinor variety. This is very surprising as the spinor variety is a very special one. One concrete version of Problem XXVIII would be to consider difference 5 and prove that we always obtain a spinor variety, up to smooth equivalence.

2.7. Punctual Hilbert schemes. The punctual Hilbert schemes (see §1.3) naturally appear when considering the elementary components of the Hilbert scheme of points. Much is known about the case of two variables and a little for the case of three variables, see for example [**17, 52, 69**]. Still, it seems that many natural questions remain open. The punctual Hilbert scheme is important for applications to enumerative problems and algebraic homotopy theory. For example, tautological counts on the smoothable component of the Hilbert scheme reduce to the curvilinear Hilbert scheme [**9, 10**]; understanding its components and upper bounds on the dimension is important here.

PROBLEM XXIX. *Say that an irreducible component* $\mathcal{Z}$ *of* $\operatorname{Hilb}_d(\mathbb{A}^n, 0)$ *is* smoothable *if* $\mathcal{Z}$ *is contained in the smoothable component of the Hilbert scheme* $\operatorname{Hilb}_d(\mathbb{A}^n)$. *What can be said about the smoothable components* $\mathcal{Z}$*? How many are there, for a given* (n, d)*?*

In contrast with the Hilbert scheme of points, as far as the author knows, there is no natural injective map from the components of $\mathrm{Hilb}_d(\mathbb{A}^n, 0)$ to the components of $\mathrm{Hilb}_{d+1}(\mathbb{A}^n, 0)$. It would be very interesting to see if the answer in Problem XXIX can sometimes decrease with increasing d.

(The following was solved in [**11**] while the present paper was in review.)

(SOLVED) PROBLEM XXX. *Do all monomial ideals in* $\mathrm{Hilb}_d(\mathbb{A}^n, 0)$ *lie in the closure of the curvilinear locus?*

As a motivation, it is known classically that all monomial ideals lie in the smoothable component of $\mathrm{Hilb}_d(\mathbb{A}^n)$, see [**57**], [**26**, Proposition 4.15]. Some examples of smoothable components other than the curvilinear component are given in [**21**, Appendix].

The problem below concerns local complete intersections. From the deformation theory point of view, the graded ones are very easy: deforming them means just deforming each equation separately. In the local case, things are more difficult, as deforming the equations can change the degree of the scheme. Things get even harder if we require that the scheme stays supported at a single point. The motivation for the problem below partially comes from motivic homotopy theory.

PROBLEM XXXI. *In the punctual Hilbert scheme* $\mathrm{Hilb}_d(\mathbb{A}^n, 0)$ *consider the open locus* $\mathrm{Hilb}_d^{\mathrm{lci}}(\mathbb{A}^n, 0)$ *of locally complete intersections supported at a point. Find an upper bound for its dimension.*

Locally complete intersections are smoothable, hence $\mathrm{Hilb}_d^{\mathrm{lci}}(\mathbb{A}^n, 0)$ is contained in $\mathrm{Hilb}_d^{\mathrm{sm}}(\mathbb{A}^n)$ and so it has dimension at most $nd-1$. However, local complete intersections are not necessarily limits of curvilinear schemes [**69**][3]. Perhaps the newly introduced method of Białynicki-Birula slicing [**80**] could be of use for this problem.

2.8. Multigraded Hilbert schemes. Fix n and d. Let $S = \Bbbk[x_0, \dots, x_n]$, note the indexing from zero. The multigraded Hilbert schemes are slightly different in flavour from the usual ones: they parameterize *homogeneous* ideals I in S with $\dim(S/I) = 1$. This is still a "zero-dimensional" situation, when we intuitively look at these ideals as defining subschemes in $\mathbb{P}^n$ rather than in $\mathbb{A}^{n+1}$. We stress however, that the ideals above are usually *not* saturated, hence there is important information lost when passing from I to $V(I) \subseteq \mathbb{P}^n$. The multigraded Hilbert scheme is recently applied very successfully in theory of tensors, for investigating the lower bounds on border rank [**19, 36**].

Let $H\colon \mathbb{N} \to \mathbb{N}$ be the Hilbert function of a tuple of d general points in $\mathbb{P}^n$, so that $H(i) = \min(d, \binom{n+i}{i})$. Let Hilb^H be the multigraded Hilbert scheme as in [**56**], so its $\Bbbk$-points correspond to homogeneous ideals $I \subseteq S$ with Hilbert function H. A bit more generally, let $H_r\colon \mathbb{N}^r \to \mathbb{N}$ be the Hilbert function of a tuple of d general points in

$$\underbrace{\mathbb{P}^n \times \ldots \times \mathbb{P}^n}_{r},$$

then Hilb^{H_r} parameterizes homogeneous ideals in $I \subseteq S$ where S is an $\mathbb{N}^r$-graded ring in $r(n+1)$ variables corresponding to the total coordinate ring of $\mathbb{P}^n \times \ldots \times \mathbb{P}^n$.

[3]On page 602 of [**69**] there is a bound $nd(1-d^{-1/n})$ attributed to Briançon-Granger without specific reference. It would be a solution to Problem XXXI. Unfortunately it seems wrong: the analogue of [**69**, Example 3] for $C(4,4,4,4)$ yields a family of too high dimension.

The multigraded Hilbert scheme Hilb^H behaves more pathologically than $\mathrm{Hilb}_d(\mathbb{A}^n)$. For $n = 1$ it is smooth irreducible [**97**], but already for $n = 2$ the multigraded Hilbert scheme is reducible and singular [**98**], while $n = 2$, naïvely speaking, corresponds to projective plane and $\mathrm{Hilb}_d(\mathbb{P}^2)$ is smooth irreducible by Fogarty's result.

The saturated ideals in Hilb^H form an open subset [**83**], whose closure is a union of irreducible components and is denoted by $\overline{\mathrm{Sat}}^H$. Inside, the saturated ideals of smooth subschemes in $\mathbb{P}^n$ form an open irreducible subset, whose closure is a distinguished component Slip^H, the *scheme of limits of ideals of points*, see [**19, 83, 98**]. The following problem is akin to smoothability and has immediate applications in the theory of border ranks of tensors:

Problem XXXII. *For a given $[I] \in \mathrm{Hilb}^H$ determine whether it lies in $\overline{\mathrm{Sat}}^H$ or in Slip^H.*

Determining this for monomial or even Borel-fixed ideals is already open. There are examples not lying in $\overline{\mathrm{Sat}}^H$, see [**83, 98**]. An extension from H to H_r would be also very interesting. See [**83**] for some results.

By a result of Pardue [**109**, Theorem 34] the scheme Hilb^H is connected. The author is not aware of any extensions of this result to the case $r > 1$.

Problem XXXIII. *Is Hilb^{H_r} connected? Does every component of Hilb^{H_r} intersect the Slip^{H_r}?*

If we remove the assumptions on H_r, that is, if we consider an arbitrary Hilbert function H_r for an arbitrary $\mathbb{N}^r$-grading, then Hilb^{H_r} may be disconnected [**120**]. The example is however complicated and the polynomial ring has 26 variables. Hence we pose the following challenge.

Problem XXXIV. *Give a "small" example of a disconnected multigraded Hilbert scheme, preferably with H_r eventually constant.*

2.9. Smoothability. Smoothability of a given ideal is investigated classically. In this section we gather problems which were not posed before explicitly, however in our opinion they are very important.

Problem XXXV. *For a subscheme $\Gamma \subseteq \mathbb{A}^n$ consider the* smoothability defect *$\delta(\Gamma)$ of Γ as the minimal possible difference* $\deg \Gamma' - \deg \Gamma$ *where $\Gamma \subseteq \Gamma'$ and Γ' is finite smoothable. What is the behavior of the function $\delta(-)$? For example, can it happen that $\delta(\Gamma) > \deg \Gamma$?*

As far as we know, nothing is known about the smoothability defect. To see that it is finite, take Γ local and such that $|\Gamma| = \{0\} \subseteq \mathbb{A}^n = \mathrm{Spec}(\Bbbk[x_1, \ldots, x_n])$. Then Γ is contained in $\Gamma' = V((x_1, \ldots, x_n)^e)$ for a large enough e. The scheme Γ' is monomial, hence smoothable [**26**, Proposition 4.15]. For a nonlocal Γ we repeat the above procedure for every point of the support of Γ. The above gives an exponential bound on the smoothability defect. However, in practice this defect seems to be very small, for example by a direct check it is equal to 1 for the $(1, 4, 3)$ example of [**26**, Proposition 5.1] and the $(1, 6, 6, 1)$ example of Emsalem-Iarrobino [**72**].

Problem XXXVI. *For a finite scheme Γ, let $\delta(\Gamma)$ be its smoothability defect. Is it true that for every $\epsilon > 0$ we have*

$$\lim_{n\to\infty} \frac{\delta(\Gamma^{\times n})}{(\deg \Gamma)^{n(1+\epsilon)}} = 0?$$

Warning: an affirmative answer in Problem XXXVI for every member of a sequence of $\Gamma := \operatorname{Spec}(\overline{A}_n)$ as defined below (following [**33**]) would imply the famous "$\omega = 2$" conjecture on the matrix multiplication exponent. This problem itself is a strengthened version of a special case of a conjecture by Strassen so *disproving* it, even on an example disconnected from the context of the exponent of matrix multiplication, would be a breakthrough for complexity theory. See [**35**] for the discussion on Strassen's conjecture in the language of tensors. Providing an example of Γ with nonzero limit for $\epsilon = 0$ is a bit weaker, but still very interesting problem.

There is a more direct link between smoothability and the matrix multiplication exponent ω. Consider a cubic form $sM_{\langle n\rangle}$ in n^2 variables $(x_{ij})_{1\leq i,j\leq n}$ given by $\operatorname{trace}([x_{ij}]^3)$, where $[x_{ij}]^3$ is the cube of a matrix $[x_{ij}]$. (The symbol $sM_{\langle n\rangle}$ stands for *symmetrized matrix multiplication tensor* [**33**].) The apolar algebra $(A_n, \mathfrak{m})$ of $sM_{\langle n\rangle}$ is graded and has Hilbert function $(1, n^2, n^2, 1)$. Let $\overline{A}_n = A_n/\mathfrak{m}^3$, then $\overline{A}_n$ has Hilbert function $(1, n^2, n^2)$. Let $\delta_n := \delta(\overline{A}_n)$ be the smoothability defect of $\overline{A}_n$, as defined above. It follows from [**33**] that δ_n grows like n^ω and that it is enough to look at surjections $B_n \twoheadrightarrow \overline{A}_n$ where $H_{B_n} = (1, *, *)$. (The details of this assertion will be given elsewhere.) Hence, to prove that $\omega = 2$ it is enough to find for every n a smoothable algebra of length around n^2 surjecting onto a given one! More modestly, it would be huge advance to find a sequence of B_n which grows slower than, say, $n^{2.3}$.

To formulate the next problem, let $\Bbbk[[t]]$ be the ring of formal power series in one variable and $\Bbbk((t))$ be the field of Laurent formal power series. For an element $f \in \Bbbk((t))$ its *valuation* is the smallest exponent $\nu(f)$ such that the coefficient in f next to $t^{\nu(f)}$ is nonzero.

Let $\Bbbk((t))^{\times d} = \Bbbk((t)) \times \ldots \times \Bbbk((t))$ be the algebra with usual coordinatewise operations. An *abstract smoothing* of A is a finite flat family $\Bbbk[[t]] \to \mathcal{A}$ with generic fiber $\mathcal{A}_t$ (geometrically) reduced and special fiber $\mathcal{A}/t\mathcal{A}$ isomorphic to A. The scheme Γ is smoothable in the usual sense if and only if an abstract smoothing exists [**22**, Theorem 1.1]. A *completely split abstract smoothing* of A is an abstract smoothing $\mathcal{A}$ of A with generic fiber $\mathcal{A}_t$ isomorphic to $\Bbbk((t))^{\times d}$.

For a finite flat $\Bbbk[[t]]$-algebra $\mathcal{A}$, the *approximation degree* of $\mathcal{A}$ is the maximum of valuations of nonzero elements appearing in the multiplication table of $\mathcal{A}$; this degree does not depend on the choice of a $\Bbbk[[t]]$-basis.

PROBLEM XXXVII. *Is it true that for every smoothable finite scheme* $\operatorname{Spec}(A)$ *of degree d there exists a completely split abstract smoothing $\mathcal{A}$ of A with approximation degree at most d? Less restrictively, does there exist a polynomial P such that for every A as above there exists an $\mathcal{A}$ with approximation degree bounded by $P(d)$? (for example, proving that the approximation degree is bounded by d^{100} would be enough!)*

An abstract smoothing is rarely completely split. For example take $A = \Bbbk[x]/(x^d)$. Then $\mathcal{A} = \Bbbk[[t]][x]/(x^d - t)$ is an abstract smoothing of A, but $\mathcal{A}_t \simeq \Bbbk((t))[x]/(x^d - t)$ is a domain, so it is not isomorphic to $\Bbbk((t))^{\times d}$. However, take

$$\mathcal{A}' = \frac{\Bbbk[[t]][x]}{(x(x-t)(x-2t)\ldots(x-(d-1)t))}.$$

Assuming that $\Bbbk$ has characteristic zero, by Chinese Remainder Theorem the generic fiber $\mathcal{A}'_t$ is indeed isomorphic to $\Bbbk((t))^{\times d}$. The approximation degree obtained from $\mathcal{A}'$ is $d-1$. Observe that in the multiplication table of $\mathcal{A}$ the largest power of t that appears is t^1, but this tells us nothing about the approximation degree since the smoothing $\mathcal{A}$ is not completely split. Completely split abstract smoothings exist for every smoothable A and can be obtained by a base change from any one-parameter family that shows smoothability of A.

Again, the motivation for the problem above is algebraic complexity theory [**23**, **74**, **94**]. In the references just given the problem is formulated differently. The author believes that the solution to Problem XXXVII yields a polynomial upper bound on the approximation degree for tensors, a classical important direction in complexity theory [**23**, **24**, **34**, **94**], but the precise argument on why it is so is nontrivial and well beyond the scope of the current work.

2.10. Moduli space of bilinear maps. Let $S = \Bbbk[x_1, \ldots, x_n]$ be the coordinate ring of $\mathbb{A}^n$. Consider the *Bilinear scheme* $\mathrm{Bilin}^{r_1,r_2}_{d_1,d_2,d_3}(\mathbb{A}^n)$ whose $\Bbbk$-points are surjections of S-modules $S^{\oplus r_1} \twoheadrightarrow M_1 = \frac{S^{\oplus r_1}}{K_1}$, $S^{\oplus r_2} \twoheadrightarrow M_2 = \frac{S^{\oplus r_2}}{K_2}$ together with a surjection of S-modules $\pi\colon M_1 \otimes_S M_2 \twoheadrightarrow M_3 = \frac{S^{\oplus r_1 r_2}}{K_3}$, where M_i is a finite S-module of degree d_i. This space can be realized as a scheme using Quot schemes.

The following is a meta-problem.

PROBLEM XXXVIII. *Investigate the geometry of this space.*

In practical terms, this means: find the components, investigate singularities, decide nonreducedness etc. Already the case $n = 2$, the case of two variables, is open. The motivation for this problem comes from the theory of tensors, but the Bilinear scheme naturally generalizes the Hilbert scheme and Quot scheme. To see this, we need to introduce a bit of notation.

Observe that a finite algebra $A = S/I$ gives rise to a surjection $\pi_A\colon A \otimes_S A \to A$ which is the multiplication in A. Every finite cyclic S-module is of the form $S^{\oplus 1}/I$ and a surjective S-linear map $\pi\colon S^{\oplus 1}/I \otimes S^{\oplus 1}/J \to S^{\oplus 1}/K$ between modules with $\dim_{\Bbbk} S/I = \dim_{\Bbbk} S/J = \dim_{\Bbbk} S/K$ exists only if $I = J = K$ and in that case it is equal to the multiplication in S/I followed by multiplying by a fixed invertible element of S/I; the latter operation is irrelevant when we pass to equivalence classes as in the Quot scheme. It follows that $\mathrm{Bilin}^{1,1}_{d_1,d_1,d_1}$ is isomorphic to $\mathrm{Hilb}_{d_1}(\mathbb{A}^n)$. More generally, the space $\mathrm{Bilin}^{r_1,r_2}_{d_1,d_1,d_1}$ admits an open subset $\mathcal{U}$ where M_1 and M_2 are cyclic. On this open subset M_3 is cyclic as well (since it is an image of the cyclic module). This open subset is smoothly equivalent to the Hilbert scheme.

For the Quot scheme, the situation is similar yet a bit more complicated, as we do not obtain exactly the Quot scheme: rather than modules, the obtained space parameterizes modules together with an algebra acting on them. Namely, a finite S-module $M = S^{\oplus r}/K$ yields a surjection $S/\operatorname{Ann}_S(M) \otimes_S M \to M$. The space $\mathrm{Bilin}^{r_1,r_2}_{d_1,d_2,d_2}$ admits an open subset $\mathcal{V}$ where M_1 is cyclic. For a point of this subset, we have $M_1 \simeq S^{\oplus 1}/I$ and the homomorphism $\pi\colon M_1 \otimes_S M_2 \to M_3$ simplifies to a surjective S-linear map $\frac{M_2}{IM_2} \to M_3$, which can only exist if $M_2 \simeq M_3$ and $IM_2 = 0$. Thus, from such a point we obtain a degree d module $M := M_2 \simeq M_3$ and an algebra $A = S/I$ such that $IM = 0$, hence M is an A-module. Let $\mathcal{Q}$ be the moduli space of pairs $M = S^{\oplus r_2}/K$ and $A = S/I$ such that $IM = 0$ and $\dim_{\Bbbk} A = d_1$, $\dim_{\Bbbk} M = d_2$. Then $\mathcal{V}$ admits a smooth map to $\mathcal{Q}$. Moreover, $\mathcal{Q}$ has a map to $\mathrm{Quot}^{r_2}_{d_2}(\mathbb{A}^n)$ which forgets about A. This map is usually not smooth. It may even be far from surjective: if d_1 is small, then the existence of I is very restrictive. If

$d_1 = d_2$, then the map $\mathcal{Q} \to \mathrm{Quot}^{r_2}_{d_2}(\mathbb{A}^n)$ is birational on the principal component, hence $\mathcal{Q}$ can be thought of as a blowup of $\mathrm{Quot}^{r_2}_{d_2}(\mathbb{A}^n)$.

In the language of [**82**], roughly speaking, the Bilinear scheme corresponds to 111-abundant tensors, the open subset $\mathcal{U}$ corresponds to 1_A- and 1_B-generic 111-abundant tensors, while $\mathcal{V}$ corresponds to 1_A-generic 111-abundant tensors, see [**82**, Proposition 5.3]. The passage between a tensor and bilinear map is straightforward: to a map $\pi\colon M_1 \otimes_S M_2 \to M_3$ one associates the tensor $\pi \in M_1^* \otimes_{\Bbbk} M_2^* \otimes_{\Bbbk} M_3$. Conversely, to a 111-abundant tensor $T \in V_1 \otimes_{\Bbbk} V_2 \otimes_{\Bbbk} V_3$ with the 111-algebra A, one associates the bilinear map $T\colon V_1^* \otimes_{\Bbbk} V_2^* \to V_3$. The spaces V_1^*, V_2^*, V_3 are A-modules in a canonical way [**82**, §5.1] and the map T is A-bilinear, thus (after fixing some generators) yields a point of the Bilinear scheme[4].

There are also points of $\mathrm{Bilin}^{r_1,r_2}_{d_2,d_2,d_2}$ in the complement of $\mathcal{V}$. Examples for $d_1 = d_2 = 5$ are given in [**82**, Theorem 1.7]. Below, we translate the tensor called $T_{\mathcal{O}_{58}}$ to the language above (see also [**82**, Example 4.6]). Here, we have $n = 4$, $d_1 = d_2 = d_3 = 5$ and $r_1 = r_2 = 2$. We now construct the modules. Consider a polynomial ring in two variables $\Bbbk[z_1, z_2]$ and the finite algebra $B = \frac{\Bbbk[z_1,z_2]}{(z_1^4, z_2^4)}$. It has Hilbert function $H_B = (1, 2, 3, 4, 3, 2, 1)$. Let B_j be the space of homogeneous elements of degree j in B. Let $S = \Bbbk[x_1, \ldots, x_4]$ act on B via the homomorphism $x_i \mapsto z_1^{i-1} z_2^{4-i}$ for $i = 1, 2, 3, 4$. Let M be equal to the $\Bbbk$-vector space $B_1 \oplus B_4$, so M has a basis $z_1, z_2, z_1^3 z_2, z_1^2 z_2^2, z_1 z_2^3$. Of course, $M \subseteq B$ is not a B-submodule. But $M \subseteq B$ *is* an S-submodule. Let $N := B_{6-1} \oplus B_{6-4} = B_2 \oplus B_5$. This is also an S-submodule of B. The multiplication in B restricted to M satisfies $M \cdot M \subseteq N$, so yields a map $\pi\colon M \otimes_S M \to N$. Choose z_1, z_2 as generators of the S-module M, so we have $M = S^{\oplus 2}/K$ and π yields a point of $\mathrm{Bilin}^{2,2}_{5,5,5}$. We stress that B is not determined by the map π: we introduced it above as an ad-hoc device to phrase the example neatly.

References

[1] *Stacks Project*, `http://math.columbia.edu/algebraic_geometry/stacks-git`, 2023. 14

[2] A. G. Aleksandrov, *Duality, derivations and deformations of zero-dimensional singularities*, Zero-dimensional schemes (Ravello, 1992), de Gruyter, Berlin, 1994, pp. 11–31. MR1292472 ↑12

[3] Allen B. Altman, Anthony Iarrobino, and Steven L. Kleiman, *Irreducibility of the compactified Jacobian*, Real and complex singularities (Proc. Ninth Nordic Summer School/NAVF Sympos. Math., Oslo, 1976), Sijthoff & Noordhoff, Alphen aan den Rijn, 1977, pp. 1–12. MR498546 ↑9

[4] American Institute of Mathematics Problem List, *Components of Hilbert schemes*, Available at `http://aimpl.org/hilbertschemes`, 2010. 5, 11, 12

[5] V. Baranovsky, *The variety of pairs of commuting nilpotent matrices is irreducible*, Transform. Groups **6** (2001), no. 1, 3–8, DOI 10.1007/BF01236059. MR1825165 ↑7

[6] Arnaud Beauville, *Variétés Kähleriennes dont la première classe de Chern est nulle* (French), J. Differential Geom. **18** (1983), no. 4, 755–782 (1984). MR730926 ↑3

[7] Kai Behrend, *Donaldson-Thomas type invariants via microlocal geometry*, Ann. of Math. (2) **170** (2009), no. 3, 1307–1338, DOI 10.4007/annals.2009.170.1307. MR2600874 ↑13

[8] Kai Behrend and Barbara Fantechi, *Symmetric obstruction theories and Hilbert schemes of points on threefolds*, Algebra Number Theory **2** (2008), no. 3, 313–345, DOI 10.2140/ant.2008.2.313. MR2407118 ↑13

[9] Gergely Bérczi, *Tautological integrals on Hilbert scheme of points I*, (2023), `https://arxiv.org/abs/2303.14807`. 14

[4] In [**82**] the spaces are named A, B, C instead of V_1, V_2, V_3.

[10] Gergely Bérczi, *Tautological integrals on Hilbert scheme of points II*, (2023), `https://arxiv.org/abs/2303.14812`. ↑14
[11] Gergely Bérczi and Jonas M. Svendsen, *Fixed point distribution on Hilbert scheme of points*, (2023), `arXiv:2306.11521`. ↑15
[12] José Bertin, *The punctual Hilbert scheme: an introduction* (English, with English and French summaries), Geometric methods in representation theory. I, Sémin. Congr., vol. 24, Soc. Math. France, Paris, 2012, pp. 1–102. MR3202701 ↑3, 4, 5
[13] Cristina Bertone, Francesca Cioffi, and Margherita Roggero, *Smoothable Gorenstein points via marked schemes and double-generic initial ideals*, Exp. Math. **31** (2022), no. 1, 120–137, DOI 10.1080/10586458.2019.1592034. MR4399113 ↑8
[14] Markus Bläser and Vladimir Lysikov, *On degeneration of tensors and algebras*, 41st International Symposium on Mathematical Foundations of Computer Science, LIPIcs. Leibniz Int. Proc. Inform., vol. 58, Schloss Dagstuhl. Leibniz-Zent. Inform., Wadern, 2016, pp. Art. No. 19, 11. MR3578455 ↑4
[15] Mats Boij, *Betti numbers of compressed level algebras*, J. Pure Appl. Algebra **134** (1999), no. 2, 111–131, DOI 10.1016/S0022-4049(97)90163-8. MR1663785 ↑11
[16] J. Briançon and A. Iarrobino, *Dimension of the punctual Hilbert scheme*, J. Algebra **55** (1978), no. 2, 536–544, DOI 10.1016/0021-8693(78)90236-3. MR523473 ↑12, 13
[17] Joël Briançon, *Description de* $H\mathrm{ilb}^n C\{x,y\}$, Invent. Math. **41** (1977), no. 1, 45–89, DOI 10.1007/BF01390164. MR457432 ↑7, 14
[18] Weronika Buczyńska and Jarosław Buczyński, *Secant varieties to high degree Veronese reembeddings, catalecticant matrices and smoothable Gorenstein schemes*, J. Algebraic Geom. **23** (2014), no. 1, 63–90, DOI 10.1090/S1056-3911-2013-00595-0. MR3121848 ↑7, 8
[19] Weronika Buczyńska and Jarosław Buczyński, *Apolarity, border rank, and multigraded Hilbert scheme*, Duke Math. J. **170** (2021), no. 16, 3659–3702, DOI 10.1215/00127094-2021-0048. MR4332674 ↑15, 16
[20] Jarosław Buczyński, Adam Ginensky, and J. M. Landsberg, *Determinantal equations for secant varieties and the Eisenbud-Koh-Stillman conjecture*, J. Lond. Math. Soc. (2) **88** (2013), no. 1, 1–24, DOI 10.1112/jlms/jds073. MR3092255 ↑8
[21] Jarosław Buczyński, Tadeusz Januszkiewicz, Joachim Jelisiejew, and Mateusz Michałek, *Constructions of k-regular maps using finite local schemes*, J. Eur. Math. Soc. (JEMS) **21** (2019), no. 6, 1775–1808, DOI 10.4171/JEMS/873. MR3945742 ↑3, 15
[22] Jarosław Buczyński and Joachim Jelisiejew, *Finite schemes and secant varieties over arbitrary characteristic*, Differential Geom. Appl. **55** (2017), 13–67, DOI 10.1016/j.difgeo.2017.08.004. MR3724212 ↑17
[23] Peter Bürgisser, *The complexity of factors of multivariate polynomials*, Found. Comput. Math. **4** (2004), no. 4, 369–396, DOI 10.1007/s10208-002-0059-5. MR2097213 ↑18
[24] Peter Bürgisser, *Correction to: The complexity of factors of multivariate polynomials*, Found. Comput. Math. **20** (2020), no. 6, 1667–1668, DOI 10.1007/s10208-020-09477-6. MR4179954 ↑18
[25] Peter Bürgisser, Michael Clausen, and M. Amin Shokrollahi, *Algebraic complexity theory*, Grundlehren der mathematischen Wissenschaften [Fundamental Principles of Mathematical Sciences], vol. 315, Springer-Verlag, Berlin, 1997. With the collaboration of Thomas Lickteig, DOI 10.1007/978-3-662-03338-8. MR1440179 ↑3, 4
[26] Dustin A. Cartwright, Daniel Erman, Mauricio Velasco, and Bianca Viray, *Hilbert schemes of 8 points*, Algebra Number Theory **3** (2009), no. 7, 763–795, DOI 10.2140/ant.2009.3.763. MR2579394 ↑7, 11, 15, 16
[27] Gianfranco Casnati, Juan Elias, Roberto Notari, and Maria Evelina Rossi, *Poincaré series and deformations of Gorenstein local algebras*, Comm. Algebra **41** (2013), no. 3, 1049–1059, DOI 10.1080/00927872.2011.636643. MR3037178 ↑9
[28] Gianfranco Casnati, Joachim Jelisiejew, and Roberto Notari, *Irreducibility of the Gorenstein loci of Hilbert schemes via ray families*, Algebra Number Theory **9** (2015), no. 7, 1525–1570, DOI 10.2140/ant.2015.9.1525. MR3404648 ↑7
[29] Gianfranco Casnati and Roberto Notari, *On some Gorenstein loci in* $\mathcal{H}ilb_6(\mathbb{P}^4_k)$, J. Algebra **308** (2007), no. 2, 493–523, DOI 10.1016/j.jalgebra.2006.09.023. MR2295072 ↑5
[30] Gianfranco Casnati and Roberto Notari, *On the Gorenstein locus of the punctual Hilbert scheme of degree 11*, J. Pure Appl. Algebra **218** (2014), no. 9, 1635–1651, DOI 10.1016/j.jpaa.2014.01.004. MR3188862 ↑7

[31] Jean-Yves Charbonnel, *Projective dimension and commuting variety of a reductive Lie algebra*, arXiv:2006.12942. 6

[32] Jan Cheah, *The cohomology of smooth nested Hilbert schemes of points*, ProQuest LLC, Ann Arbor, MI, 1994. Thesis (Ph.D.)–The University of Chicago. MR2716513 ↑5

[33] Luca Chiantini, Jonathan D. Hauenstein, Christian Ikenmeyer, Joseph M. Landsberg, and Giorgio Ottaviani, *Polynomials and the exponent of matrix multiplication*, Bull. Lond. Math. Soc. **50** (2018), no. 3, 369–389, DOI 10.1112/blms.12147. MR3829726 ↑17

[34] Matthias Christandl, Fulvio Gesmundo, Vladimir Lysikov, and Vincent Steffan, *Partial degeneration of tensors*, arXiv:2212.14095. 18

[35] Austin Conner, Fulvio Gesmundo, Joseph M. Landsberg, Emanuele Ventura, and Yao Wang, *Towards a geometric approach to Strassen's asymptotic rank conjecture*, Collect. Math. **72** (2021), no. 1, 63–86, DOI 10.1007/s13348-020-00280-8. MR4204580 ↑17

[36] Austin Conner, Hang Huang, and Joseph M. Landsberg, *Bad and good news for Strassen's laser method: Border rank of* Perm_3 *and strict submultiplicativity*, Foundations of Computational Mathematics (2022). 15

[37] Don Coppersmith and Shmuel Winograd, *Matrix multiplication via arithmetic progressions*, J. Symbolic Comput. **9** (1990), no. 3, 251–280, DOI 10.1016/S0747-7171(08)80013-2. MR1056627 ↑4

[38] A. Craw, *An introduction to Hilbert schemes of points on ADE singularities*, https://people.bath.ac.uk/ac886/pubs/Hilbert_schemes_of_points_on_ADE_singularities.pdf. 9

[39] Alastair Craw, Søren Gammelgaard, Ádám Gyenge, and Balázs Szendrői, *Punctual Hilbert schemes for Kleinian singularities as quiver varieties*, Algebr. Geom. **8** (2021), no. 6, 680–704, DOI 10.14231/ag-2021-021. MR4371544 ↑9, 10

[40] Theodosios Douvropoulos, Joachim Jelisiejew, Bernt Ivar Utstøl Nødland, and Zach Teitler, *The Hilbert scheme of 11 points in* $\mathbb{A}^3$ *is irreducible*, Combinatorial algebraic geometry, Fields Inst. Commun., vol. 80, Fields Inst. Res. Math. Sci., Toronto, ON, 2017, pp. 321–352. MR3752506 ↑7, 10

[41] Alan H. Durfee, *Fifteen characterizations of rational double points and simple critical points*, Enseign. Math. (2) **25** (1979), no. 1-2, 131–163. MR543555 ↑9

[42] J. Elias and M. E. Rossi, *Isomorphism classes of short Gorenstein local rings via Macaulay's inverse system*, Trans. Amer. Math. Soc. **364** (2012), no. 9, 4589–4604, DOI 10.1090/S0002-9947-2012-05430-4. MR2922602 ↑5

[43] Geir Ellingsrud and Manfred Lehn, *Irreducibility of the punctual quotient scheme of a surface*, Ark. Mat. **37** (1999), no. 2, 245–254, DOI 10.1007/BF02412213. MR1714770 ↑7

[44] Geir Ellingsrud and Stein Arild Strømme, *On the homology of the Hilbert scheme of points in the plane*, Invent. Math. **87** (1987), no. 2, 343–352, DOI 10.1007/BF01389419. MR870732 ↑7

[45] Elden Elmanto, Marc Hoyois, Adeel A. Khan, Vladimir Sosnilo, and Maria Yakerson, *Modules over algebraic cobordism*, Forum Math. Pi **8** (2020), e14, 44, DOI 10.1017/fmp.2020.13. MR4190058 ↑3

[46] Jacques Emsalem, *Géométrie des points épais* (French, with English summary), Bull. Soc. Math. France **106** (1978), no. 4, 399–416. MR518046 ↑11

[47] Barbara Fantechi, Lothar Göttsche, Luc Illusie, Steven L. Kleiman, Nitin Nitsure, and Angelo Vistoli, *Fundamental algebraic geometry*, Mathematical Surveys and Monographs, vol. 123, American Mathematical Society, Providence, RI, 2005. Grothendieck's FGA explained, DOI 10.1090/surv/123. MR2222646 ↑4

[48] John Fogarty, *Algebraic families on an algebraic surface*, Amer. J. Math. **90** (1968), 511–521, DOI 10.2307/2373541. MR237496 ↑6, 7

[49] Terence Gaffney, *Multiple points, chaining and Hilbert schemes*, Amer. J. Math. **110** (1988), no. 4, 595–628, DOI 10.2307/2374643. MR955290 ↑8

[50] Franco Giovenzana, Luca Giovenzana, Michele Graffeo, and Paolo Lella, *A counterexample to the parity conjecture*, arXiv:2305.18191. 13

[51] A. Gorsky, N. Nekrasov, and V. Rubtsov, *Hilbert schemes, separated variables, and D-branes*, Comm. Math. Phys. **222** (2001), no. 2, 299–318, DOI 10.1007/s002200100503. MR1859601 ↑3

[52] Michel Granger, *Géométrie des schémas de Hilbert ponctuels* (French, with English summary), Mém. Soc. Math. France (N.S.) **8** (1983), 84. MR716783 ↑14

[53] Robert M. Guralnick, *A note on commuting pairs of matrices*, Linear and Multilinear Algebra **31** (1992), no. 1-4, 71–75, DOI 10.1080/03081089208818123. MR1199042 ↑8

[54] Mark Haiman, *Hilbert schemes, polygraphs and the Macdonald positivity conjecture*, J. Amer. Math. Soc. **14** (2001), no. 4, 941–1006, DOI 10.1090/S0894-0347-01-00373-3. MR1839919 ↑3

[55] Mark Haiman, *Vanishing theorems and character formulas for the Hilbert scheme of points in the plane*, Invent. Math. **149** (2002), no. 2, 371–407, DOI 10.1007/s002220200219. MR1918676 ↑3

[56] Mark Haiman and Bernd Sturmfels, *Multigraded Hilbert schemes*, J. Algebraic Geom. **13** (2004), no. 4, 725–769, DOI 10.1090/S1056-3911-04-00373-X. MR2073194 ↑4, 15

[57] Robin Hartshorne, *Connectedness of the Hilbert scheme*, Inst. Hautes Études Sci. Publ. Math. **29** (1966), 5–48. MR213368 ↑15

[58] Robin Hartshorne, *Deformation theory*, Graduate Texts in Mathematics, vol. 257, Springer, New York, 2010, DOI 10.1007/978-1-4419-1596-2. MR2583634 ↑3, 4, 12

[59] John Holbrook and Matjaž Omladič, *Approximating commuting operators*, Linear Algebra Appl. **327** (2001), no. 1-3, 131–149, DOI 10.1016/S0024-3795(00)00286-X. MR1823346 ↑8

[60] Roser Homs, Joachim Jelisiejew, Mateusz Michałek, and Tim Seynnaeve, *Bounds on complexity of matrix multiplication away from Coppersmith-Winograd tensors*, J. Pure Appl. Algebra **226** (2022), no. 12, Paper No. 107142, 16, DOI 10.1016/j.jpaa.2022.107142. MR4432327 ↑4

[61] Yi-Ning Hsiao and Andras Szenes, *A note on the E-polynomials of a stratification of the Hilbert scheme of points*, Representation theory, mathematical physics, and integrable systems, Progr. Math., vol. 340, Birkhäuser/Springer, Cham, [2021] ©2021, pp. 317–344, DOI 10.1007/978-3-030-78148-4_11. MR4391352 ↑9

[62] Xiaowen Hu, *On singular Hilbert schemes of points: local structures and tautological sheaves*, arXiv:2101.05236. 13, 14

[63] Mark E. Huibregtse, *Some elementary components of the Hilbert scheme of points*, Rocky Mountain J. Math. **47** (2017), no. 4, 1169–1225, DOI 10.1216/RMJ-2017-47-4-1169. MR3689951 ↑8

[64] Mark E. Huibregtse, *More elementary components of the Hilbert scheme of points*, (2022), https://arxiv.org/abs/2102.00494. 8

[65] Marc Härkönen, Jonas Hirsch, and Bernd Sturmfels, *Making Waves*, Matematica **2** (2023), no. 3, 593–615, DOI 10.1007/s44007-023-00056-z. MR4651750 ↑11

[66] A. Iarrobino, *The number of generic singularities*, Rice Univ. Stud. **59** (1973), no. 1, 49–51. MR345967 ↑6, 7

[67] A. Iarrobino, *Hilbert scheme of points: overview of last ten years*, Algebraic geometry, Bowdoin, 1985 (Brunswick, Maine, 1985), Proc. Sympos. Pure Math., vol. 46, Amer. Math. Soc., Providence, RI, 1987, pp. 297–320, DOI 10.1090/pspum/046.2/927986. MR927986 ↑5, 6, 12

[68] A. Iarrobino, *Reducibility of the families of* 0*-dimensional schemes on a variety*, Invent. Math. **15** (1972), 72–77, DOI 10.1007/BF01418644. MR301010 ↑7, 10

[69] Anthony Iarrobino, *Deforming complete intersection Artin algebras. Appendix: Hilbert function of* $\mathbf{C}[x, y]/I$, Singularities, Part 1 (Arcata, Calif., 1981), Proc. Sympos. Pure Math., vol. 40, Amer. Math. Soc., Providence, RI, 1983, pp. 593–608. MR713096 ↑14, 15

[70] Anthony Iarrobino, *Compressed algebras: Artin algebras having given socle degrees and maximal length*, Trans. Amer. Math. Soc. **285** (1984), no. 1, 337–378, DOI 10.2307/1999485. MR748843 ↑7, 10

[71] Anthony A. Iarrobino, *Associated graded algebra of a Gorenstein Artin algebra*, Mem. Amer. Math. Soc. **107** (1994), no. 514, viii+115, DOI 10.1090/memo/0514. MR1184062 ↑9

[72] A. Iarrobino and J. Emsalem, *Some zero-dimensional generic singularities; finite algebras having small tangent space*, Compositio Math. **36** (1978), no. 2, 145–188. MR515043 ↑7, 8, 16

[73] Anthony Iarrobino and Vassil Kanev, *Power sums, Gorenstein algebras, and determinantal loci*, Lecture Notes in Mathematics, vol. 1721, Springer-Verlag, Berlin, 1999. Appendix C by Iarrobino and Steven L. Kleiman, DOI 10.1007/BFb0093426. MR1735271 ↑5, 11

[74] Christian Ikenmeyer, *De-bordering symmetric border rank*, https://agates.mimuw.edu.pl/images/warsaw_ikenmeyer.pdf. 18

[75] Nathan Owen Ilten, *Versal deformations and local Hilbert schemes*, J. Softw. Algebra Geom. **4** (2012), 12–16, DOI 10.2140/jsag.2012.4.12. MR2947667 ↑12

[76] Joachim Jelisiejew, *Classifying local Artinian Gorenstein algebras*, Collect. Math. **68** (2017), no. 1, 101–127, DOI 10.1007/s13348-016-0183-1. MR3591467 ↑5

[77] Joachim Jelisiejew, *VSPs of cubic fourfolds and the Gorenstein locus of the Hilbert scheme of 14 points on* $\mathbb{A}^6$, Linear Algebra Appl. **557** (2018), 265–286, DOI 10.1016/j.laa.2018.08.002. MR3848270 ↑11

[78] Joachim Jelisiejew, *Elementary components of Hilbert schemes of points*, J. Lond. Math. Soc. (2) **100** (2019), no. 1, 249–272, DOI 10.1112/jlms.12212. MR3999690 ↑8

[79] Joachim Jelisiejew, *Pathologies on the Hilbert scheme of points*, Invent. Math. **220** (2020), no. 2, 581–610, DOI 10.1007/s00222-019-00939-5. MR4081138 ↑12, 14

[80] Joachim Jelisiejew and Hanieh Keneshlou, *On construction of k-regular maps to Grassmannians via algebras of socle dimension two*, To appear in Annali della Scuola Normale Superiore di Pisa, `arXiv:2112.14106`. 7, 8, 15

[81] J. Jelisiejew, M. Kool, and R. F. Schmiermann, *Behrend's function is not constant on* $\mathrm{Hilb}^n(\mathbb{C}^3)$, (2023), `https://doi.org/10.48550/arXiv.2311.05408`. 13

[82] Joachim Jelisiejew, J. M. Landsberg, and Arpan Pal, *Concise tensors of minimal border rank*, Math. Ann. (2023), DOI https://doi.org/10.1007/s00208-023-02569-y. 4, 19

[83] Joachim Jelisiejew and Tomasz Mańdziuk, *Limits of saturated ideals*, (2022), `arXiv:2210.13579`. 16

[84] Joachim Jelisiejew and Klemen Šivic, *Components and singularities of Quot schemes and varieties of commuting matrices*, J. Reine Angew. Math. **788** (2022), 129–187, DOI 10.1515/crelle-2022-0018. MR4445543 ↑6, 7, 10, 11

[85] Jesse Leo Kass, *The compactified jacobian can be nonreduced*, Bull. Lond. Math. Soc. **47** (2015), no. 4, 686–692, DOI 10.1112/blms/bdv036. MR3375936 ↑6

[86] Sheldon Katz, *The desingularization of* $\mathrm{Hilb}^4\mathbf{P}^3$ *and its Betti numbers*, Zero-dimensional schemes (Ravello, 1992), de Gruyter, Berlin, 1994, pp. 231–242. MR1292488 ↑14

[87] Steven L. Kleiman and Jan O. Kleppe, *Macaulay duality and its geometry*, `arXiv:2210.10934`. 8, 10, 11

[88] Hans Kleppe, *Deformation of schemes defined by vanishing of Pfaffians*, J. Algebra **53** (1978), no. 1, 84–92, DOI 10.1016/0021-8693(78)90207-7. MR498556 ↑7

[89] Jan O. Kleppe and Rosa M. Miró-Roig, *Unobstructedness and dimension of families of codimension 3 ACM algebras*, Algebra, geometry and their interactions, Contemp. Math., vol. 448, Amer. Math. Soc., Providence, RI, 2007, pp. 141–164, DOI 10.1090/conm/448/08663. MR2389240 ↑7

[90] Michael Kunte, *Gorenstein modules of finite length*, Math. Nachr. **284** (2011), no. 7, 899–919, DOI 10.1002/mana.200810173. MR2815960 ↑11

[91] J. M. Landsberg and Mateusz Michałek, *Abelian tensors* (English, with English and French summaries), J. Math. Pures Appl. (9) **108** (2017), no. 3, 333–371, DOI 10.1016/j.matpur.2016.11.004. MR3682743 ↑4

[92] J. M. Landsberg, *Tensors: geometry and applications*, Graduate Studies in Mathematics, vol. 128, American Mathematical Society, Providence, RI, 2012, DOI 10.1090/gsm/128. MR2865915 ↑4

[93] François Le Gall, *Powers of tensors and fast matrix multiplication*, ISSAC 2014—Proceedings of the 39th International Symposium on Symbolic and Algebraic Computation, ACM, New York, 2014, pp. 296–303, DOI 10.1145/2608628.2608664. MR3239939 ↑4

[94] Thomas Lehmkuhl and Thomas Lickteig, *On the order of approximation in approximative triadic decompositions of tensors*, Theoret. Comput. Sci. **66** (1989), no. 1, 1–14, DOI 10.1016/0304-3975(89)90141-2. MR1018840 ↑18

[95] Paolo Lella, *`MarkedSchemes.m2` - A M2 package for marked schemes and Groebner strata*, `http://www.paololella.it/publications/lm2/`. 12

[96] Paolo Lella and Margherita Roggero, *Rational components of Hilbert schemes*, Rend. Semin. Mat. Univ. Padova **126** (2011), 11–45, DOI 10.4171/RSMUP/126-2. MR2918197 ↑12

[97] Diane Maclagan and Gregory G. Smith, *Smooth and irreducible multigraded Hilbert schemes*, Adv. Math. **223** (2010), no. 5, 1608–1631, DOI 10.1016/j.aim.2009.10.003. MR2592504 ↑16

[98] T. Mańdziuk, *Identifying limits of ideals of points in the case of projective space*, Linear Algebra Appl. **634** (2022), 149–178, DOI 10.1016/j.laa.2021.11.003. MR4341635 ↑16

[99] D. Maulik, N. Nekrasov, A. Okounkov, and R. Pandharipande, *Gromov-Witten theory and Donaldson-Thomas theory. I*, Compos. Math. **142** (2006), no. 5, 1263–1285, DOI 10.1112/S0010437X06002302. MR2264664 ↑13

[100] Guerino Mazzola, *Generic finite schemes and Hochschild cocycles*, Comment. Math. Helv. **55** (1980), no. 2, 267–293, DOI 10.1007/BF02566686. MR576606 ↑7

[101] Luca Migliorini, *Homfly polynomials from the Hilbert schemes of a planar curve [after D. Maulik, A. Oblomkov, V. Shende, ...]*, Astérisque **422, Séminaire Bourbaki. Vol. 2018/2019. Exposés 1151–1165** (2020), Exp. No. 1160, 355–389, DOI 10.24033/ast. MR4224640 ↑9

[102] T. S. Motzkin and Olga Taussky, *Pairs of matrices with property L. II*, Trans. Amer. Math. Soc. **80** (1955), 387–401, DOI 10.2307/1992996. MR86781 ↑7

[103] Hiraku Nakajima, *Lectures on Hilbert schemes of points on surfaces*, University Lecture Series, vol. 18, American Mathematical Society, Providence, RI, 1999, DOI 10.1090/ulect/018. MR1711344 ↑9

[104] Nham V. Ngo and Klemen Šivic, *On varieties of commuting nilpotent matrices*, Linear Algebra Appl. **452** (2014), 237–262, DOI 10.1016/j.laa.2014.03.032. MR3201099 ↑8

[105] Alexei Oblomkov and Vivek Shende, *The Hilbert scheme of a plane curve singularity and the HOMFLY polynomial of its link*, Duke Math. J. **161** (2012), no. 7, 1277–1303, DOI 10.1215/00127094-1593281. MR2922375 ↑9

[106] Kevin C. O'Meara, John Clark, and Charles I. Vinsonhaler, *Advanced topics in linear algebra*, Oxford University Press, Oxford, 2011. Weaving matrix problems through the Weyr form. MR2849857 ↑7, 8

[107] R. Pandharipande, *A tour of the geometry of points in affine space*, (2022), DMV Jahrestagung lecture (2022). 13

[108] R. Pandharipande and R. P. Thomas, *Curve counting via stable pairs in the derived category*, Invent. Math. **178** (2009), no. 2, 407–447, DOI 10.1007/s00222-009-0203-9. MR2545686 ↑13

[109] Keith Pardue, *Deformation classes of graded modules and maximal Betti numbers*, Illinois J. Math. **40** (1996), no. 4, 564–585. MR1415019 ↑16

[110] Irena Peeva and Mike Stillman, *Toric Hilbert schemes*, Duke Math. J. **111** (2002), no. 3, 419–449, DOI 10.1215/S0012-7094-02-11132-6. MR1885827 ↑5

[111] Bjorn Poonen, *The moduli space of commutative algebras of finite rank*, J. Eur. Math. Soc. (JEMS) **10** (2008), no. 3, 817–836, DOI 10.4171/JEMS/131. MR2421162 ↑12

[112] Ritvik Ramkumar and Alessio Sammartano, *On the parity conjecture for Hilbert schemes of points on threefolds*, `arXiv:2302.02204`. 13

[113] Ritvik Ramkumar and Alessio Sammartano, *On the tangent space to the Hilbert scheme of points in* $\mathbf{P}^3$, Trans. Amer. Math. Soc. **375** (2022), no. 9, 6179–6203, DOI 10.1090/tran/8657. MR4474889 ↑13

[114] Ritvik Ramkumar and Alessio Sammartano, *Rational Singularities of Nested Hilbert Schemes*, International Mathematics Research Notices (2023). 9

[115] Kristian Ranestad and Frank-Olaf Schreyer, *The variety of polar simplices*, Doc. Math. **18** (2013), 469–505. MR3084557 ↑14

[116] Miles Reid, *The Du Val singularities* A_n, D_n, E_6, E_7, E_8, `https://homepages.warwick.ac.uk/~masda/surf/more/DuVal.pdf`. 9, 10

[117] Jørgen Vold Rennemo, *Homology of Hilbert schemes of points on a locally planar curve*, J. Eur. Math. Soc. (JEMS) **20** (2018), no. 7, 1629–1654, DOI 10.4171/JEMS/795. MR3807309 ↑9

[118] Andrea T. Ricolfi, *A sign that used to annoy me, and still does*, J. Geom. Phys. **195** (2024), Paper No. 105032, DOI 10.1016/j.geomphys.2023.105032. MR4660630 ↑13

[119] Tim Ryan and Gregory Taylor, *Irreducibility and singularities of some nested Hilbert schemes*, J. Algebra **609** (2022), 380–406, DOI 10.1016/j.jalgebra.2022.05.037. MR4457391 ↑7

[120] Francisco Santos, *Non-connected toric Hilbert schemes*, Math. Ann. **332** (2005), no. 3, 645–665, DOI 10.1007/s00208-005-0643-5. MR2181765 ↑16

[121] Matthew Satriano and Andrew P. Staal, *Small elementary components of Hilbert schemes of points*, Forum Math. Sigma **11** (2023), Paper No. e45, 36, DOI 10.1017/fms.2023.42. MR4599219 ↑8, 12

[122] Edoardo Sernesi, *Deformations of algebraic schemes*, Grundlehren der mathematischen Wissenschaften [Fundamental Principles of Mathematical Sciences], vol. 334, Springer-Verlag, Berlin, 2006. MR2247603 ↑3, 4, 5

[123] I. R. Shafarevich, *Degeneration of semisimple algebras*, Comm. Algebra **29** (2001), no. 9, 3943–3960, DOI 10.1081/AGB-100105983. Special issue dedicated to Alexei Ivanovich Kostrikin. MR1857022 ↑3

[124] I. R. Shafarevich, *Deformations of commutative algebras of class* 2 (Russian), Algebra i Analiz **2** (1990), no. 6, 178–196; English transl., Leningrad Math. J. **2** (1991), no. 6, 1335–1351. MR1092534 ↑8, 11

[125] Klemen Šivic, *On varieties of commuting triples III*, Linear Algebra Appl. **437** (2012), no. 2, 393–460, DOI 10.1016/j.laa.2011.08.015. MR2921710 ↑8

[126] Lei Song, *On the universal family of Hilbert schemes of points on a surface*, J. Algebra **456** (2016), 348–354, DOI 10.1016/j.jalgebra.2016.03.005. MR3484148 ↑9

[127] Samuel Stark, *On the Quot scheme* $\mathit{Quot}^l(\mathcal{E})$, (2021), `https://arxiv.org/abs/2107.03991`. 6

[128] Jan Stevens, *Deformations of singularities*, Lecture Notes in Mathematics, vol. 1811, Springer-Verlag, Berlin, 2003, DOI 10.1007/b10723. MR1976016 ↑14

[129] Stein Arild Strømme, *Elementary introduction to representable functors and Hilbert schemes*, Parameter spaces (Warsaw, 1994), Banach Center Publ., vol. 36, Polish Acad. Sci. Inst. Math., Warsaw, 1996, pp. 179–198. MR1481486 ↑3, 4

[130] Bernd Sturmfels, *Four counterexamples in combinatorial algebraic geometry*, J. Algebra **230** (2000), no. 1, 282–294, DOI 10.1006/jabr.1999.7950. MR1774768 ↑13

[131] Michał Szachniewicz, *Non-reducedness of the Hilbert schemes of few points*, 2021, arXiv:2109.11805. 12

[132] Robert Szafarczyk, *New elementary components of the Gorenstein locus of the Hilbert scheme of points*, Comm. Algebra **51** (2023), no. 6, 2688–2704, DOI 10.1080/00927872.2023.2171425. MR4563459 ↑11, 12

[133] André Weil, *Scientific works. Collected papers. Vol. I (1926–1951)* (French), Springer-Verlag, New York-Heidelberg, 1979. MR537937 ↑3

[134] Hermann Weyl, *Generalized Riemann matrices and factor sets*, Ann. of Math. (2) **37** (1936), no. 3, 709–745, DOI 10.2307/1968485. MR1503306 ↑3

[135] Virginia Vassilevska Williams, *Multiplying matrices faster than Coppersmith-Winograd [extended abstract]*, STOC'12—Proceedings of the 2012 ACM Symposium on Theory of Computing, ACM, New York, 2012, pp. 887–898, DOI 10.1145/2213977.2214056. MR2961552 ↑4

[136] Xudong Zheng, *Irreducibility of the Hilbert schemes of points on surfaces with Kleinian singularities*, Comm. Algebra **51** (2023), no. 1, 98–115, DOI 10.1080/00927872.2022.2089993. MR4525283 ↑9

Institute of Mathematics, University of Warsaw, Banacha 2, 02-097 Warsaw, Poland
Email address: `j.jelisiejew@uw.edu.pl`

Contemporary Mathematics
Volume **805**, 2024
https://doi.org/10.1090/conm/805/16123

Commuting Jordan types: A survey

Leila Khatami

Abstract. In this paper, we survey the progress in the problem of finding the maximum commuting nilpotent orbit that intersects the centralizer of a given nilpotent matrix.

1. Introduction

Let k be an infinite field and n be a positive integer. We denote the set of all $n \times n$ matrices with entries in k by $\mathcal{M}at_n(\mathsf{k})$, and the set of nilpotent matrices in $\mathcal{M}at_n(\mathsf{k})$ by $\mathcal{N}ilp_n(\mathsf{k})$. For a matrix $A \in \mathcal{N}ilp_n(\mathsf{k})$, let $\mathcal{N}_A$ be the nilpotent commutator of A, that is $\mathcal{N}_A = \{B \in \mathcal{N}ilp_n(\mathsf{k}) \mid [A, B] = 0\}$.

By Jordan decomposition theorem, there is a bijection between the set of partitions of n and $\mathrm{GL}_n(\mathsf{k})$-orbits of $\mathcal{N}ilp_n(\mathsf{k})$. For a nilpotent matrix $A \in \mathcal{N}ilp_n(\mathsf{k})$, we refer to the partition of n determined by the size of Jordan blocks in the Jordan canonical form of A as the *Jordan type* of A and we denote it by P_A. It is well known that the nilpotent commutator of a nilpotent matrix is an irreducible algebraic variety (for example see [**5**, Lemma 2.3]). This in particular implies that the nilpotent commutator $\mathcal{N}_A$ contains an open dense subset over which the Jordan type is constant. In the past two decades, many authors have studied the characteristics and properties of the generic Jordan type in the nilpotent commutator of a nilpotent matrix. Many efforts were motivated by proving a conjecture of P. Oblak in which the the generic Jordan type in the nilpotent commutator of a nilpotent matrix is explicitly determined by the Jordan type of the matrix itself. The conjecture was finally proved in general by R. Basili in her paper [**6**], published in 2022. In this survey, we provide a general overview of those results. A wide variety of algebraic and combinatorial tools are used to establish the results that we survey in this paper. In the interest of brevity, we leave out technical details and keep the focus of this paper on the results themselves.

We should note that the study of commuting nilpotent matrices in the past two decades has not been limited to the results surveyed in this paper. For example, see the work of V. Baranovsky, R. Basili and A. Iarrobino in [**4, 5, 7**] where the study of pairs of commuting nilpotent matrices has been related to the punctual Hilbert scheme of a plane. Pairs of commuting nilpotent orbits have also been studied by A. Premet, G. McNinch and D. Panyushev in the broader context of Lie algebras (see [**25, 30, 31**]). Triples (or more generally d-tuples) of commuting

---

2020 *Mathematics Subject Classification.* Primary 15A27, Secondary 05E40, 06A11, 13E10, 15A21.

nilpotent matrices have also been studied by R. Guralnick, B. Sethuranam, K. Šivic, W. Haboush and D. Hyeon, and others (see [**13**, **14**, **26**, **27**, **32**–**35**]).

Although the problem of explicitly determining the generic Jordan type in the nilpotent commutator of a nilpotent matrix has been resolved, there are several related problems that remain open. Determining Jordan types of nilpotent matrices with the same fixed generic commuting Jordan type, as well as the equation of loci of such matrices have been studied and answered for the case that the fixed generic Jordan type has two parts (see [**16**]), but the general case is still wide open (with the exception of limited partial results established for the 3-part case). The general question of finding pairs of Jordan types corresponding to commuting nilpotent matrices remains open, as well, and only partial results in special cases are established (for example see [**9**, **29**]). Another related longstanding open problem is the study of possible Jordan types of the multiplication maps $m_\ell : A \to A$, where A is an Artinian algebra and ℓ is in A. In the past decade, there has been significant progress in this area and the problem continues to be studied extensively. For example see [**1**–**3**, **11**, **17**–**19**, **24**]. We refer the reader to Section 6 for more details and additional references about these open problems and future directions.

2. Preliminaries

Throughout this paper, we write a partition as a weakly decreasing sequence of positive integers. If a partition includes ℓ parts of size k, in writing P, we often include k^ℓ (instead of writing ℓ copies of k). For example, we write $(3^2, 1^3)$ for $(3, 3, 1, 1, 1)$.

Definition 2.1. Let n be a positive integer and P and Q be partitions of n. We say partitions P and Q *commute* if there exist commuting nilpotent matrices A and B in $\mathcal{N}ilp_n(\mathsf{k})$ with Jordan types P and Q, respectively.

The Jordan type of a nilpotent matrix can be determined by the ranks of its powers.

Lemma 2.2. ([**7**, Lemma 1.1]) *Suppose that $A \in \mathcal{N}ilp_n(\mathsf{k})$. For $i \geq 1$, let $d_i = \operatorname{rank} A^{i-1} - \operatorname{rank} A^i$. Then the partition $d = (d_1, d_2, \dots)$ and the Jordan type P_A of A are conjugate partitions.*

Example 2.3. Let J_4 be the 4×4 nilpotent Jordan block. Obviously the Jordan type of J_P is (4). Now consider $B = J_4^2$. It is easy to see that $\operatorname{rank} B = 2$, while $B^2 = 0$. Thus, the Jordan type of B is a conjugate of $(2, 2)$, which is incidentally $(2, 2)$ itself. This in particular shows that partitions (4) and $(2, 2)$ commute. Note that although partitions $(2, 2)$ and (4) commute, Jordan matrix $J_{(2,2)}$, the nilpotent Jordan matrix corresponding to the partition $(2, 2)$, and J_4 do not commute.

Suppose that matrices A and B commute. Then for every matrix A' in the $\mathrm{GL}_n(\mathsf{k})$-orbit of A, there exists a matrix B' in the $\mathrm{GL}_n(\mathsf{k})$-orbit of B such that A' and B' commute (if $A' = P^{-1}AP$, simply let $B' = P^{-1}BP$). However, as the example above also shows, this does not imply that every matrix in the orbit of A must commute with all matrices in the orbit of B.

We can extend the approach we took in Example 2.3 to easily find Jordan types of powers of a single Jordan block. This will immediately give rise to a family of partitions of n that commute with partition (n).

Let J_n be the $n \times n$ nilpotent Jordan block. Then multiplying any $n \times n$ matrix A by J_n shifts the components of each column of A one position up, with a zero in the last position (we work with upper triangular Jordan blocks). In particular, for $1 \leq k \leq n$, the matrix J_n^k has 1's on its k-diagonal and 0's everywhere else. Consequently,

$$\operatorname{rank} J_n^k = \begin{cases} n-k, & 1 \leq k \leq n, \\ 0, & \text{otherwise.} \end{cases}$$

Therefore, if non-negative integers x and r are such that $0 \leq r < k$ and $n = kx + r$, then $(\operatorname{rank} \left(J_n^k\right)^i)_{i \geq 0} = (n, n-k, n-2k, \ldots, n-kx, 0)$. Thus the difference sequence, which is conjugate to the Jordan type of J_n^k is (k^x, r). Therefore, the Jordan type of J_n^k is the partition $\left((x+1)^r, x^{k-r}\right)$. Note that this is the unique partition of n consisting of exactly k "almost equal" parts.

Definition 2.4. A partition is called *almost rectangular* if its largest and smallest parts differ by at most 1. For a positive integer n and $1 \leq k \leq n$, the almost rectangular partition of n with k parts is denoted by $[n]^k$. If r denotes the remainder of dividing n by k, then the almost rectangular partition $[n]^k$ consists of r (possibly 0) parts of size $\lfloor \frac{n}{k} \rfloor + 1$ and $k - r$ parts of size $\lfloor \frac{n}{k} \rfloor$.

Every partition P can be decomposed into a union (concatenation) of its almost rectangular subpartitions. We denote the minimum number of non overlapping almost rectangular subpartitions required for such a decomposition of P by r_P.

Example 2.5. Partitions $[6]^2 = (3,3)$ and $[6]^4 = (2^2, 1^2)$ are both almost rectangular partitions of $n = 6$.

Now consider $P = (5, 3, 2^2, 1^2)$. Then we can write

$$P = (\underline{5}, \underline{3, 2^2}, \underline{1^2}) = ([5]^1, [7]^3, [2]^2) \text{ or } P = (\underline{5}, \underline{3}, \underline{2^2, 1^2}) = ([5]^1, [3]^1, [6]^4).$$

The partition can not be decomposed into fewer almost rectangular subpartitions. Therefore, $r_P = 3$.

3. The generic commuting partition

Definition 3.1. Let n be a positive integer and $\mathcal{P}(n)$ be the set of all partitions of n. We define the map $\mathcal{Q} : \mathcal{P}(n) \to \mathcal{P}(n)$ to send each partition P to the Jordan type of a generic element in the irreducible variety $\mathcal{N}_{J_P}$, where J_P is the nilpotent Jordan matrix with Jordan type P. In other words, $\mathcal{Q}(P)$ is the generic Jordan type that commutes with P.

In this section, we give an overview of the properties of the map $\mathcal{Q}$ above.

Recall that the *dominance order* is a partial order defined on the set of all partitions of n as follows. For partitions $P = (p_1, p_2, \ldots)$ and $Q = (q_1, q_2, \ldots)$ in $\mathcal{P}(n)$, we say $Q \geq P$ if and only if for all k,

$$\sum_{1 \leq i \leq k} q_i \geq \sum_{1 \leq i \leq k} p_i.$$

Equivalently, $Q \geq P$ if and only if for all k, $\operatorname{rank} {J_Q}^k \geq \operatorname{rank} {J_P}^k$.

Using semicontinuity of the ranks of powers of a matrix and the definition of $\mathcal{Q}(P)$ as the Jordan type of a generic element in the nilpotent commutator of J_P, we get the following statement.

PROPOSITION 3.2. *Let n be a positive integer and P be a partition of n. If partition Q commutes with P, then $\mathcal{Q}(P) \geq Q$.*

In other words, $Q(P)$ is the maximum partition (in the dominance order) among Jordan types of matrices in the nilpotent commutator of the Jordan matrix J_P (or any other matrix with Jordan type P).

In [**7**], R. Basili and A. Iarrobino characterize partitions that are stable under the action of $\mathcal{Q}$. In [**30**, Theorem 2.1], Panyushev independently proved the same result in a more general context.

THEOREM 3.3. *([**7**, Theorem 1.12]) For a partition P, $\mathcal{Q}(P) = P$ if and only if parts of P differ pairwise by at least 2.*

For a pair of commuting nilpotent matrices A and B in $\mathcal{N}ilp_n(\mathsf{k})$, consider the subalgebra $\mathsf{k}[A, B]$ of $\mathcal{M}at_n(\mathsf{k})$ that is generated by A and B. Then $\mathsf{k}[A, B]$ is a commutative Artinian k-algebra. Let $H(\mathsf{k}[A, B])$ denote the Hilbert function of this algebra and assume that $\operatorname{char} \mathsf{k}$ is either 0 or is greater than the socle degree of $H(\mathsf{k}[A, B])$. In [**7**, Theorem 2.16], R. Basili and A. Iarrobino prove that if $\dim_{\mathsf{k}} \mathsf{k}[A, B] = n$, then the Jordan type of the multiplication by the general linear form in $\mathsf{k}[A, B]$, as well as its associated graded algebra, is the conjugate of the partition obtained from $H(\mathsf{k}[A, B])$. On the other hand, in [**23**, Corollary 5], T. Košir and P. Oblak show that if B is a generic element in $\mathcal{N}_A$, then the algebra $\mathsf{k}[A, B]$ is Gorenstein of vector space dimension n and therefore also complete intersection. Equivalently, in this case the action of $\mathsf{k}[A, B]$ is both cyclic and acyclic. Using Macaulay's classical characterization of Hilbert functions of complete intersection Artinian algebras, T. Košir and P. Oblak then prove the following statement.

THEOREM 3.4. *([**23**, Theorem 6]) Assume that* $\operatorname{char} \mathsf{k} = 0$ *or* $\operatorname{char} \mathsf{k} > n$. *Then the map $\mathcal{Q}$ defined in Definition* 3.1 *is idempotent, i.e., for every partition P, $\mathcal{Q}(\mathcal{Q}(P)) = \mathcal{Q}(P)$.*

In [**5**], R. Basili determines the number of parts of $\mathcal{Q}(P)$ by calculating the rank of a generic matrix in the nilpotent commutator of J_P. Also see [**8**, Theorem 2.17] for another proof.

THEOREM 3.5. *([**5**, Proposition 2.4]) Partition $\mathcal{Q}(P)$ has r_P parts.*

Note that an immediate consequence of this result is the fact that a partition P of n commutes with the maximal partition (n) if and only if P is almost rectangular.

In [**28**, Theorem 12], P. Oblak gives an explicit formula for the largest part of $\mathcal{Q}(P)$. Although her result was proved over a field of characteristic 0, it was later generalized to an arbitrary infinite field by A. Iarrobino and the author in [**15**, Corollary 3.10]. In [**22**], the author gives an explicit formula for the smallest part of $\mathcal{Q}(P)$ over an arbitrary infinite field.

These partial results in particular determined the partition $\mathcal{Q}(P)$ for every partition P with $r_P \leq 3$. They also confirmed a conjecture of P. Oblak that was finally proved by R. Basili in [**6**] in full generality after being partially proved in [**15**] and [**21**]. These results are closely tied to a recursive process introduced by P. Oblak. In the next section, we discuss the process, as well as an algorithm to explicitly determine $\mathcal{Q}(P)$ from P.

4. Oblak process

In this section, we review a recursive process, proposed by P. Oblak, which she conjectured could be used to explicitly determine the generic commuting partition $\mathcal{Q}(P)$ in terms of P. As we noted earlier, this conjecture was partially confirmed through the work of several authors and was finally proved in general by R. Basili in [**6**]. The material presented in this section is based on P. Oblak's original ideas but the notations and the set-up are mainly based on [**8**], [**15**], [**22**] and [**21**].

The poset $\mathcal{D}_P$. Let n be a positive integer and P be a partition of n. We write $P = (p_1^{n_1}, p_2^{n_2}, \dots, p_t^{n_t})$, where $p_1 > \cdots > p_t$ and $n_i > 0$, for all i. We define the directed graph $\mathcal{D}_P$ as follows.

There are n vertices in $\mathcal{D}_P$. The vertices are arranged in rows such that each row corresponds to a part of P. The rows are arranged in a weakly decreasing order from top to bottom of $\mathcal{D}_P$. In other words, there are n_1 rows of length p_1 followed by n_2 rows of length p_2 below them, etc. We label each vertex by a triple (u, p_i, k), where for each i, $1 \leq u \leq p_i$ and $1 \leq k \leq n_i$. The vertex (u, p_i, k) corresponds to the u-th vertex in the k-th length-p_i row of $\mathcal{D}_P$. For each i, the vertices in the n_i rows of length p_i are labeled in a way that the first (respectively, last) component of the triple is increasing when we go from left to right (from top to bottom, respectively).

For $1 \leq i < t$, there are four families of edges in $\mathcal{D}_P$.

- For each $1 \leq u \leq p_{i+1}$, there is an edge from $(u, p_i, 1)$, the u-th vertex in the top row corresponding to p_i, to (u, p_{i+1}, n_i), the u-th vertex in the bottom row corresponding to p_{i+1} (dashed arrows in Figure 1).
- For each $1 \leq u \leq p_{i+1}$, there is an edge from $(u, p_{i+1}, 1)$, the u-th vertex in the top row corresponding to p_{i+1} to $(u + p_i - p_{i+1}, p_i, n_i)$, the $u + (p_i - p_{i+1})$-th vertex in the bottom row corresponding to p_i (dotted arrows in Figure 1).
- For each $1 \leq u \leq p_i$ and each $1 \leq k < n_i$, there is an edge from (u, p_i, k), the u-th vertex in the k-th row corresponding to p_i, to $(u, p_i, k-1)$ the u-th vertex in the $(k-1)$-st row corresponding to p_i (solid arrows in Figure 1).
- If p_i is an isolated part, i.e, $p_{i-1} - p_i > 1$ and $p_i - p_{i+1} > 1$, then for every $1 \leq u < p_i$, the directed graph $\mathcal{D}_P$ also includes an edge from $(u, p_i, 1)$, the u-th vertex in the top row corresponding to p_i, to $(u+1, p_i, n_i)$, the $(u+1)$-st vertex in the bottom row corresponding to p_i (dash-dotted arrows in arrows in Figure 1). Note that if $n_i = 1$, then these edges go from left to right in the same row (the only row of $\mathcal{D}_P$ corresponding to p_i).

In Figure 1, we illustrate the poset $\mathcal{D}_P$ for $P = (6, 4^2, 3^2, 2^2, 1)$. We have only labeled a few vertices of the poset by their corresponding triples to illustrate the labeling logic without overcrowding the figure.

The directed graph $\mathcal{D}_P$ is in fact the covering edge graph of a partially ordered set (poset), which we also denote by $\mathcal{D}_P$. For elements v and v' in the poset, $v > v'$ if and only if there is a directed path from vertex v to vertex v' in the directed graph $\mathcal{D}_P$. The partial order is defined over a basis of k^n in close connection with a maximal nilpotent subalgebra of the centralizer of J_P. The generic Jordan type in this algebra is $\mathcal{Q}(P)$, as well. We refer the reader to [**8**] and [**21**] for details. Roughly speaking, the edges in $\mathcal{D}_P$ correspond to non-zero entries in a generic (upper triangular) matrix in the nilpotent commutator of J_P. As a simple

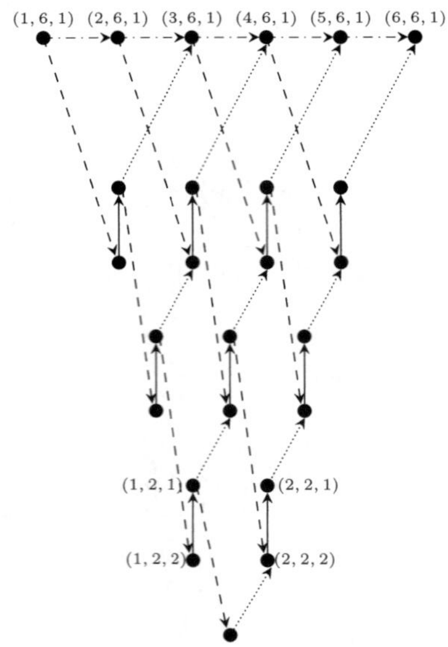

FIGURE 1. The directed graph $\mathcal{D}_P$ corresponding to the partition $P = (6, 4^2, 3^2, 2^2, 1)$.

$$B = \left(\begin{array}{ccc|c} 0 & a & 0 & b \\ 0 & 0 & a & 0 \\ 0 & 0 & 0 & 0 \\ \hline 0 & 0 & c & 0 \end{array}\right)$$

Generic matrix in the
Nilpotent commutator of $J_{(3,1)}$

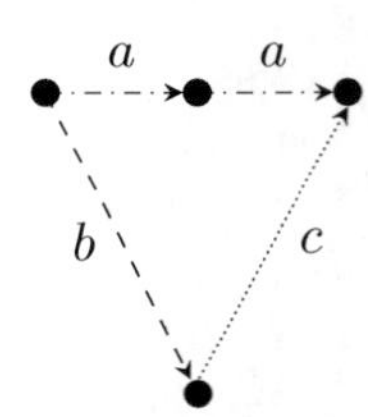

$\mathcal{D}_{(3,1)}$

FIGURE 2. A generic matrix in the nilpotent commutator of J_P, as well as the covering edge graph of poset $\mathcal{D}_P$ for $P = (3, 1)$ and the poset

example, consider partition $P = (3, 1)$. Calculations show that a generic element in the nilpotent commutator of J_P is a matrix B of the form given in Figure 2, where a, b, c are general enough. The Figure also illustrates $\mathcal{D}_P$ with each edge labeled by the corresponding entry in B.

Recall that a totally ordered subset of a poset is called a *chain.* In [**12**], Gansner proves that in a finite poset S, the Jordan type of a generic element in the incidence algebra of S, is obtained by the difference sequence of the sequence of $\{c_k\}_k$, where c_k is the maximal cordiality of a union of k chains in the poset. We denote this partition by $\lambda(S)$ (see [**10**]).

As our example above illustrates, each element of the nilpotent commutator of J_P is a matrix in the incidence algebra of $\mathcal{D}_P$. However, additional conditions on entries of the matrix must be imposed in order for the commuting condition to be satisfied. Consequently, the partition $\mathcal{Q}(P)$ is dominated by $\lambda(\mathcal{D}_P)$ (see [**15**]). P. Oblak's recursive process gives an algorithm to determine $\mathcal{Q}(P)$ from the poset $\mathcal{D}_P$ by studying special chains in the poset. These chains correspond to almost rectangular subpartitions of P.

DEFINITION 4.1. Let n be a positive integer and P be a partition of n. For a positive integer p, let $N_P(p)$ denote the (potentially zero) number of parts of P of size p. Suppose that positive integer p is such that $N_P(p) > 0$, i.e., P has at least one part of size p. Let $R_P(p) = \left(p^{N_P(p)}, (p-1)^{N_P(p-1)}\right)$ be the almost rectangular subpartition of P with greatest part of size p. Note that P may not include any parts of size $p-1$. If that is the case, then $N_P(p-1) = 0$ and $R_P(p)$ will consist of $N_P(p)$ parts of size p.

In $\mathcal{D}_P$, we define a U-chain to be a chain that consists of all vertices in $\mathcal{D}_P$ that correspond to an almost rectangular subpartition of P, as well as the first and the last vertex in any row above that almost rectangular subpartition (see Figure 3 for an example). For a positive integer p, we refer to the U-chain corresponding to the almost rectangular subpartition $R_P(p)$ by $U_P(p)$. In other words, $U_P(p)$ is defined as the set consisting of all vertices of $\mathcal{D}_P$ corresponding to $R_P(p)$, as well as the first and the last vertex in every row of $\mathcal{D}_P$ that is longer than p (see Figure 3 and Example 4.2).

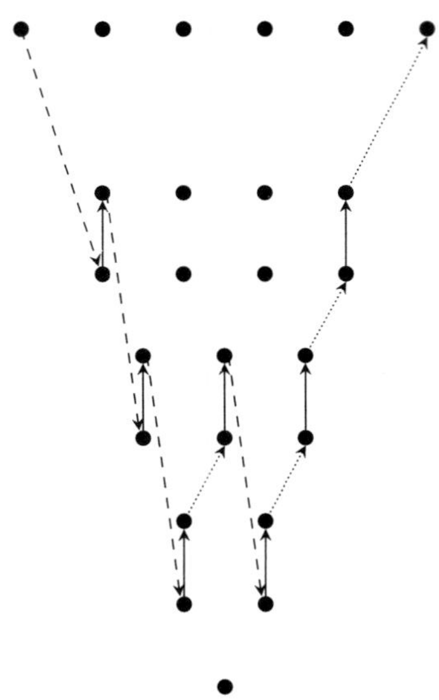

FIGURE 3. The U-chain $U_P(3)$ corresponding to the almost rectangular subpartition $R_P(3) = (3^2, 2^2)$ of $P = (6, 4^2, 3^2, 2^2, 1)$.

Evidently, the cardinality of $U_P(p)$, which we denote by $\mathfrak{u}_P(p)$, is $\mathfrak{u}_P(p) = pN_P(p) + (p-1)N_P(p-1) + 2\sum_{j>p} N_P(j)$. Here the summand $pN_P(p) + (p-1)N_P(p-1)$ is the size of the almost rectangular subpartition $R_P(p)$, while $\sum_{j>p} N_P(j)$ is the number of parts of P that are greater than p.

EXAMPLE 4.2. Figure 3, illustrates the U-chain $U_P(3)$ for $P = (6, 4^2, 3^2, 2^2, 1)$. Here the almost rectangular subpartition $R_P(3)$ is $(3^2, 2^2)$ and

$$\mathfrak{u}_P(3) = 3(2) + 2(2) + 2(2+1) = 16.$$

For the same partition $P = (6, 4^2, 3^2, 2^2, 1)$, there are other U-chains in $\mathcal{D}_P$ corresponding to almost rectangular subpartitions $R_P(6) = (6)$, $R_P(4) = (4^2, 3^2)$, $R_P(2) = (2^2, 1)$, and even $R_P(1) = (1)$. We have

$$\begin{aligned} \mathfrak{u}_P(6) &= \ 6(1) = 6, \\ \mathfrak{u}_P(4) &= \ 4(2) + 3(2) + 2(1) = 16, \\ \mathfrak{u}_P(3) &= \ 3(2) + 2(2) + 2(2+1) = 16, \\ \mathfrak{u}_P(2) &= \ 2(2) + 1(1) + 2(2+2+1) = 15, \text{ and} \\ \mathfrak{u}_P(1) &= \ 1(1) + 2(2+2+2+1) = 15. \end{aligned}$$

Oblak process. The Oblak process for a partition P is a recursive process to obtain $\mathcal{Q}(P)$ by finding a U-chain in $\mathcal{D}_P$ with maximum possible cardinality, removing it from $\mathcal{D}_P$ to obtain a new partition, and repeating the process until all vertices of $\mathcal{D}_P$ are accounted for.

Let $P = (p_1^{n_1}, \ldots, p_t^{n_t})$ such that $p_1 > \cdots > p_t$ and $n_i > 0$ for $1 \leq i \leq t$. Assume that $p \in \{p_1, \ldots, p_t\}$. Consider the partition P_1 determined by the number of remaining vertices in $\mathcal{D}_P - U_P(p)$. In other words, P_1 is obtained from P by removing the almost rectangular subpartition $R_P(p)$, keeping the subpartition of P that is below $R_P(p)$ unchanged, and shortening each part of P that is above $R_P(p)$ by 2. If we write $P_1 = (q_1^{m_1}, \ldots, q_t^{m_t})$, then for $1 \leq i \leq t$, we have

$$m_i = \begin{cases} 0, & \text{if } p_i \in \{p, p-1\}, \\ n_i, & \text{otherwise}, \end{cases} \quad \text{and} \quad q_i = \begin{cases} p_i - 2, & \text{if } p_i > p, \\ p_i, & \text{if } p_i < p - 1. \end{cases}$$

Suppose that $\mathcal{D}_P = U_1 \cup \cdots \cup U_r$, where U_i is the maximum U-chain associated with the partition obtained in the i-th step of the process. P. Oblak conjectured that the partition $(|U_1|, \ldots, |U_r|)$ is in fact the generic commuting partition $\mathcal{Q}(P)$. As we saw in Example 4.2, there may be more than one U-chain with maximum cardinality in each poset. Initially, P. Oblak addressed this issue by imposing an additional condition on the particular maximum U-chain that must be chosen in each step of the process. However, in [**21**], the author proves that the restriction is not necessary and the partition described above is indeed independent of the choice of the maximum U-chains and is uniquely determined by P. Over the years since P. Oblak made her conjecture, results by several authors confirmed the conjecture (see [**28**], [**23**], [**8**], [**15**], [**21**], and [**22**]). The conjecture was finally proved in full generality by R. Basili in [**6**, Theorem 1.2].

EXAMPLE 4.3. Revisiting the example $P = (6, 4^2, 3^3, 2^2, 1)$, as we saw, there are two maximum U-chains in $\mathcal{D}_P$, namely $U_P(4)$, corresponding to the almost rectangular subpartition $(4^2, 3^2)$, and $U_P(3)$, corresponding to the almost rectangular subpartition $(3^2, 2^2)$. Both these chains have cardinality 16.

Assume that we start the Oblak process by removing $U_P(4)$ from $\mathcal{D}_P$. This eliminates parts of sizes 4 and 3 in P, shortens the part 6 of P (the only one greater than 4) by 2, and keeps parts 2 and 1 (those smaller than 3) unchanged. This leads to a new partition $P_1 = (4, 2^2, 1)$. Next, we perform the Oblak step for this new partition. Here, the only maximum U-chain is the one corresponding to the almost rectangular subpartition $(2^2, 1)$. The cardinality of this U-chain is $2(2) + 1(1) + 2(1) = 7$. Removing the U-chain from $\mathcal{D}_{P_1}$, we get partition $P_2 = (2)$, in which the maximum U-chain (which is the whole partition) has cardinality 2. Thus the process gives rise to the partition $(16, 7, 2)$, which is indeed the generic partition commuting with P. See Figure 4, where for simplicity, we have not

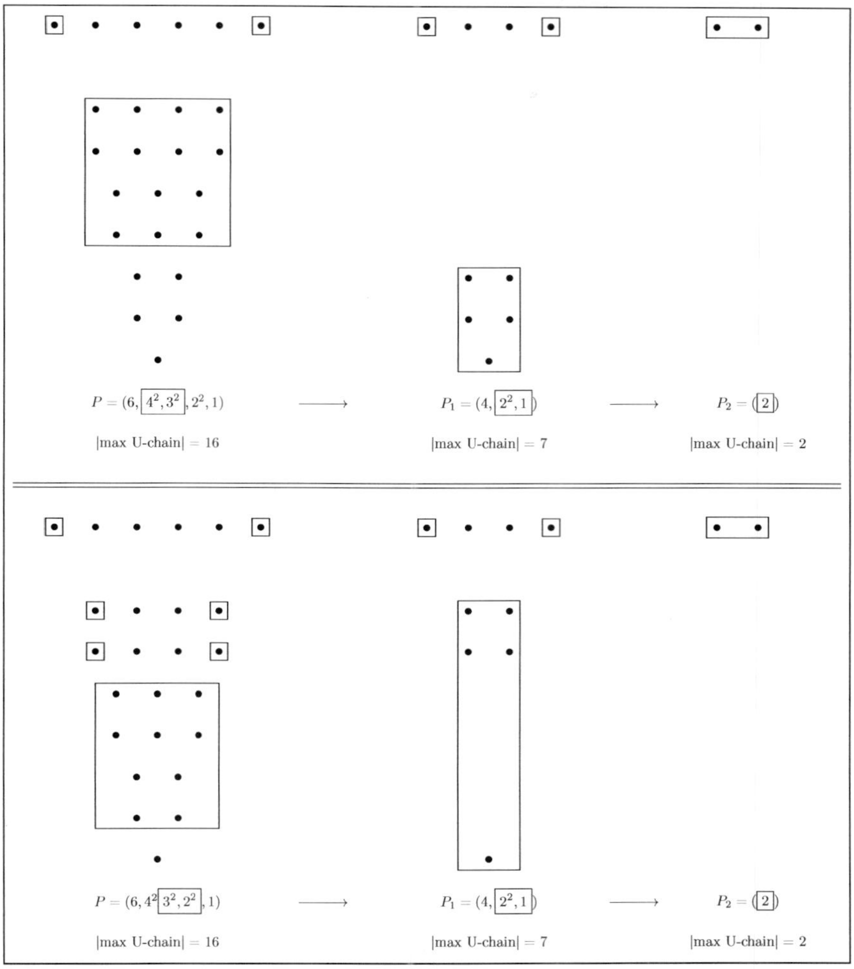

FIGURE 4. Performing Oblak recursive process, in two different ways, to find the generic commuting Jordan type $\mathcal{Q}(P) = (16, 7, 2)$ for partition $P = (6, 4^2, 3^2, 2^2, 1)$

included directed edges and have simply boxed vertices that correspond to the maximum U-chain selected in each step of the process. It is worth noting that the U-chain selected in each step of Oblak process is not necessarily a chain in the original poset. For example, in the example represented in Figure 4, the maximum U-chain for P_1 is not a chain in the original poset $\mathcal{D}_P$. Recall that two elements in the poset of P are comparable if and only if there is a directed path between their corresponding vertices in the directed graph $\mathcal{D}_P$. As Figure 1 illustrates, in the poset of $P = (6, 4^2, 3^2, 2^2, 1)$, vertices $(2, 6, 1)$ and $(1, 2, 1)$ are not comparable while they are both in the maximum U-chain selected in P_1 (the second step of the process).

There is an alternative way to perform Oblak process for P, as well. We can start the process by removing the maximum U-chain $U_P(3)$ from $\mathcal{D}_P$. Then continue the process as illustrated in Figure 4. We can observe that in this example, we arrive at partition $(16, 7, 2)$, regardless of the choice of the maximum U-chain in the Oblak process.

5. Partitions with the same generic commuting partition

In previous sections, we gave an overview of results about the map $\mathcal{Q}$. In this section, we review the studies focusing on the inverse map $\mathcal{Q}^{-1}$. Recall that by Theorems 3.3 and 3.4, a partition Q is in the image of the map $\mathcal{Q}$ if and only if parts of Q differ pairwise by at least 2, i.e., Q is stable. Fixing a stable partition Q, what is $\mathcal{Q}^{-1}(Q)$, the set of all partition with the same generic commuting partition Q?

In [**29**], P. Oblak conjectured that if $Q = (u, u-r)$, where $r \geq 2$ and $u - r \geq 1$, then $\mathcal{Q}^{-1}(Q)$ consists of $(u-r) \times (r-1)$ partitions. She proved the conjecture for $2 \leq r \leq 5$. In [**16**], A. Iarrobino, B. Van Steirteghem, R. Zhao, and the author prove the conjecture in general. Additionally, we showed that partitions in $\mathcal{Q}^{-1}(Q)$ can be arranged in a $(r-1) \times (u-r)$ table such that the partition in the k-th row and ℓ-th column of the table has $k + \ell$ parts. In [**16**], we also made the following general conjecture.

CONJECTURE 5.1. *(The Box Conjecture) Let n be a positive integer and $Q = (q_1, \dots, q_k)$ be a partition of n such that for $q_i - q_{i+1} \geq 2$, for $1 \leq i < k$. Set*

$$s_i = \begin{cases} q_i - q_{i+1} - 1, & \text{if } 1 \leq i < k, \\ q_k, & \text{if } i = k. \end{cases}$$

Then $\mathcal{Q}^{-1}(Q)$ consists of $\prod_{1 \leq i \leq k} s_i$ partitions. These partitions can be arranged in a $s_1 \times \cdots \times s_k$ "box", $\mathcal{B}(Q) = \{P_{i_1,\dots,i_k} \mid 1 \leq i_j \leq s_j\}$, such that $P_{i_1,\dots,i_k}$ has $\sum_{1 \leq j \leq k} i_j$ parts.

In [**16**], the conjecture is proved for every stable partition $Q = (u+s, u, u-r)$, such that $s \leq 4$.

6. Open problems

We conclude this survey by stating a few active related directions and open problems for future studies.

One direction, still focusing on the generic commuting partition, is to tackle the Box Conjecture (Conjecture 5.1), which remains open in general. Given the nature of the statement, it is reasonable to hope for an inductive argument. In fact, partial results given in [**16**] for the case $k = 3$ solve the problem by "decomposing" the corresponding 3-dimensional box into multiple 2-dimensional tables (sheets). However, the subtleties of the map $\mathcal{Q}(P)$, which are also captured in Oblak process, make this approach quite involved. So a general solution to the problems may require novel tools. The loci of nilpotent matrices whose Jordan types are in $\mathcal{Q}^{-1}(Q)$, for a stable partition Q with 2 have been studied by M. Boij, A. Iarrobino, B. Van Steirteghem and the author. A complete solution to that problem and its expansion to the case of Q with more than two parts remains open.[1]

[1]While this paper was in print, J. Irving, T. Košir, and M. Mastnak gave a beautiful proof of the Box conjecture, as well as an independent proof of Oblak's conjecture. In their proof, they utilize Shayman's results on invariant subspaces, and Burge's encoding of partitions as binary words. See [**20**].

Another related direction is to look beyond the generic commuting partition $\mathcal{Q}(P)$ for a given P and study all Jordan types that commute with P. One set of partial results in this direction are given in [**29**], where P. Oblak shows the following.

- A partition $P \in \mathcal{P}(n)$ commutes with all partitions in $\mathcal{P}(n)$, if and only if $n \leq 3$, or $P = (2^k, 1^\ell)$, where $k, \ell \geq 0$.
- Assume that P and Q are both partitions of n with exactly two parts. Then P commutes with Q if and only if $P = Q$ or n is even and $P = \left(\frac{n}{2}, \frac{n}{2}\right)$ and $Q = \left(\frac{n}{2}+1, \frac{n}{2}-1\right)$.

The study of commuting partitions can also be tied to another active area of research. Namely, the study of Jordan types $P_{A,\ell}$ of the multiplications map $m_\ell : A \to A$, where A is a Artinian algebra and ℓ is in A. In the past decade, significant progress has been made in this longstanding open problem, especially in the graded case (for example, see [**1–3, 11, 18, 19, 24**]). The local case has been studied as well (for example see [**17**]). We note that for a pair of commuting nilpotent matrices A and B, the Artinian algebra $\mathsf{k}[A, B]$ is in general not graded.

Acknowledgment

The author would like to express her heartfelt gratitude to Tony Iarrobino for initially introducing her to this problem, which has led to numerous engaging discussions and productive collaborations over the years. The author would also like to thank the referee for careful reading of the paper and helpful comments.

References

[1] Nasrin Altafi, *Jordan types with small parts for Artinian Gorenstein algebras of codimension three*, Linear Algebra Appl. **646** (2022), 54–83, DOI 10.1016/j.laa.2022.03.013. MR4403081

[2] Nasrin Altafi, Anthony Iarrobino, and Leila Khatami, *Complete intersection Jordan types in height two*, J. Algebra **557** (2020), 224–277, DOI 10.1016/j.jalgebra.2020.04.015. MR4092365

[3] Nasrin Altafi, Anthony Iarrobino, Leila Khatami, and Joachim Yaméogo, *Number of generators of ideals in Jordan cells of the family of graded Artinian algebras of height two*, J. Pure Appl. Algebra **227** (2023), no. 12, Paper No. 107419, 43, DOI 10.1016/j.jpaa.2023.107419. MR4601075

[4] V. Baranovsky, *The variety of pairs of commuting nilpotent matrices is irreducible*, Transform. Groups **6** (2001), no. 1, 3–8, DOI 10.1007/BF01236059. MR1825165

[5] Roberta Basili, *On the irreducibility of commuting varieties of nilpotent matrices*, J. Algebra **268** (2003), no. 1, 58–80, DOI 10.1016/S0021-8693(03)00388-0. MR2004480

[6] Roberta Basili, *On the maximal nilpotent orbit that intersects the centralizer of a matrix*, Transform. Groups **27** (2022), no. 1, 1–30, DOI 10.1007/s00031-022-09704-0. MR4400714

[7] Roberta Basili and Anthony Iarrobino, *Pairs of commuting nilpotent matrices, and Hilbert function*, J. Algebra **320** (2008), no. 3, 1235–1254, DOI 10.1016/j.jalgebra.2008.03.006. MR2427639

[8] Roberta Basili, Anthony Iarrobino, and Leila Khatami, *Commuting nilpotent matrices and Artinian algebras*, J. Commut. Algebra **2** (2010), no. 3, 295–325, DOI 10.1216/JCA-2010-2-3-295. MR2728146

[9] Duško Bogdanić, Alen łDurić, Sara Koljančić, Polona Oblak, and Klemen Šivic, *Jordan structures of nilpotent matrices in the centralizer of a nilpotent matrix with two jordan blocks of the same size*, `arXiv:2212.08425`.

[10] Thomas Britz and Sergey Fomin, *Finite posets and Ferrers shapes*, Adv. Math. **158** (2001), no. 1, 86–127, DOI 10.1006/aima.2000.1966. MR1814900

[11] Barbara Costa and Rodrigo Gondim, *The Jordan type of graded Artinian Gorenstein algebras*, Adv. in Appl. Math. **111** (2019), 101941, 27, DOI 10.1016/j.aam.2019.101941. MR4011109

[12] Emden R. Gansner, *Acyclic digraphs, Young tableaux and nilpotent matrices*, SIAM J. Algebraic Discrete Methods **2** (1981), no. 4, 429–440, DOI 10.1137/0602046. MR634367

[13] Robert M. Guralnick and B. A. Sethuraman, *Commuting pairs and triples of matrices and related varieties*, Linear Algebra Appl. **310** (2000), no. 1-3, 139–148, DOI 10.1016/S0024-3795(00)00065-3. MR1753173

[14] William J. Haboush and Donghoon Hyeon, *Conjugacy classes of commuting nilpotents*, Trans. Amer. Math. Soc. **372** (2019), no. 6, 4293–4311, DOI 10.1090/tran/7782. MR4009390

[15] Anthony Iarrobino and Leila Khatami, *Bound on the Jordan type of a generic nilpotent matrix commuting with a given matrix*, J. Algebraic Combin. **38** (2013), no. 4, 947–972, DOI 10.1007/s10801-013-0433-1. MR3119366

[16] Anthony Iarrobino, Leila Khatami, Bart Van Steirteghem, and Rui Zhao, *Nilpotent matrices having a given Jordan type as maximum commuting nilpotent orbit*, Linear Algebra Appl. **546** (2018), 210–260, DOI 10.1016/j.laa.2018.02.007. MR3771880

[17] Anthony Iarrobino, Pedro Macias Marques, and Chris McDaniel, *Artinian algebras and Jordan type*, J. Commut. Algebra **14** (2022), no. 3, 365–414, DOI 10.1216/jca.2022.14.365. MR4492997

[18] Anthony Iarrobino, Pedro Macias Marques, and Chris McDaniel, *Free extensions and Jordan type*, J. Algebra **549** (2020), 346–364, DOI 10.1016/j.jalgebra.2020.01.003. MR4052278

[19] Anthony Iarrobino, Chris McDaniel, and Alexandra Seceleanu, *Connected sums of graded Artinian Gorenstein algebras and Lefschetz properties*, J. Pure Appl. Algebra **226** (2022), no. 1, Paper No. 106787, 52, DOI 10.1016/j.jpaa.2021.106787. MR4262073

[20] J. Irving, T. Košir, and M. Mastnak, *A Proof of the Box Conjecture for Commuting Pairs of Matrices*, `arXiv:2212.08425`.

[21] Leila Khatami, *The poset of the nilpotent commutator of a nilpotent matrix*, Linear Algebra Appl. **439** (2013), no. 12, 3763–3776, DOI 10.1016/j.laa.2013.09.045. MR3133456

[22] Leila Khatami, *The smallest part of the generic partition of the nilpotent commutator of a nilpotent matrix*, J. Pure Appl. Algebra **218** (2014), no. 8, 1496–1516, DOI 10.1016/j.jpaa.2013.12.002. MR3175035

[23] Tomaž Košir and Polona Oblak, *On pairs of commuting nilpotent matrices*, Transform. Groups **14** (2009), no. 1, 175–182, DOI 10.1007/s00031-008-9045-6. MR2480857

[24] Chris McDaniel, Shujian Chen, Anthony Iarrobino, and Pedro Macias Marques, *Free extensions and Lefschetz properties, with an application to rings of relative coinvariants*, Linear Multilinear Algebra **69** (2021), no. 2, 305–330, DOI 10.1080/03081087.2019.1598930. MR4219281

[25] George McNinch, *On the centralizer of the sum of commuting nilpotent elements*, J. Pure Appl. Algebra **206** (2006), no. 1-2, 123–140, DOI 10.1016/j.jpaa.2005.04.016. MR2220085

[26] Michael G. Neubauer and B. A. Sethuraman, *Commuting pairs in the centralizers of 2-regular matrices*, J. Algebra **214** (1999), no. 1, 174–181, DOI 10.1006/jabr.1998.7703. MR1684884

[27] Nham V. Ngo and Klemen Šivic, *On varieties of commuting nilpotent matrices*, Linear Algebra Appl. **452** (2014), 237–262, DOI 10.1016/j.laa.2014.03.032. MR3201099

[28] Polona Oblak, *The upper bound for the index of nilpotency for a matrix commuting with a given nilpotent matrix*, Linear Multilinear Algebra **56** (2008), no. 6, 701–711, DOI 10.1080/03081080701571083. MR2457695

[29] Polona Oblak, *On the nilpotent commutator of a nilpotent matrix*, Linear Multilinear Algebra **60** (2012), no. 5, 599–612, DOI 10.1080/03081087.2011.618837. MR2916845

[30] Dmitri I. Panyushev, *Two results on centralisers of nilpotent elements*, J. Pure Appl. Algebra **212** (2008), no. 4, 774–779, DOI 10.1016/j.jpaa.2007.07.003. MR2363491

[31] Alexander Premet, *Nilpotent commuting varieties of reductive Lie algebras*, Invent. Math. **154** (2003), no. 3, 653–683, DOI 10.1007/s00222-003-0315-6. MR2018787

[32] B. A. Sethuraman and Klemen Šivic, *Jet schemes of the commuting matrix pairs scheme*, Proc. Amer. Math. Soc. **137** (2009), no. 12, 3953–3967, DOI 10.1090/S0002-9939-09-10029-1. MR2538555

[33] Klemen Šivic, *On varieties of commuting triples*, Linear Algebra Appl. **428** (2008), no. 8-9, 2006–2029, DOI 10.1016/j.laa.2007.11.004. MR2401636

[34] Klemen Šivic, *On varieties of commuting triples II*, Linear Algebra Appl. **437** (2012), no. 2, 461–489, DOI 10.1016/j.laa.2011.08.014. MR2921711

[35] Klemen Šivic, *On varieties of commuting triples III*, Linear Algebra Appl. **437** (2012), no. 2, 393–460, DOI 10.1016/j.laa.2011.08.015. MR2921710

Department of Mathematics, Union College, 807 Union St, Schenectady, NY 12308, USA

Email address: `khatamil@union.edu`

Contemporary Mathematics
Volume **805**, 2024
https://doi.org/10.1090/conm/805/16138

A brief survey of Specht ideals and their perfection in the two-rowed case

Chris McDaniel

ABSTRACT. We give a brief survey of results on general Specht ideals, then we focus on the two-rowed case, focusing on issues of primary decompositions and perfection, the dependence of these properties on field characteristic, and a surprising connection to Lefschetz properties.

1. Introduction

Specht polynomials $F_T \in R = \mathbb{F}[x_1, \ldots, x_n]$ are generalizations of Vandermonde polynomials and are indexed by tableaux T on an integer partition $\lambda \vdash n$. The $\mathbb{F}$-linear span of Specht polynomials over all tableaux on λ is called the Specht module of λ, and defines a representation of the symmetric group $\mathfrak{S}_n$. In fact, if the characteristic of $\mathbb{F}$ is 0, then these Specht modules $\{V_\lambda \mid \lambda \vdash n\}$ constitute a complete list of irreducible representations of $\mathfrak{S}_n$; see [**19**]. The R-linear span of Specht polynomials is a symmetric homogeneous ideal I_λ called the Specht ideal of λ.

The study of Specht ideals, from the point of view of commutative algebra, seems to have been initiated by Yanagawa in his paper [**26**], although these ideals have appeared in the literature prior to this, e.g. [**1**], [**3**], [**5**], [**6**], [**12**]. Since the appearance of [**26**], there have been several works published on various aspects of Specht ideals, e.g. [**13**], [**16**], [**17**], [**22**], [**23**], [**25**], . In this short article, we shall collect some of these general commutative algebraic results dealing with prime decompositions, radicals, and perfection. We shall then focus on the two-rowed case and the results of [**13**], namely perfection of two-rowed Specht ideals and some related ideals, and the dependence of this property on the field characteristic. We shall also discuss some interesting connections between two-rowed Specht ideals and Lefschetz properties of certain quadratic monomial complete intersections.

This paper is meant as a brief survey and guide to the literature of Specht ideals, but it is by no means meant to be comprehensive. One topic that has been omitted is free resolutions of Specht ideals; the reader may find such results, both established and conjectural, in the papers of Shibata-Yanagawa [**22, 23**] and Griffeth-Sam-Berkesch [**5**]. In this paper, for the sake of brevity, we have chosen to reference the literature for proofs rather than give them here.

2020 *Mathematics Subject Classification.* Primary 13A50, 13C14, 20C30.
The author thanks both anonymous referees for their thoughtful comments and their careful readings of this manuscript.

2. Generalities on Specht Ideals

2.1. Preliminaries. A partition $\lambda = (\lambda_1, \dots, \lambda_r)$ is a weakly decreasing sequence of non-negative integers $\lambda_1 \geq \cdots \geq \lambda_r \geq 0$. The length of the partition is the number of parts, denoted $\ell(\lambda) = r$, and the size of the partition is the sum of its parts, denoted $|\lambda| = \lambda_1 + \cdots + \lambda_r$; if $|\lambda| = n$ we sometimes write $\lambda \vdash n$ to mean that λ is a partition of n. We identify a partition with its Young diagram, consisting of r-rows of left-justified boxes with λ_i boxes in row i. A filling of the boxes of λ with the numbers $[n] = \{1, \dots, n\}$ where each number is used only once is called a tableau T on λ or a tableau T of shape λ. The set of all tableaux of shape λ is denoted by $\mathrm{Tab}(\lambda)$. A tableau T is called standard if its row entries are increasing from left to right and its column entries are increasing from top to bottom. The set of all standard tableaux of shape λ is denoted by $\mathrm{STab}(\lambda)$.

Let $\mathbb{F}$ be any field, and let $R = R^n_{\mathbb{F}} = \mathbb{F}[x_1, \dots, x_n]$. Given a subset $S \subset [n]$, define the polynomial $\Delta_S \in R$ by

$$\Delta_S = \prod_{i<j \in S} (x_i - x_j).$$

Given a partition $\lambda \vdash n$ and a tableau $T \in \mathrm{Tab}(\lambda)$, let $\mathrm{Col}(T)$ denote the set partition of $[n]$ given by the columns of T, and define the Specht polynomial of T by

$$F_T = \prod_{C \in \mathrm{Col}(T)} \Delta_C = \prod_{C \in \mathrm{Col}(T)} \prod_{i<j \in C} (x_i - x_j).$$

The Specht module V_λ (over $\mathbb{F}$) is defined to be $\mathbb{F}$-linear span of its Specht polynomials, i.e.

$$V_\lambda = \sum_{T \in \mathrm{Tab}(\lambda)} F_T \cdot \mathbb{F} = \langle F_T \mid T \in \mathrm{Tab}(\lambda) \rangle .$$

The Specht ideal I_λ (over $\mathbb{F}$) is defined to be the R-linear span of (or the ideal generated by) its Specht polynomials, i.e.

$$I_\lambda = \sum_{T \in \mathrm{Tab}(\lambda)} F_T \cdot R = (F_T \mid T \in \mathrm{Tab}(\lambda)) .$$

With the natural action of the symmetric group $\mathfrak{S}_n$ on R defined on the variables by $\sigma.x_i = x_{\sigma(i)}$, both V_λ and I_λ define $\mathfrak{S}_n$ representations (of course I_λ is not finite dimensional, but it is finite dimensional in each graded component). The following fact was evidently first proved by Specht [**24**], and is by now well known among representation theorists and combinatorialists; a proof can be found for example in [**19**, Theorem 1.1]. See also [**7**, Proposition 2] or [**21**, Theorem 2.5.2] for a more combinatorial flavor.

FACT 2.1. Over any field $\mathbb{F}$, the Specht polynomials corresponding to standard tableaux form a minimal generating set of V_λ and hence also of I_λ, i.e.

$$V_\lambda = \langle F_T \mid T \in \mathrm{STab}(\lambda) \rangle , \text{ and } I_\lambda = (F_T \mid T \in \mathrm{STab}(\lambda)) .$$

2.2. Vanishing Sets and Minimal Prime Divisors of Specht Ideals. Denote by $\mathcal{P}_n$ the set of all integer partitions of n. The dominance order on $\mathcal{P}_n$ is a partial order defined by $\lambda \trianglelefteq \mu$ if $\sum_{i=1}^k \lambda_i \leq \sum_{i=1}^k \mu_i$ for all k. The following fact, due to Moustrou-Riener-Verdure [**16**, Theorem 1], says that the poset $(\mathcal{P}_n, \trianglelefteq)$ is isomorphic to the poset of Specht ideals ordered by inclusion.

FACT 2.2. Let $\lambda, \mu \in \mathcal{P}_n$ be two partitions of size n. Then $\lambda \trianglelefteq \mu$ if and only if $I_\lambda \subseteq I_\mu$.

Related to the notion of an integer partition is a set partition. Given a set $[n] = \{1, \dots, n\}$ a set partition is a sequence $\Lambda = (\Lambda_1, \dots, \Lambda_r)$ of disjoint subsets of $[n]$ whose union is $[n]$; keeping in line with our convention for integer partitions, we shall always assume that the parts of Λ are arranged by cardinality, in weakly decreasing order, i.e. $|\Lambda_1| \geq \cdots \geq |\Lambda_r|$. Let P_n denote the set of set partitions of $[n]$; it is partially ordered by refinement, i.e. $\Lambda \leq \Theta$ if and only if every block of Λ belongs to some block of Θ. To a set partition $\Lambda \vdash [n]$, we may associate a linear prime ideal of R, denoted by P_Λ, defined by

$$P_\Lambda = \sum_{k=1}^{r} P_{\Lambda_k} = (x_i - x_j \mid i, j \in \Lambda_k,\ 1 \leq k \leq r).$$

The linear subspaces cut out from the linear prime ideals P_Λ form a lattice under reverse inclusion which is the intersection lattice of the braid arrangement. The next fact says that the intersection lattice of the braid arrangement coincides exactly with the poset $(\mathsf{P}_n, \leq)$; for a proof we refer the reader to the book of Orlik-Terao [**18**, Proposition 2.9].

FACT 2.3. Given two set partitions $\Lambda, \Theta \vdash [n]$, we have $\Lambda \leq \Theta$ if and only if $P_\Lambda \subseteq P_\Theta$.

There is a natural surjective map of posets $\phi \colon (\mathsf{P}_n, \leq) \to (\mathcal{P}_n, \trianglelefteq)$ defined by $\phi(\Lambda_1, \dots, \Lambda_r) = (|\Lambda_1|, \dots, |\Lambda_r|)$; that it is a map of posets can be reformulated as the implication:

$$\Lambda \leq \Theta \ \Rightarrow\ \phi(\Lambda) \trianglelefteq \phi(\Theta).$$

The integer partition $\phi(\Lambda)$ is called the type of the set partition Λ.

For an ideal $J \subset R$, let $\mathcal{V}(J) = \{a \in \mathbb{F}^n \mid F(a) = 0,\ \forall F \in J\}$, the vanishing set of J. Given an element $a = (a_1, \dots, a_n) \in \mathbb{F}^n$, define its orbit type $\mu(a) = (\mu_1, \dots, \mu_s) \in \mathcal{P}_n$ if up to permutation of coordinates we have

$$a = \left(\underbrace{a_1, \dots, a_1}_{\mu_1}, \underbrace{a_2, \dots, a_2}_{\mu_2}, \dots, \underbrace{a_s, \dots, a_s}_{\mu_s} \right);$$

In other words, a has orbit type μ if and only if $a \in \mathcal{V}(P_\Lambda)$ for some $\Lambda \in \mathsf{P}_n$ of type $\phi(\Lambda) = \mu$. On the other hand, $a \in \mathcal{V}(I_\lambda)$ if and only if for every tableau $T \in \operatorname{Tab}(\lambda)$, there exists indices i, j in the same column of T such that $a_i = a_j$. We can reformulate this condition combinatorially: given $a \in \mathbb{F}^n$ with orbit type $\mu = (\mu_1, \dots, \mu_s)$, $a \in \mathcal{V}(I_\lambda)$ if and only if there are no fillings of the boxes of the Young diagram of λ with the numbers $[s] = \{1, \dots, s\}$ with rows weakly increasing and columns strictly increasing; such fillings are called semi-standard Young tableaux of shape λ. A standard Young tableau is also a semi-standard Young tableau, but a semi-standard Young tableau may have repeated indices, and hence it need not even be a Young tableaux as defined above. The ordered integer s-tuple $\mu = (\mu_1, \dots, \mu_s)$ is called the content of T, a semi-standard Young tableaux of shape λ, if T is filled with μ_1 1's and μ_2 2's, and so on. The number of semi-standard Young tableaux of shape λ and content μ is called the Kostka number for the pair (λ, μ), and is denoted by $K_{\lambda\mu}$; see [**14**, page 101]. In fact, one can show that $K_{\lambda\mu} = 0$ if and only if $\mu \not\trianglelefteq \lambda$, e.g. [**14**, Theorem 6.5] or [**7**, Exercise 2, pg. 26]. The following result, also due to Moustrou-Riener-Verdure [**16**, Corollary 1], may be regarded as a geometric interpretation of this fact.

FACT 2.4. For any partition $\lambda \in \mathcal{P}_n$ we have

$$\mathcal{V}(I_\lambda) = \bigcup_{\phi(\Lambda) \not\trianglelefteq \lambda} \mathcal{V}(P_\Lambda)$$

or equivalently

$$\sqrt{I_\lambda} = \bigcap_{\phi(\Lambda) \not\trianglelefteq \lambda} P_\Lambda.$$

Fact 2.4 shows that the vanishing set of a Specht ideal is a union of linear spaces, and more specifically, that the minimal associated prime divisors of the Specht ideal I_λ are the linear prime ideals corresponding to set partitions whose type is not dominated by λ and that are minimal with respect to refinement. For example, the partition $\lambda = (3,2,1) \in \mathcal{P}_6$ has types $\mu_1 = (3,3)$, $\mu_2 = (4,1,1)$ that are not dominated by λ and that are minimal under refinement, and hence

$$\begin{aligned}\sqrt{I_\lambda} &= \bigcap_{\substack{\Lambda_1 \in \mathrm{P}_6 \\ \phi(\Lambda_1)=\mu_1}} P_{\Lambda_1} \cap \bigcap_{\substack{\Lambda_2 \in \mathrm{P}_6 \\ \phi(\Lambda_2)=\mu_2}} P_{\Lambda_2} \\ &= \bigcap_{\sigma \in \mathfrak{S}_6} \left(x_{\sigma(1)} - x_{\sigma(2)}, x_{\sigma(1)} - x_{\sigma(3)}, x_{\sigma(4)} - x_{\sigma(5)}, x_{\sigma(4)} - x_{\sigma(6)}\right) \\ &\quad \cap \bigcap_{\sigma \in \mathfrak{S}_6} \left(x_{\sigma(1)} - x_{\sigma(2)}, x_{\sigma(1)} - x_{\sigma(3)}, x_{\sigma(1)} - x_{\sigma(4)}\right);\end{aligned}$$

this computation shows that the Specht ideal I_λ is mixed, meaning that it has minimal associated prime divisors of different heights. It turns out that most Specht ideals are mixed, as shown by the following result of Yanagawa [**26**, Corollary 2.5].

FACT 2.5. For any partition $\lambda = (\lambda_1, \ldots, \lambda_r) \in \mathcal{P}_n$, its Specht ideal I_λ is unmixed if and only if either $\lambda_2 = 1$ or $\lambda_{r-1} = \lambda_1$.

Note that $\lambda_2 = 1$ if and only if λ is a hook-shaped partition of the form $\lambda = (n-k, 1, \ldots, 1) = (n-k, 1^k)$. In this case, the Specht ideal is generated by the maximal $(k+1) \times (k+1)$ minor determinants of the generalized Vandermonde matrix

$$M_{k,n} = \begin{pmatrix} 1 & \cdots & 1 \\ x_1 & \cdots & x_n \\ \vdots & \ddots & \vdots \\ x_1^k & \cdots & x_n^k \end{pmatrix};$$

such determinantal ideals have been studied by Fröberg-Shapiro [**6**], and more recently by Watanabe-Yanagawa [**25**]. One can show that the partitions not dominated by the hook shape $\lambda = (n-k, 1^k)$ that are minimal with respect to refinement are those partitions in $\mathcal{P}_n$ with exactly k parts, and since $\mathrm{ht}(P_\Lambda) = n - \#(\text{parts of } \Lambda)$ it follows that these associated primes all have the same height $n-k$.

If $\lambda_{r-1} = \lambda_1$, then we call the partition λ almost rectangular, and one can show that in this case, there is a unique partition that is not dominated by λ and is minimal with respect to refinement, namely the hook shape $\mu = (h = \lambda_1 + 1, 1, \ldots, 1)$, and we have

$$\sqrt{I_\lambda} = \bigcap_{\sigma \in \mathfrak{S}_n} \left(x_{\sigma(1)} - x_{\sigma(2)}, \ldots, x_{\sigma(1)} - x_{\sigma(h)} \right).$$

The vanishing set $\mathcal{V}(I_\lambda)$ can be identified with the set of points in $\mathbb{F}^n$ in which some h of the coordinates are equal; this is colloquially referred to as the h-equals set and has shown up in combinatorics [**12**], representation theory [**1, 3**], and even in connection with physics [**5**]. A special case occurs when $r = 2$, where $\lambda = (n-k, k)$ is called a *two-rowed partition.*

2.3. Radicals of Specht Ideals. It was conjectured by Yanagawa in his 2020 paper [**26**] that all Specht ideals are radical, although it was later discovered that this result had been obtained for almost rectangular partitions some forty years earlier by graph theorists Li-Li [**12**]. Watanabe-Yanagawa [**25**] have shown that Specht ideals for hook shaped partitions are always radical. The general case that follows appears in the beautiful short 2021 paper of Murai-Ohsugi-Yanagawa [**17**, Theorem 1.1], although they attribute it to an unpublished manuscript of Haiman-Woo.

FACT 2.6. For each partition $\lambda \in \mathcal{P}_n$, the Specht ideal I_λ is radical, i.e. $I_\lambda = \sqrt{I_\lambda}$. Moreover the Specht polynomials $\{F_T \mid T \in \mathrm{Tab}(\mu),\ \mu \trianglelefteq \lambda\}$ are a universal Gröbner basis for I_λ.

2.4. Perfection of Specht Ideals. Recall that a homogeneous ideal I in a polynomial ring R is called *perfect* if the grade (height) of I equals its projective dimension, or equivalently, if its quotient R/I is Cohen-Macaulay. Note that a necessary condition for an ideal to be perfect is that the ideal is unmixed. Hence Fact 2.5 already precludes many Specht ideals from being perfect (i.e. all except the hook shapes and almost rectangles). The following fact is due to Yanagawa [**26**, Proposition 2.8].

FACT 2.7. If I_λ is perfect, then λ must be one of the following three types:

(1) $\lambda = (n-k, 1^k)$ (hook shape)
(2) $\lambda = (n-k, k)$ (two rowed almost rectangular)
(3) $\lambda = (d, d, 1)$ (special three rowed almost rectangular)

It is tempting to conjecture that the converse holds, namely that the Specht ideal for partitions of the above types are always perfect. In fact, Watanabe-Yanagawa [**25**, Theorem 1.1] proved it for hook shapes:

FACT 2.8. For any field $\mathbb{F}$, if $\lambda = (n-k, 1^k)$ is hook-shaped, then its Specht ideal I_λ is perfect.

Using deep results from representation theory of rational Cherednik algebras, Etingof-Gorsky-Losev [**3**, Proposition 3.11] classified the Cohen-Macaulay h-equals sets over the field of complex numbers:

FACT 2.9. Given integers $h \leq n$, the h-equals set (over $\mathbb{C}$)

$$X_{n,h} = \bigcup_{\sigma \in \mathfrak{S}_n} \sigma.\{v \in \mathbb{C}^n \mid x_1(v) = \cdots = x_h(v)\}$$

is Cohen-Macaulay if and only if $h = 2$ or $2h > n$.

The homogeneous coordinate ring of $X_{n,h}$ is $R/\sqrt{I_\lambda} = R/I_\lambda$ where λ is the almost rectangular partition with $\lambda_1 = h-1$. Hence for $h=2$, $I_\lambda = \left(\prod_{1 \leq i<j \leq n}(x_i - x_j)\right)$ is the principal ideal generated by the determinant of the Vandermonde matrix, and for $2h > n$, it follows that either $\lambda = (h-1, h-1, 1)$, the special three rowed almost rectangle, in case $2h = n+1$, or that $\lambda = (h-1, n-h+1)$, the two-rowed almost rectangle, for $2h > n+1$. From Fact 2.9, Yanagawa [**26**, Theorem 2.6] has proved the following:

FACT 2.10. If $\mathbb{F}$ has characteristic zero, and $\lambda = (d, d, 1)$ or $\lambda = (n-k, k)$, then I_λ is perfect.

It turns out however that Fact 2.10 can fail for fields of positive characteristic. For the two-rowed case, McDaniel-Watanabe [**13**, Theorem 1.2] have obtained the following result using only elementary methods and avoiding rational Cherednik algebras altogether:

FACT 2.11. Fix integers n, k satisfying $k \leq n-k$, let $\lambda = (n-k, k)$, and let $\mathbb{F}$ be any field of characteristic $p \geq 0$.

(1) If $p = 0$ or $p \geq k$, then I_λ is perfect.
(2) If $p = k-1$ then I_λ is imperfect.

According to Fact 2.11, if $\lambda = (n-2, 2)$, then I_λ is perfect over any field, a fact already observed by Yanagawa [**26**]. On the other hand, if $\lambda = (n-3, 3)$ then I_λ is imperfect over any field of characteristic $p = 2$, and similarly for any prime satisfying $p+1 \leq n-p-1$, if $\lambda = (n-p-1, p+1)$ then I_λ is not perfect over any field of characteristic p.

3. Two-Rowed Specht Ideals and Friends

For this section, we focus on the two-rowed case, and results from [**13**], centered around Fact 2.11. We remark that the notation used here may differ slightly from that used in [**13**].

3.1. Two-Rowed Specht Ideals. A partition with exactly two parts $\lambda = (n-k, k)$ is called a two-rowed partition. Specht polynomials of two-rowed partitions are square-free, and thus occupy a special place among other Specht polynomials. A tableau T on shape λ has the form

$T =$

<table>
<tr><td>i_1</td><td>$\cdots$</td><td>i_k</td><td>i_{k+1}</td><td>$\cdots$</td><td>i_{n-k}</td></tr>
<tr><td>j_1</td><td>$\cdots$</td><td>j_k</td></tr>
</table>

and its Specht polynomial is

$$F_T = \prod_{p=1}^{k} (x_{i_p} - x_{j_p}).$$

The following facts can be deduced from Fact 2.1, Fact 2.4, and Fact 2.6.

FACT 3.1. Let $\lambda = (n-k, k)$ be a two-rowed partition for some positive integers n, k satisfying $2k \leq n$. Then

(1) $I_\lambda = (F_T \mid T \in \mathrm{STab}(\lambda))$
(2) $\dim(V_\lambda) = \binom{n}{k} - \binom{n}{k-1}$.
(3) $I_\lambda = \bigcap_{\sigma \in \mathfrak{S}_n} \sigma.(x_1 - x_2, \ldots, x_1 - x_{n-k+1}) = \bigcap_{\substack{\Lambda \in \mathrm{P}_n \\ \phi(\Lambda)=(n-k+1,1^{k-1})}} P_\Lambda$

3.2. Shifted Two-Rowed Specht Ideals. By a shifted or skew diagram, $\lambda \setminus \mu$ we mean the diagram obtained from that of λ by removing the sub-diagram μ (we will assume in this setting that $\mu_i \leq \lambda_i$ for $1 \leq i \leq r$). A (respectively, standard) tableau T on a skew shape $\lambda \setminus \mu$ is defined in an analogous way, and we shall denote the set of such tableaux $\mathrm{Tab}(\lambda \setminus \mu)$ (respectively, $\mathrm{STab}(\lambda \setminus \mu)$).

Given a two-rowed partition $\lambda = (n-k, k)$ and an integer $r \leq n - 2k$, and given a tableau $T \in \mathrm{Tab}(\lambda)$

$$T = \begin{array}{|c|c|c|c|c|c|c|c|} \hline i_1 & \cdots & i_k & i_{k+1} & \cdots & i_{k+r} & \cdots & i_{n-k} \\ \hline j_1 & \cdots & j_k \\ \cline{1-3} \end{array} \tag{3.1}$$

define its r-shifted two-rowed Specht polynomial

$$F_{T,r} = \prod_{p=1}^{k} (x_{i_p} - x_{j_p}) \cdot \prod_{q=1}^{r} x_{i_{k+q}}.$$

Thus $F_{T,r}$ is a product of a two-rowed Specht polynomial and a square-free monomial of degree r in a disjoint subset of variables; in particular it is has a square-free monomial expansion. Define the r-shifted (or skew) shape $\lambda(r) = (n, n-r) \setminus (n - 2k - r, 0)$, and note that there is a one-to-one corrspondence between $\mathrm{Tab}(\lambda)$ and $\mathrm{Tab}(\lambda(r))$ where $T \in \mathrm{Tab}(\lambda)$ in (3.1) maps to $T(r) \in \mathrm{Tab}(\lambda(r))$ defined by

$$T(r) = \begin{array}{cccccccccc} \cline{4-9} & & & \multicolumn{1}{|c|}{i_1} & \multicolumn{1}{c|}{\cdots} & \multicolumn{1}{c|}{i_k} & \multicolumn{1}{c|}{i_{k+1}} & \multicolumn{1}{c|}{\cdots} & \multicolumn{1}{c|}{i_{k+r}} \\ \hline \multicolumn{1}{|c|}{i_{k+r+1}} & \multicolumn{1}{c|}{\cdots} & \multicolumn{1}{c|}{i_{n-k}} & \multicolumn{1}{c|}{j_1} & \multicolumn{1}{c|}{\cdots} & \multicolumn{1}{c|}{j_k} \\ \cline{1-6} \end{array} \tag{3.2}$$

Identifying $F_{T(r)}$ with $F_{T,r}$, let $V_{\lambda(r)}$ denote the F-linear span of the r-shifted Specht polynomials, called the r-shifted Specht module, and let $I_{\lambda(r)}$ denote their R-linear span, i.e. the ideal generated by r-shifted Specht polynomials, called the r-shifted Specht ideal:

$$V_{\lambda(r)} = \langle F_{T(r)} \mid T(r) \in \mathrm{Tab}(\lambda(r)) \rangle \quad I_{\lambda(r)} = (F_{T(r)} \mid T(r) \in \mathrm{Tab}(\lambda(r))).$$

In the following result, which appears in [**13**], we use the notation $(x_1,\dots,x_n)^{\langle k\rangle}$ to denote the ideal generated by all square-free monomials of degree k, called the complete square-free monomial ideal of order k.

FACT 3.2. Let $\lambda=(n-k,k)\in\mathcal{P}_n$ be the two-rowed partition, and $\lambda(r)$ its r-shifted counterpart.

(1) $I_{\lambda(r)}=\big(F_{T(r)}\mid T(r)\in\mathrm{STab}(\lambda(r))\big)$ minimally generated by standard r-shifted tableaux.
(2) $\dim(V_{\lambda(r)})=\binom{n}{r}-\binom{n}{k-1}$
(3) $I_{\lambda(r)}=I_\lambda\cap(x_1,\dots,x_n)^{\langle k+r\rangle}$.

It follows from Fact 3.2(1) that $I_{\lambda(r)}$ is radical for all $r\geq 0$ ($\lambda(0)=\lambda$), and also that $I_{\lambda(r)}$ is mixed (hence also imperfect) for all $r>1$. In fact one can further show [**13**, Corollary 2.7] that under the natural projection map $\Phi\colon R^{n+1}\to R^n$, defined by $\Phi(x_i)=x_i$ for $1\leq i\leq n$ and $\Phi(x_{n+1})=0$, the two-rowed Specht ideal $I_{\hat\lambda}$ where $\hat\lambda=(n-k,k+1)\in\mathcal{P}_{n+1}$, is mapped isomorphically onto the 1-shifted Specht ideal $I_{\lambda(1)}$. We therefore have the following result, equivalent to Fact 2.11:

FACT 3.3. Let $\lambda=(n-k,k)$ be a two-rowed partition for some positive integers n,k satisfying $2k+1\leq n$, let $\lambda(1)=(n-k+1,k)\setminus(1)$ be the 1-shifted shape, let $I_{\lambda(1)}$ be its shifted Specht ideal, and let $\mathbb{F}$ be any field of characteristic $p\geq 0$.

(1) If $p=0$ or $p\geq k+1$, then $I_{\lambda(1)}$ is perfect.
(2) If $p=k$, then $I_{\lambda(1)}$ is imperfect.

In characteristics $p=0$ or $p\geq n+1$, the r-shifted Specht module $V_{\lambda(r)}$ turns out to be a skew representation of $\mathfrak{S}_n$ studied by James-Peel [**10**]. This identification is surprisingly related to the strong Lefschetz property; see Fact 3.7.

3.3. Specht-Monomial Ideals. The *Specht-monomial ideal* for a two-rowed partition $\lambda=(n-k,k)$ is defined to be the sum of the Specht ideal I_λ and the complete square-free monomial ideal of order $k+1$, i.e.

$$I(n,k)=I_\lambda+(x_1,\dots,x_n)^{\langle k+1\rangle}.$$

In other words the Specht monomial ideal is generated by Specht polynomials of degree k and square-free monomials of degree $k+1$. It is easy to see that the Specht-monomial ideal is not radical; in fact it contains the square of every square-free monomial of order k and one can show

$$\sqrt{I(n,k)}=(x_1,\dots,x_n)^{\langle k\rangle}.$$

For the following results, which comprise [**13**, Theorem 1.8], define the $n-1$ linear forms $y_i=x_n-x_i$ for $1\leq i\leq n$, and set $(y_1,\dots,y_{n-1})^{\langle k\rangle}$ be the ideal generated by the square-free monomials in those variables of degree k. Also set $I(n-1,k-1)=I_{(n-k,k-1)}+(x_1,\dots,x_{n-1})^{\langle k\rangle}$, which we may view as an ideal in R^n, generated by polynomials in R^{n-1}.

FACT 3.4. The following are equivalent.

(1) $p=0$ or $p\geq k+1$.
(2) $I(n,k)$ has no embedded associated prime divisors.
(3) $I(n,k)=I(n-1,k-1)\cap\Big((y_1,\dots,y_{n-1})^{\langle k\rangle}+(x_n^2)\Big)$.

It follows from Fact 3.4(3) that a primary decomposition for the Specht-monomial ideal $I(n,k)$ is

$$I(n,k) = \bigcap_{\sigma \in \mathfrak{S}_n} \left(x_{\sigma(1)} - x_{\sigma(2)}, \dots, x_{\sigma(1)} - x_{\sigma(n-k+1)}, x_{\sigma(1)}^2 \right),$$

as long as $p = 0$ or $p \geq k+1$. Concerning perfection we have the following result:

FACT 3.5. Fix any integers n, k satisfying $2k+1 \leq n$, and any field $\mathbb{F}$ of characteristic $p \geq 0$. Then $I(n,k)$ is perfect if and only if $p = 0$ or $p \geq k+1$.

According to Fact 3.4, imperfection of a Specht-monomial ideal can be detected by the existence of an embedded prime divisor, which evidently appears when the characteristic is positive but too small. Contrast with Specht ideals which are radical over any field (hence no embedded prime divisors), where imperfection is much more subtle and harder to detect!

3.4. Connection to Lefschetz Properties. Square-free polynomials in the variables $x_1, \dots, x_n$ can be viewed as elements of the Artinian monomial complete intersection

$$A = \frac{\mathbb{F}[x_1, \dots, x_n]}{(x_1^2, \dots, x_n^2)},$$

where the square-free monomials are an $\mathbb{F}$-basis. Let $L = \times(x_1 + \cdots + x_n)$ be the multiplication operator on A and let $D = \frac{\partial}{\partial x_1} + \cdots + \frac{\partial}{\partial x_n}$ be the partial differentiation operator on A. Then letting $H(a) = (n - 2\deg(a)) \cdot a$ be the semi-simple operator on A, one can show that $\{L, D, H\}$ is an $\mathfrak{sl}_2(\mathbb{F})$-triple on A. For $0 \leq k \leq \lfloor \frac{n}{2} \rfloor$, we say that A satisfies the weak Lefschetz property in degree k (WLP_k) if the multiplication map $L\colon A_{k-1} \to A_k$ is injective, or equivalently if the differentiation map $D\colon A_k \to A_{k-1}$ is surjective. The connection to Specht polynomials is that if $\lambda = (n-k, k)$ then $V_\lambda \subset \ker(D) \cap A_k$, with equality if and only if A has WLP_k. The following result appears in [**13**, Lemma 3.4]:

FACT 3.6. The following are equivalent:
(1) $p = 0$ or $p \geq k+1$
(2) A has WLP_k
(3) $\ker(D) \cap A_k = V_{(n-k,k)}$.

Equivalence of (1) and (2) in Fact 3.6 can be deduced from general results of Kustin-Vraciu [**11**]; see also [**15**] and [**9**] for further discussion of Lefschetz properties.

If $p = k$, then the elementary symmetric polynomial in $p+1$ variables $e_p(x_1, \dots, x_{p+1})$ lies in $\ker(D) \cap A_k$ because $D(e_p) = pe_{p-1} \equiv 0 \mod p$, but on the other hand, as a square-free polynomial in n-variables, $E_p^n(x_1, \dots, x_n) = e_p(x_1, \dots, x_{p+1})$ cannot lie in $V_{(n-k,k)}$ because otherwise we must have $E(1, \dots, 1) \equiv 0 \mod p$, but $E(1, \dots, 1) = p + 1 \equiv 1 \mod p$. In fact one can show that as a homogeneous polynomial in R^n over a field of characteristic p, we always have

$$E_p^n \in \bigcap_{\sigma \in \mathfrak{S}_n} \left(x_{\sigma(1)} - x_{\sigma(2)}, \dots, x_{\sigma(1)} - x_{\sigma(n-p+1)}, x_{\sigma(1)}^2 \right) \setminus I(n,p).$$

Thus, according to Fact 3.4, one might say that the polynomial E_p^n "witnesses" the imperfection of $I(n,p)$ in characteristic p.

The algebra A satisfies the strong Lefschetz property (SLP) means that powers of the multiplication map $L^j\colon A_k \to A_{k+j}$ have full rank for all $j,k \geq 0$, or equivalently that the maps $L^{d-2k}\colon A_k \to A_{n-k}$ are isomorphisms for all $0 \leq k \leq \lfloor \frac{n}{2} \rfloor$. The subspace $P_k = \ker(D) \cap A_k$ is sometimes called the primitive subspace of A with respect to L. The following result is [**13**, Lemma 3.6].

FACT 3.7. The following are equivalent.

(1) $p = 0$ or $p \geq n+1$.
(2) A has the SLP.
(3) For all integers $0 \leq k \leq r \leq n-k$, the r-shifted Specht modules $V_{(n-k,k)(r)}$ satisfy

$$V_{(n-k,k)(r)} = \ker(L^{n-k-r+1}) \cap A_r = \bigoplus_{i=k}^{r} L^{r-i}(P_i) \cong \bigoplus_{i=k}^{r} V_{(n-i,i)}[r-i].$$

Equivalence of (1) and (2) in Fact 3.7 is evidently due to Ikeda; see [**9**, Proposition 3.67]. Fact 3.7(3) is related to the primitive decomposition of A, and identifies our shifted Specht modules as the skew representations of $\mathfrak{S}_n$ studied by James-Peel [**10**].

3.5. Principal Perfection Systems. Our notion of principal perfection systems is inspired directly from the principal radical systems invented by Hochster-Eagon in their seminal paper [**8**]. The following elementary facts from commutative algebra are at the basis of our method.

FACT 3.8. Let $I \subset R$ be a homogeneous ideal in a polynomial ring (over any field), and let $x \in R \setminus I$ be any homogeneous polynomial not in I. Then

(1) If x is a non-zero divisor of I, then $I + (x)$ is perfect if and only if I is perfect.
(2) If x is a zero divisor of I, and both $(I : x)$ and $I + (x)$ are perfect of the same grade g, then I is also perfect of grade g.

A proof of Fact 3.8 is given in [**13**, Lemma 5.1]. It is interesting to note that a strikingly similar fact holds if "perfect" is replaced by "radical" everywhere in Fact 3.8; see [**13**, Lemma 4.1]. The following fact appears in [**8**, Theorem 18].

FACT 3.9. Let $I, J \subset R$ be homogeneous ideals in a polynomial ring (over any field).

(1) If I, J, and $I \cap J$ are perfect of the same grade, then $I + J$ is perfect.
(2) If I and J are perfect of grade g and $I + J$ is perfect of grade $g+1$, then $I \cap J$ is perfect.

The conditions on grade in (2) in both Fact 3.8 and Fact 3.9 cannot be dropped and is illustrated by essentially the same example. Take $I = (x^2, y)$ and $J = (x)$. Then $I + J = (x, y)$ is perfect, but $I \cap J = (x^2, xy)$ is not, and $(I \cap J) + (x) = (x)$ and $(I \cap J : x) = (x, y)$ are both perfect whereas $(I \cap J)$ is not.

DEFINITION 3.10. A family $\mathcal{F}$ of homogeneous ideals in a polynomial ring is called a principal perfection system if for every $I \in \mathcal{F}$, one of the following conditions holds:

(1) I is $\mathcal{F}$-maximal, i.e. maximal with respect to inclusion in $\mathcal{F}$,
(2) there exists $x \in R \setminus I$ such that $(I : x), I + (x) \in \mathcal{F}$,

(3) there exists ideals $J_1, J_2 \in \mathcal{F}$ such that $I = J_1 \cap J_2$ and $J_1 + J_2 \in \mathcal{F}$,
(4) there exists an ideal $J \in \mathcal{F}$ such that $I \cap J, I + J \in \mathcal{F}$.

Ideals satisfying either (2) or (3) are called *$\mathcal{F}$-interior ideals*; the set of all $\mathcal{F}$-interior ideals is denoted $\mathcal{F}^\circ \subset \mathcal{F}$. Ideals of $\mathcal{F}$ that are not $\mathcal{F}$-interior are called *$\mathcal{F}$-boundary ideals* and their set is denoted by $\partial\mathcal{F} \subset \mathcal{F}$.

The following result is a direct consequence of Fact 3.8 and Fact 3.9.

FACT 3.11. If $\mathcal{F}$ is a principal perfection system such that all $\mathcal{F}$-boundary ideals are perfect, then every ideal of $\mathcal{F}$ is perfect.

Using the machine in Fact 3.11, we obtain the results Fact 2.11, Fact 3.3, and Fact 3.5. Let $\lambda = (n-k,k) \in \mathcal{P}_n$ and $\hat{\lambda} = (n-k,k+1) \in \mathcal{P}_{n+1}$.

FACT 3.12. If $\mathrm{char}(\mathbb{F}) = p$ and $n, k \in \mathbb{Z}$ are positive integers satisfying $2k + 1 \leq n$, and either $p = 0$ or $p \geq k+1$, then there exists a principal perfection system $\mathcal{F}(n,k)$ in R^{n+1} in which the the Specht ideal $I_{(n-k,k+1)}$, the shifted Specht ideal $I_{(n-k,k)[1]}$, and the Specht-monomial ideal $I(n,k)$ are $\mathcal{F}(n,k)$-interior ideals, and the $\mathcal{F}(n,k)$-boundary ideals are either monomial ideals of the form $(x_1,\ldots,x_n)^{\langle k+1\rangle}$, $(x_1,\ldots,x_{n-1})^{\langle k\rangle}$, or $(y_1,\ldots,y_{n-1})^{\langle k\rangle}+(x_n^2)$, or else either Specht ideals $I_{(n-k,k)}$ or Specht-monomial ideals $I(n-1,k-1)$ with smaller parameters.

Unfortunately, the implication in Fact 3.11 is strict, hence in the case $0 < p < k+1$ where the $\mathcal{F}$-boundary ideals are imperfect, we cannot deduce the same for the $\mathcal{F}$-interior ideals.

4. Concluding Remarks and Open Problems

The following was essentially conjectured by Yanagawa [**26**]:

CONJECTURE 4.1. (1) *If $\lambda = (n-k,k)$ is a two-rowed partition, and if $0 < p < k$ then the Specht ideal $I_{(n-k,k)}$ is not perfect.*
(2) *If $\lambda = (d,d,1)$ is a special three-rowed partition, then the Specht ideal I_λ is perfect if and only if $p = 0$ or $p \geq d+1$*

Fact 2.11 together with Conjecture 4.1(1) would prove that the two-rowed Specht ideal $I_{(n-k,k)}$ is perfect if and only if $p = 0$ or $p \geq k$, which was Yanagawa's original conjecture for two-rowed partitions. For special three rowed partitions $\lambda = (d,d,1)$, Fact 2.10 and Fact 2.7 prove Conjecture 4.1 in the case $p = 0$, but in positive characteristic, the problem is open.

Our method of principal perfection systems in the two-rowed case, while completely elementary, relies on knowing an ideal *both* in terms of its generating set *and* in terms of its primary decomposition, and therein lies the difficulty. One cannot help but wonder if there might be an alternative, simpler proof of Fact 2.11 or even of Conjecture 4.1 awaiting discovery. One idea is to exploit Fact 2.6 and use the theory of Gröbner bases. For example, the following result is due to Conca-Varbaro [**2**, Theorem 1.3]:

FACT 4.2. Let $I \subset R$ be a homogeneous ideal in a polynomial ring over any field $\mathbb{F}$ and $<$ be any monomial ordering. Assume that $in_<(I)$ is a square-free monomial ideal. Then I is perfect if and only if $in_<(I)$ is perfect.

Fact 4.2 applies to two-rowed Specht ideals since two-rowed Specht polynomials are square-free and form a Gröbner basis by Fact 2.6. Given a two-rowed Specht ideal I_λ its initial ideal $in_<(I_\lambda)$ (with respect to the lexicographic ordering say) is the Stanley-Reisner ideal of a simplicial complex Δ_λ. Perfection of $in_<(I_\lambda)$ can thus be determined by Reisner's criterion:

FACT 4.3. The square-free monomial ideal $in_<(I_\lambda)$ is perfect over $\mathbb{F}$ if and only if for every face $F \in \Delta_\lambda$, its link

$$\operatorname{link}(F, \Delta_\lambda) = \{G \in \Delta_\lambda \mid F \cup G \in \Delta_\lambda,\ F \cap G = \emptyset\}$$

has reduced cohomology groups satisfying

$$\tilde{H}^i(\operatorname{link}(F, \Delta_\lambda), \mathbb{F}) = 0, \quad \forall i \neq \dim(\Delta_\lambda) - |F|.$$

PROBLEM 4.4. For $\lambda = (n-k, k)$ and $p = \operatorname{char}(\mathbb{F})$, prove (or disprove) that Δ_λ satisfies Reisner's criterion if and only if $p = 0$ or $p \geq k$.

References

[1] A. Brookner, D. Corwin, P. Etingof, and S. V. Sam, *On Cohen-Macaulayness of S_n-invariant subspace arrangements*, Int. Math. Res. Not. IMRN **7** (2016), 2104–2126, DOI 10.1093/imrn/rnv200. MR3509948

[2] A. Conca and M. Varbaro, *Square-free Gröbner degenerations*, Invent. Math. **221** (2020), no. 3, 713–730, DOI 10.1007/s00222-020-00958-7. MR4132955

[3] P. Etingof, E. Gorsky, and I. Losev, *Representations of rational Cherednik algebras with minimal support and torus knots*, Adv. Math. **277** (2015), 124–180, DOI 10.1016/j.aim.2015.03.003. MR3336085

[4] D. R. Grayson and M. E. Stillman: *Macaulay2, a software system for research in algebraic geometry.*

[5] C. B. Zamaere, S. Griffeth, and S. V. Sam, *Jack polynomials as fractional quantum Hall states and the Betti numbers of the $(k+1)$-equals ideal*, Comm. Math. Phys. **330** (2014), no. 1, 415–434, DOI 10.1007/s00220-014-2010-4. MR3215587

[6] R. Fröberg and B. Shapiro, *On Vandermonde varieties*, Math. Scand. **119** (2016), no. 1, 73–91, DOI 10.7146/math.scand.a-24185. MR3548867

[7] W. Fulton, *Young tableaux*, London Mathematical Society Student Texts, vol. 35, Cambridge University Press, Cambridge, 1997. With applications to representation theory and geometry. MR1464693

[8] M. Hochster and J. A. Eagon, *Cohen-Macaulay rings, invariant theory, and the generic perfection of determinantal loci*, Amer. J. Math. **93** (1971), 1020–1058, DOI 10.2307/2373744. MR302643

[9] T. Harima, T. Maeno, H. Morita, Y. Numata, A. Wachi, and J. Watanabe, *The Lefschetz properties*, Lecture Notes in Mathematics, vol. 2080, Springer, Heidelberg, 2013, DOI 10.1007/978-3-642-38206-2. MR3112920

[10] G. D. James and M. H. Peel, *Specht series for skew representations of symmetric groups*, J. Algebra **56** (1979), no. 2, 343–364, DOI 10.1016/0021-8693(79)90342-9. MR528580

[11] A. R. Kustin and A. Vraciu, *The weak Lefschetz property for monomial complete intersection in positive characteristic*, Trans. Amer. Math. Soc. **366** (2014), no. 9, 4571–4601, DOI 10.1090/S0002-9947-2014-05944-8. MR3217693

[12] S.-Y. R. Li and W. C. W. Li, *Independence numbers of graphs and generators of ideals*, Combinatorica **1** (1981), no. 1, 55–61, DOI 10.1007/BF02579177. MR602416

[13] C. McDaniel and J. Watanabe, *Principal radical systems, Lefschetz properties, and perfection of Specht ideals of two-rowed partitions*, Nagoya Math. J. **247** (2022), 690–730, DOI 10.1017/nmj.2021.17. MR4480098

[14] I. G. Macdonald, *Symmetric functions and Hall polynomials*, 2nd ed., Oxford Classic Texts in the Physical Sciences, The Clarendon Press, Oxford University Press, New York, 2015. With contribution by A. V. Zelevinsky and a foreword by Richard Stanley; Reprint of the 2008 paperback edition [MR1354144]. MR3443860

[15] J. C. Migliore, R. M. Miró-Roig, and U. Nagel, *Monomial ideals, almost complete intersections and the weak Lefschetz property*, Trans. Amer. Math. Soc. **363** (2011), no. 1, 229–257, DOI 10.1090/S0002-9947-2010-05127-X. MR2719680

[16] P. Moustrou, C. Riener, and H. Verdure, *Symmetric ideals, Specht polynomials and solutions to symmetric systems of equations*, J. Symbolic Comput. **107** (2021), 106–121, DOI 10.1016/j.jsc.2021.02.002. MR4224847

[17] S. Murai, H. Ohsugi, and K. Yanagawa, *A note on the reducedness and Gröbner bases of Specht ideals*, Comm. Algebra **50** (2022), no. 12, 5430–5434, DOI 10.1080/00927872.2022.2085288. MR4484722

[18] P. Orlik and H. Terao, *Arrangements of hyperplanes*, Grundlehren der mathematischen Wissenschaften [Fundamental Principles of Mathematical Sciences], vol. 300, Springer-Verlag, Berlin, 1992, DOI 10.1007/978-3-662-02772-1. MR1217488

[19] M. H. Peel, *Specht modules and symmetric groups*, J. Algebra **36** (1975), no. 1, 88–97, DOI 10.1016/0021-8693(75)90158-1. MR374253

[20] G. A. Reisner, *Cohen-Macaulay quotients of polynomial rings*, Advances in Math. **21** (1976), no. 1, 30–49, DOI 10.1016/0001-8708(76)90114-6. MR407036

[21] B. E. Sagan, *The symmetric group*, 2nd ed., Graduate Texts in Mathematics, vol. 203, Springer-Verlag, New York, 2001. Representations, combinatorial algorithms, and symmetric functions, DOI 10.1007/978-1-4757-6804-6. MR1824028

[22] K. Shibata and K. Yanagawa, *Elementary construction of minimal free resolutions of the Specht ideals of shapes* $(n-2,2)$ *and* $(d,d,1)$, J. Algebra Appl. **22** (2023), no. 9, Paper No. 2350199, 26, DOI 10.1142/S0219498823501992. MR4614721

[23] K. Shibata and K. Yanagawa, *Regularity of Cohen-Macaulay Specht ideals*, J. Algebra **582** (2021), 73–87, DOI 10.1016/j.jalgebra.2021.04.022. MR4256906

[24] W. Specht, *Die irreduziblen Darstellungen der symmetrischen Gruppe* (German), Math. Z. **39** (1935), no. 1, 696–711, DOI 10.1007/BF01201387. MR1545531

[25] J. Watanabe and K. Yanagawa, *Vandermonde determinantal ideals*, Math. Scand. **125** (2019), no. 2, 179–184, DOI 10.7146/math.scand.a-114906. MR4031045

[26] K. Yanagawa, *When is a Specht ideal Cohen-Macaulay?*, J. Commut. Algebra **13** (2021), no. 4, 589–608, DOI 10.1216/jca.2021.13.589. MR4366840

Department of Mathematics, Endicott College, Beverly, Massachusetts 01915
Email address: cmcdanie@endicott.edu

Articles

Contemporary Mathematics
Volume **805**, 2024
https://doi.org/10.1090/conm/805/16125

A note on Artin Gorenstein algebras with Hilbert function $(1, 4, k, k, 4, 1)$

Nancy Abdallah

ABSTRACT. We study the free resolutions of some Artin Gorenstein algebras of Hilbert function $(1, 4, k, k, 4, 1)$ and we prove that all such algebras have the strong Lefschetz property if they have the weak Lefschetz property. In the case $k = 4$ we prove that the Hilbert function alone fixes the Betti table. For higher k stronger conditions on the algebras are needed to fix the Betti table. In particular, if the algebra is a complete intersection or if it is defined by an equigenerated ideal then the Betti table is unique.

1. Introduction

Let $S = \mathbb{K}[x_1, \ldots, x_n]$ be the standard graded polynomial ring over a field $\mathbb{K}$ of characteristic zero, and I a nondegenerate homogeneous ideal (containing no linear form) such that $A = S/I$ is an Artinian algebra. The Hilbert function of A is the function $H(A, i) : \mathbb{Z} \to \mathbb{Z}$, $H(A, i) = \dim_{\mathbb{K}} A_i$. These dimensions are given by the H-vector of A, $H(A) = (1, h_1, h_2, \ldots, h_s)$ where $h_i = H(A, i) > 0$. The integer s is called socle degree of A (the highest integer i for which $\dim_{\mathbb{K}} A_i \neq 0$).

Consider the ring $R = \mathbb{K}[X_1, \ldots, X_n]$. S acts on R by differentiation

$$x_i(X_j) = \frac{\partial X_j}{\partial X_i} = \delta_{ij}.$$

Define $I_F = \operatorname{ann}_S(F)$ for a homogeneous polynomial F of degree s. Macaulay shows in [**15**], that S/I_F is Gorenstein of socle degree s, and that for every Artin Gorenstein (AG) algebra $A = S/I$ of socle degree s, $I = I_F$ for some homogenous polynomial $F \in R_s$. F is called the Macaulay dual generator of I and I the Macaulay inverse system of F.

For a finitely $\mathbb{Z}$-graded S-module M, the Hilbert Syzygy Theorem [**9**] ensures the existence of a minimal graded finite free resolution, which is an exact sequence

$$0 \longrightarrow F_i \longrightarrow F_{i-1} \longrightarrow \cdots \longrightarrow F_0 \longrightarrow M \longrightarrow 0,$$

for $i \leq n$ and $F_i \simeq \oplus_j S(-j)^{b_{i,j}}$ with $b_{i,j} \in \mathbb{Z}$. This information can be captured by the *Betti table*, an array whose entry in position (i, j) (reading over and down) is

2020 *Mathematics Subject Classification.* Primary 13E10, 13F55, 13D40, 13C13, 13D02.
Key words and phrases. Artinian algebra, Gorenstein algebra, Free resolutions, Lefschetz property.

$b_{i,i+j}$ (see [**9**]). With this indexing, the bottom row of the Betti table encodes the regularity of M. When $M = S/I$ is Artin Gorenstein, the regularity of M coincides with its socle degree.

EXAMPLE 1.1. *For $F = y_1y_2y_3 \in \mathbb{K}[y_1, y_2, y_3]$, we have $I_F = \langle x_1^2, x_2^2, x_3^2\rangle$ and the minimal free resolution is given by the Koszul complex*

$$0 \longrightarrow S(-6) \longrightarrow S(-4)^3 \longrightarrow S(-2)^3 \longrightarrow S \longrightarrow S/I_F \longrightarrow 0,$$

which in Betti table notation is written as

```
+--------------+
|       0 1 2 3|
|total: 1 3 3 1|
|    0: 1 . . .|
|    1: . 3 . .|
|    2: . . 3 .|
|    3: . . . 1|
+--------------+
```

For codimension 4 regularity (socle degree) 3 Artin Gorenstein (AG) algebras, the H-vector is $(1, 4, 4, 1)$ and there are three possible Betti tables (see [**3**]). For AG algebras with H-vector $(1, 4, k, 4, 1)$, $k = 4, \ldots, 10$ (codimenion 4 regularity 4), there are 16 possible betti tables (see [**18**]). In both cases the corresponding algebras have the weak Lefschetz property (see [**10**]). In this note, we study the structure of AG algebras with regularity 5. In this case the H-vector is $(1, 4, k, k, 4, 1)$, $k = 4, \ldots, 10$. We prove first that for these algebras the weak Lefschetz property implies the strong Lefschetz property, which was noticed in [**3**] by looking at the Jordan type for a general linear form. In [**3**], the authors conjectured also that there are 36 possible Betti tables for such AG algebras. Among these 36 Betti tables, there is only one that corresponds to the case $k = 4$, and one that corresponds to a complete intersection. The main results of this paper are Theorems 3.4 and 4.2 where we prove that the conjecture holds for these two cases.

In Section 2, we recall some definitions and results about Lefschetz properties and we investigate these properties in the case of algebras with codimension 4 and regularity 5. AG algebras of H-vector $(1, 4, 4, 4, 4, 1)$ are considered in Section 3. Section 4 is devoted to the study of complete intersections. We conclude in Section 5 with some open problems.

2. Lefschetz Properties

DEFINITION 2.1. *Let A be an Artinian $\mathbb{Z}$-graded algebra. We say that*

(1) *A has the weak Lefschetz property (WLP) if there exists a linear form $\ell \in A_1$ such that the multiplication map $\ell : A_i \to A_{i+1}$ has maximal rank for all i. ℓ is called a weak Lefschetz element of A.*
(2) *A has the strong Lefschetz property (SLP), if there is a linear form ℓ such that the multiplication map $\ell^k : A_i \to A_{i+k}$ has maximal rank for all i and k. ℓ is called a strong Lefschetz element A.*

When A is Gorenstein, A has SLP if there is a linear form ℓ such that the multiplication map $\ell^{s-2i} : A_i \to A_{s-i}$ has maximal rank (bijective) for all $i = 1, \ldots, \lfloor \frac{s}{2} \rfloor$, where s is the socle degree of A.

It is clear from the definition that if A has SLP then A has WLP. In codimension 2, all standard graded Artinian algebras have SLP provided that the field has characteristic zero or $p > s$ ([**13**]). In codimension 3, all complete intersections

(CI) have WLP ([**13**]), and in [**5**], the authors conjectured that all codimension 3 AG algebras have WLP. In a recent paper ([**2**]), we prove that all codimension 3 AG algebras with Sperner number (maximal value of H-vector) at most 6, as well as algebras of H-vector $(1,3,4,5,\ldots,(s-1),s^k,(s-1),\ldots,3,1)$, have SLP. Our computation in [**3**] shows that for codimension 4 AG algebras with socle degree 5, the SLP holds when the Sperner number is at most 6, and the bound is sharp.

We recall now the definition of Hessians and its relation with strong Lefschetz properties.

DEFINITION 2.2. *([**16**], Definition 3.1) Let F be a polynomial in R and $A = S/\operatorname{ann}_S F$ be its associated AG algebra. Let $\mathcal{B}_i = \{\alpha_m^{(i)}\}_m$ be a $\mathbb{K}$-basis of A_i. The i-th Hessian matrix of F with respect to $\mathcal{B}_i$ is defined by*

$$(\operatorname{Hess}^i(F))_{u,v} = (\alpha_u^{(i)}\alpha_v^{(i)} \circ F).$$

We write $\operatorname{hess}^i(F)$ for the determinant of $\operatorname{Hess}^i(F)$ which is independent of the basis $\mathcal{B}_i$ up to a non-zero constant multiple. For $i=1$, $\operatorname{Hess}^1(F)$ coincides with the usual Hessian.

The following result due to T. Maeno and J. Watanabe gives a strong connection between the Hessian matrices and the Strong Lefschetz propeties (see [**16**]).

THEOREM 2.3. *Let $A = S/\operatorname{ann}_S F$ be an AG algebra. $\ell = a_1x_1+\cdots+a_nx_n \in A_1$ is a strong Lefschetz element of A if and only if $\operatorname{hess}^i(F)(a_1,\ldots,a_n) \neq 0$ for all $i = 1,\ldots,\lfloor \frac{s}{2} \rfloor$. In particular, the multiplication map $\ell^{s-2i} : A_i \to A_{s-i}$ is bijective if and only if $\operatorname{hess}^i(F)(a_1,\ldots,a_n) \neq 0$.*

The following theorem due to P. Gordan and M. Noether in [**11**] was reproved in [**14**, Theorem 1.2], [**19**], [**6**].

THEOREM 2.4 (Gordan-Noether). *If F is a form expressed in at most four variables and the characteristic of $\mathbb{K}$ is zero, then the Hessian determinant $\operatorname{hess}(F)$ is identically zero if and only if F is a cone: that is, if F is annihilated by a linear form (element of S_1) in the Macaulay duality.*

In [**3**], our computation showed that for AG algebras with H-vector $(1,4,k,k,4,1)$, $k=4,\ldots,10$, whenever WLP holds, SLP holds as well. The following proposition proves that the claim holds.

PROPOSITION 2.5. *Let $A = S/I$ be an AG algebra where I is a nondegenerate homogeneous ideal and $H(A) = (1,4,k,k,4,1)$, $k = 4,\ldots,10$. If A has WLP then A has SLP.*

PROOF. The first Hessian is non zero by Gordan Noether (Theorem 2.4) and the second Hessian is non zero by the WLP property (the middle map is bijective). The result follows by Theorem 2.3. □

3. Artin Gorenstein algebras of H-vector $(1,4,4,4,4,1)$

We consider in this section *AG* algebras with H-vector $(1,4,4,4,4,1)$. In [**3**], we conjectured that there is only one Betti table for this H-vector and that all these AG algebras have SLP. We prove in this section that both claims hold.

PROPOSITION 3.1. *Every AG algebra with H-vector $(1,4,4,4,4,1)$ has SLP.*

PROOF. The fact that such an algebra has WLP follows from in [**1**, Proposition 3.3]. By Proposition 2.5, SLP holds as well. $\square$

We review the theorems of Macaulay and Gotzmann (§7.2 of [**17**]). For a graded algebra S/I with Hilbert function h_i, write

$$h_i = \binom{a_i}{i} + \binom{a_{i-1}}{i-1} + \cdots \text{ and } h_i^{\langle i\rangle} = \binom{a_i+1}{i+1} + \binom{a_{i-1}+1}{i} + \cdots,$$

where $a_i > a_{i-1} > \cdots$. Then we have

THEOREM 3.2 (Macaulay). *In the setting above,*

$$h_{i+1} \leq h_i^{\langle i\rangle}.$$

THEOREM 3.3 (Gotzmann). *If I is generated in a single degree t and equality holds in Macaulay's formula in the first degree t, then*

$$h_{t+j} = \binom{a_t+j}{t+j} + \binom{a_{t-1}+j-1}{t+j-1} + \cdots$$

We now prove our first main result.

THEOREM 3.4. *An AG algebra A with Hilbert function* $(1,4,4,4,4,1)$ *has Betti table*

```
+-----------------+
|      0 1  2 3 4|
|total: 1 9 16 9 1|
|    0: 1 .  . . .|
|    1: . 6  8 3 .|
|    2: . .  . . .|
|    3: . .  . . .|
|    4: . 3  8 6 .|
|    5: . .  . . 1|
+-----------------+
```

PROOF. We know that the Betti table of a Gorenstein algebra is symmetric, and since $h_2(A) = 4$, the ideal I has 6 quadrics, therefore $b_{12} := b_{1,2} = 6 = b_{37}$. To avoid many indices we write the Betti table as below (we ommit the row of total dimensions for simplicity).

```
+--------------+
|    0 1  2 3 4|
| 0: 1 .  . . .|
| 1: . 6  b c .|
| 2: . d  e f .|
| 3: . f  e d .|
| 4: . c  b 6 .|
| 5: . .  . . 1|
+--------------+
```

Denote by I_j the degree j component of the ideal I. The Hilbert number $h_j(A)$ is given by $h_j(A) = \binom{3+j}{j} - \dim_{\mathbb{K}}(I_j)$. For $j = 3$ and $j = 4$ we have, $h_3(A) = 20 - 6\cdot 4 - d + b = 4$ and $h_4(A) = 35 - 6\cdot 10 - 4d - f + 4b + e - c = 4$. This gives the relations

$$b = d + 8 \tag{1}$$

and

$$c - e + f = 3 \tag{2}$$

Consider now the algebra $A = S/J_2$ where J_2 is the ideal generated by the quadrics of I. Then $h_i(S/J_2) = h_i(A)$ for $i \leq 2$ and the first two rows of the Betti tables of A and S/J_2 coincide. Since $h_2(S/J_2) = 4 = \binom{3}{2} + \binom{1}{1}$, by Theorem 3.2, $h_3(S/J_2) \leq h_2^{<2>}(S/J_2) = \binom{4}{3} + \binom{2}{2} = 5$. On the other hand, $h_3(S/J_2) = 20 - 6 \cdot 4 + b \leq 5$ and $b \leq 9$. Together with Equation 1, we get $b \in \{8, 9\}$. We want to prove that the case $b = 9$ cannot occur.

Suppose $b = 9$, i.e. $h_3(S/J_2) = h_2^{<2>}(S/J_2)$. By Theorem 3.3, $h_4(S/J_2) = \binom{3+2}{2+2} + \binom{1+2-1}{2+2-1} = 5$, but $h_4(S/J_2) = 35 - 6 \cdot 10 + 9 \cdot 4 + b_{24}(S/J_2) - c$. Therefore $c = b_{24}(S/J_2) + 6 \geq 6$.

We study now the Betti tables of S/J_3 and S/J_4 where J_3 and J_4 are the ideals generated by the generators of I of degree at most 3 (respectively 4), and we prove that $c \in \{2, 3\}$ which contradicts the inequality $c \geq 6$ in the case $b = 9$.

For $j = 3, 4$ the first j rows of the Betti table of S/J_j coincide with those of A, and $h_i(S/J_j) = h_i(A)$ for $i \leq j$. Then, $h_3(S/J_3) = 4 = \binom{4}{3}$ and by Macaulay

$$h_3^{<3>} = 5 \geq h_4(S/J_3) = 35 - 60 - 4d + 4b + e - c = 7 + e - c.$$

Therefore, $c \geq 2 + e \geq 2$. By Equation 2, we get $f \leq 1$. On the other hand, $h_4(S/J_4) = 4 = \binom{4}{4} + \binom{3}{3} + \binom{2}{2} + \binom{1}{1}$, and again by Macaulay,

$$h_4^{<4>}(S/J_4) = 4 \geq h_5(S/J_4) = 56 - 120 - 10d - 4f + 10b + 4e + e - 4c - f.$$

Using Equations 1 and 2, we get $c \leq 3$, a contradiction. Hence, $b = 8$ and $d = 0$.

We have now, $b = 8$, $c \in \{2, 3\}$, $d = 0$, $e = f + c - 3$ and $f \in \{0, 1\}$. It is enough to prove that $f = 1$ cannot occur.

Since $d = 0$, the ideal I has no degree three generators. This implies that S/J_2 and A share the first three rows of the Betti table, and that $h_4(S/J_2) = 35 - 60 + 32 + e - c$. If $f = 1$, $e - c = -2$, and $h_4(S/J_2) = 5 = \binom{5}{4}$, then

$$h_4^{<4>}(S/J_2) = 6 \geq h_5(S/J_2) = 56 - 120 + 80 + 4e + b_{25}(S/J_2) - 4c - 1,$$

which gives $b_{25}(S/J_2) \leq -1$, a contradiction. The remaining case is the desired case which occurs by a computation using Macaulay2. □

EXAMPLE 3.5. *The ideal* $I = (x_1x_3 - x_2x_4, x_2^2, x_2x_3, x_3^2, x_3x_4, x_4^2, x_1^4x_2, x_1^4x_4, x_1^5)$ *defines an AG algebra of H-vector* $(1, 4, 4, 4, 4, 1)$ *and Betti table as in Theorem* 3.4.

4. Complete intersection

In the previous section, we saw that certain Hilbert functions lead to strong constraints on the possible Betti table for an AG ring. Hence it is natural to ask if there are other conditions which strongly constrain the Betti table, and in this section, we consider the case where I is a CI. We prove that given codimension $c = 4$ and the regularity $r = 5$ of an AG algebra, the complete intersection condition fixes the Hilbert function and the Betti table.

To prove our theorem, we prove first that, for the same codimension and socle degree, there is exactly one Betti table for $A = S/I$ where I is an equigenerated ideal.

PROPOSITION 4.1. *Let $A = S/I$ be an AG algebra with socle degree 5. If I is equigenerated then $H(A) = (1, 4, 10, 10, 4, 1)$ and the Betti table is given by*

```
+--------------------+
|       0  1  2  3 4|
|total: 1 10 18 10 1|
|    0: 1  .  .  . .|
|    1: .  .  .  . .|
|    2: . 10  9  . .|
|    3: .  .  9 10 .|
|    4: .  .  .  . .|
|    5: .  .  .  . 1|
+--------------------+
```

PROOF. If I is equigenerated, the Betti table has one of the two forms

```
    +-------------+           +-------------+
    |    0 1 2 3 4|           |    0 1 2 3 4|
    | 0: 1 . . . .|           | 0: 1 . . . .|
    | 1: . a b . .|           | 1: . . . . .|
(1) | 2: . . e . .|       (2) | 2: . d e . .|
    | 3: . . e . .|           | 3: . . e d .|
    | 4: . . b a .|           | 4: . . . . .|
    | 5: . . . . 1|           | 5: . . . . 1|
    +-------------+           +-------------+
```

We prove first that Table (1) cannot occur. Suppose A has Betti table (1). Since the H-vector of A is $(1, 4, k, k, 4, 1)$, $k = 4, \ldots, 10$, then I has $a = 10 - k$ quadratic generators.We compute $h_i(A)$ for $i = 3, 4, 5$. $h_i(A) = \binom{3+i}{i} - \dim_{\mathbb{K}} I_i$ where I_i is the ith graded component of the graded ideal I. For $i = 3$, $h_3(A) = 20 - (4a - b) = k$; for $i = 4$, $h_4(A) = 35 - (10a - 4b - e) = 4$; and for $i = 5$, $h_5(A) = 56 - (20a - 10b - 4e - e) = 1$. Therefore, k, a, b, e are nonnegative integer solutions to the system

$$\begin{cases} k &= 10 - a \\ k &= 20 - 4a + b \\ 4 &= 35 - 10a + 4b + e \\ 1 &= 56 - 20a + 10b + 5e \end{cases}$$

The only integer solution of this system is $(k, a, b, e) = (6, 4, 2, 1)$ which gives a complete intersection, but the Betti table of a CI with 4 quadrics is

```
+----------------+
|       0 1 2 3 4|
|total: 1 4 6 4 1|
|    0: 1 . . . .|
|    1: . 4 . . .|
|    2: . . 6 . .|
|    3: . . . 4 .|
|    4: . . . . 1|
+----------------+
```

and the socle degree in this case is 4, a contradiction.

We know now that the Betti table of A is of the form of Table (2). Since I has no quadrics, $H(A) = (1, 4, 10, 10, 4, 1)$ and therefore I is generated by 10 cubics, so $d = 10$. Now since $h_4(A) = 35 - \dim I_4 = 35 - (4d - e) = 4$ then $e = 9$ and we get the desired Betti table. $\square$

We can now prove our second main theorem.

THEOREM 4.2. *Let A be a complete intersection with H-vector $(1, 4, k, k, 4, 1)$. Then the Betti table of A is*

```
      +----------------+
      |       0 1 2 3 4|
      |total: 1 4 6 4 1|
      |    0: 1 . . . .|
(3)   |    1: . 3 . . .|
      |    2: . 1 3 . .|
      |    3: . . 3 1 .|
      |    4: . . . 3 .|
      |    5: . . . . 1|
      +----------------+
```

In particular, the H-vector is $(1, 4, 7, 7, 4, 1)$.

PROOF. Since A is a CI, I has at most four quadratic generators. Then k cannot be 4 or 5. If $k = 6$, then I is an equigenerated ideal generated by 4 quadrics. This is not possible by Proposition 4.1. If $k = 10$, then I has ten cubic generators and is not a CI. We are left with the cases $k = 7, 8, 9$.

Suppose $k = 9$. The ideal I is generated by one quadric ($b_{12} = 1$), and three other generators of degree 3, 4 or 5. Since it is not possible to have syzygies on one generator, $b_{23} = b_{34} = 0$. Therefore $h_3(A) = \binom{6}{3} - \dim I_3 = 20 - (4 \cdot 1 + b_{13}) = 9$, and $b_{13} = 7$ which does not give a CI.

Suppose now that $k = 8$. In this case I has two quadratic generators and two generators of degree 3, 4 or 5 (i.e., $b_{13} + b_{14} + b_{15} = 2$). There is at most one linear syzygy on two quadrics. Then, b_{23} is either 0 or 1. Therefore, $h_3(A) = 20 - (4 \cdot 2 + b_{13} - b_{23}) = 8$, and $b_{13} = 4 - b_{23}$ is either 3 or 4; a contradiction since $b_{13} \leq 2$.

Hence, we proved that $H(A) = (1, 4, 7, 7, 4, 1)$, so $b_{12} = 3$ and $b_{13}+b_{14}+b_{15} = 1$. If $b_{13} = 0$, $h_3(A) = 20 - (4 \cdot 3 - b_{23}) = 7$ and $b_{23} = -1$, a contradiction. Therefore, the Betti table of A is of the form

```
+--------------+
|    0 1  2 3 4|
| 0: 1 .  . . .|
| 1: . 3  b . .|
| 2: . 1  e . .|
| 3: . .  e 1 .|
| 4: . .  b 3 .|
| 5: . .  . . 1|
+--------------+
```

Now $h_3(A) = 20-(12+1-b) = 7$ and $b = 0$; $h_4(A) = 35-(3 \cdot 10+1 \cdot 4-e) = 4$ and $e = 3$. We get the desired table which is realizable (see the following example). $\square$

EXAMPLE 4.3. *Let $I = (x_1^2, x_2^2, x_3^2, x_4^3)$. $A = S/I$ is a CI of H-vector $H(A) = (1, 4, 7, 7, 4, 1)$ and Betti table as in Table (3). The Macaulay dual generator is $F = X_1 X_2 X_3 X_4^2$.*

REMARK 4.4. *It was pointed out by Anthony Iarrobino that a part of the proof can be simplified using* [**7**, Theorem 2.3.11]*: for a CI in codimension 4, if $D = (d_1, d_2, d_3, d_4)$ are the degrees generators of I with $2 \leq d_1 \leq d_2 \leq d_3 \leq d_4$, then $s + 4 = \Sigma_i d_i$. For the case $(1, 4, k, k, 4, 1)$, we have then only one choice for D: $D = (2, 2, 2, 3)$. This forces k to be 7 and gives the second and fourth column of the Betti table.*

REMARK 4.5. *Note that for the H-vector $(1,4,7,7,4,1)$ it is conjectured in* [**3**] *that there are 9 possible Betti tables, and that SLP holds for all AG algebras with the Betti tables below.*

```
+-----------------+-----------------+-----------------+-----------------+
|       0 1  2 3 4|       0 1  2 3 4|       0 1  2 3 4|       0 1  2 3 4|
|total: 1 4  6 4 1|total: 1 6 10 6 1|total: 1 6 10 6 1|total: 1 7 12 7 1|
|    0: 1 .  . . .|    0: 1 .  . . .|    0: 1 .  . . .|    0: 1 .  . . .|
|    1: . 3  . . .|    1: . 3  2 . .|    1: . 3  2 . .|    1: . 3  . . .|
|    2: . 1  3 . .|    2: . 3  3 . .|    2: . 2  4 1 .|    2: . 1  6 3 .|
|    3: . .  3 1 .|    3: . .  3 3 .|    3: . 1  4 2 .|    3: . 3  6 1 .|
|    4: . .  . 3 .|    4: . .  2 3 .|    4: . .  . 3 .|    4: . .  . 3 .|
|    5: . .  . . 1|    5: . .  . . 1|    5: . .  . . 1|    5: . .  . . 1|
+-----------------+-----------------+-----------------+-----------------+
|       0 1  2 3 4|       0 1  2 3 4|       0 1  2 3 4|       0 1  2 3 4|
|total: 1 7 12 7 1|total: 1 9 16 9 1|total: 1 9 16 9 1|total: 1 6 10 6 1|
|    0: 1 .  . . .|    0: 1 .  . . .|    0: 1 .  . . .|    0: 1 .  . . .|
|    1: . 3  2 . .|    1: . 3  2 . .|    1: . 3  3 1 .|    1: . 3  3 1 .|
|    2: . 3  4 . .|    2: . 3  6 3 .|    2: . 4  5 1 .|    2: . 4  7 3 .|
|    3: . 1  4 1 .|    3: . 3  6 3 .|    3: . 1  5 4 .|    3: . 3  7 4 .|
|    4: . .  2 3 .|    4: . .  2 3 .|    4: . 1  3 3 .|    4: . 1  3 3 .|
|    5: . .  . . 1|    5: . .  . . 1|    5: . .  . . 1|    5: . .  . . 1|
+-----------------+-----------------+-----------------+-----------------+
```

For AG algebras with the table

```
+-----------------+
|       0 1  2 3 4|
|total: 1 8 14 8 1|
|    0: 1 .  . . .|
|    1: . 3  1 . .|
|    2: . 2  6 3 .|
|    3: . 3  6 2 .|
|    4: . .  1 3 .|
|    5: . .  . . 1|
+-----------------+
```

there are examples where SLP holds and other examples where WLP fails.

5. Open problems

We conclude this note with some open problems. Let $\mathrm{Gor}(H)$ be the set of all AG with H-vector H. In [**8**] S.J. Diesel showed that in three variables and over an algebraically closed field the variety $\mathrm{Gor}(H)$ is irreducible. In four variables, Boij showed in [**4**] that $\mathrm{Gor}(H)$ can consist of several components.

PROBLEM 5.1. *It would be interesting to investigate the loci in* $\mathrm{Gor}(H)$ *corresponding to the possible betti tables. For example, if we order Betti tables as a poset with $\beta \geq \beta'$ if $b_{i,j} \geq b'_{i,j}$ for all i,j, then for the H-vector $(1,4,5,5,4,1)$ there is a strict inclusion of the Betti tables whereas for the $(1,4,6,6,4,1)$ this is not the case.*

The Jordan type of a linear form ℓ acting on an Artinian algebra A is the partition describing the Jordan block decomposition of the matrix representing the multiplication by a linear form between the graded pieces of $A_i \xrightarrow{\cdot\ell} A_{i+1}$. In [**12**] the authors show that A has SLP if and only if the partition $P_{A,\ell}$ is conjugate to the H-vector. In [**3**], by a computation with Macaulay2 we see that the Jordan type of a CI with Betti table as in (3) is $(6,4,4,4,2,2,2)$ for a general linear form, which means that such algebras have SLP.

PROBLEM 5.2. *Show that all codimension 4 CI with socle degree 5 have SLP.*

EXAMPLE 5.3. *The algebra* $A = S/(x_1^2, x_2^2, x_3x_4, x_3^3 - x_4^3)$ *is a CI. For* $\ell = x_1 + x_2 + x_3 + x_4$ *the Jordan type is* $(6,4,4,4,2,2,2,2)$ *which means that* A *has SLP.*

Acknowledgments

The author would like to thank the referee, Anthony Iarrobino and Hal Schenck for their useful remarks, and Anthony Iarrobino, Pedro Macias Marques, Maria Evelina Rossi and Jean Vallès for organizing the conference AMS-EMS-SMF special session on Deformation of Artinian algebras and Jordan Type in Grenoble.

References

[1] N. Abdallah, N. Altafi, P. De Poi, L. Fiorindo, A. Iarrobino, P. M. Marques, E. Mezzetti, R. M. Miró-Roig, L. Nicklasson. Hilbert functions and Jordan type of Perazzo Artinian algebras, *Springer INdAM Series (to appear)*, arXiv:2303.16768, (2023).

[2] N. Abdallah, N. Altafi, A. Iarrobino, A. Seceleanu, and J. Yaméogo, *Lefschetz properties of some codimension three Artinian Gorenstein algebras*, J. Algebra **625** (2023), 28–45, DOI 10.1016/j.jalgebra.2023.03.005. MR4567483

[3] N. Abdallah and H. Schenck, *Free resolutions and Lefschetz properties of some Artin Gorenstein rings of codimension four*, J. Symbolic Comput. **121** (2024), Paper No. 102257, 13, DOI 10.1016/j.jsc.2023.102257. MR4634957

[4] M. Boij, *Components of the space parametrizing graded Gorenstein Artin algebras with a given Hilbert function*, Pacific J. Math. **187** (1999), no. 1, 1–11, DOI 10.2140/pjm.1999.187.1. MR1674301

[5] M. Boij, J. Migliore, R. M. Miró-Roig, U. Nagel, and F. Zanello, *On the weak Lefschetz property for Artinian Gorenstein algebras of codimension three*, J. Algebra **403** (2014), 48–68, DOI 10.1016/j.jalgebra.2014.01.003. MR3166064

[6] D. Bricalli, F. F. Favale, and G. P. Pirola, *A theorem of Gordan and Noether via Gorenstein rings*, Selecta Math. (N.S.) **29** (2023), no. 5, Paper No. 74, 25, DOI 10.1007/s00029-023-00882-7. MR4651997

[7] W. Bruns and J. Herzog, *Cohen-Macaulay rings*, Cambridge Studies in Advanced Mathematics, vol. 39, Cambridge University Press, Cambridge, 1993. MR1251956

[8] S. J. Diesel, *Irreducibility and dimension theorems for families of height* 3 *Gorenstein algebras*, Pacific J. Math. **172** (1996), no. 2, 365–397. MR1386623

[9] D. Eisenbud, *Commutative algebra*, Graduate Texts in Mathematics, vol. 150, Springer-Verlag, New York, 1995. With a view toward algebraic geometry, DOI 10.1007/978-1-4612-5350-1. MR1322960

[10] R. Gondim, *On higher Hessians and the Lefschetz properties*, J. Algebra **489** (2017), 241–263, DOI 10.1016/j.jalgebra.2017.06.030. MR3686979

[11] P. Gordan and M. Nöther, *Ueber die algebraischen Formen, deren Hesse'sche Determinante identisch verschwindet* (German), Math. Ann. **10** (1876), no. 4, 547–568, DOI 10.1007/BF01442264. MR1509898

[12] T. Harima, T. Maeno, H. Morita, Y. Numata, A. Wachi, and J. Watanabe, *The Lefschetz properties*, Lecture Notes in Mathematics, vol. 2080, Springer, Heidelberg, 2013, DOI 10.1007/978-3-642-38206-2. MR3112920

[13] T. Harima, J. C. Migliore, U. Nagel, and J. Watanabe, *The weak and strong Lefschetz properties for Artinian K-algebras*, J. Algebra **262** (2003), no. 1, 99–126, DOI 10.1016/S0021-8693(03)00038-3. MR1970804

[14] C. Lossen, *When does the Hessian determinant vanish identically? (On Gordan and Noether's proof of Hesse's claim)*, Bull. Braz. Math. Soc. (N.S.) **35** (2004), no. 1, 71–82, DOI 10.1007/s00574-004-0004-0. MR2057045

[15] F. S. Macaulay, *Some Properties of Enumeration in the Theory of Modular Systems*, Proc. London Math. Soc. (2) **26** (1927), 531–555, DOI 10.1112/plms/s2-26.1.531. MR1576950

[16] T. Maeno and J. Watanabe, *Lefschetz elements of Artinian Gorenstein algebras and Hessians of homogeneous polynomials*, Illinois J. Math. **53** (2009), no. 2, 591–603. MR2594646

[17] H. Schenck, *Computational algebraic geometry*, London Mathematical Society Student Texts, vol. 58, Cambridge University Press, Cambridge, 2003, DOI 10.1017/CBO9780511756320. MR2011360

[18] H. Schenck, M. Stillman, and B. Yuan, *Calabi-Yau threefolds in $\mathbb{P}^n$ and Gorenstein rings*, Adv. Theor. Math. Phys. **26** (2022), no. 3, 765–792, DOI 10.4310/atmp.2022.v26.n3.a7. MR4557610

[19] J. Watanabe and M. de Bondt, *On the theory of Gordan-Noether on homogeneous forms with zero Hessian (improved version)*, Polynomial rings and affine algebraic geometry, Springer Proc. Math. Stat., vol. 319, Springer, Cham, [2020] ©2020, pp. 73–107, DOI 10.1007/978-3-030-42136-6_3. MR4113942

Department of Engineering, University of Borås, Borås, Sweden
Email address: `nancy.abdallah@hb.se`
URL: `https://nancyabdallah.info`

Contemporary Mathematics
Volume **805**, 2024
https://doi.org/10.1090/conm/805/16126

A note on Jordan types and Jordan degree types

Nasrin Altafi

ABSTRACT. We discuss whether the Jordan degree type encodes more information about graded Artinian Gorenstein algebras than the Jordan type for linear forms. It's known that in codimension two, the Jordan type determines the Jordan degree type. We provide examples showing that this is no longer the case in higher codimensions.

1. Introduction

For a standard graded Artinian algebra $A = K[x_1, \dots, x_n]/I$ and a linear form $\ell \in A_1$ the *Jordan type* $P_{\ell,A}$ is the partition of $\dim_K A$ determining the Jordan block decomposition of the multiplication map by ℓ on A. Jordan type determines the weak and strong Lefschetz properties of Artinian algebras. A graded Artinian algebra A is said to satisfy the weak Lefschetz property (WLP) if multiplication map by a linear form on A has maximal rank in every degree. If this holds for all powers of a linear form the algebra A is said to have the strong Lefschetz property (SLP). It is known that an Artinian algebra A has the WLP if there is a linear form ℓ where the number of parts in $P_{\ell,A}$ is equal to the Sperner number of A, the maximum value of the Hilbert function h_A. Also A has the SLP if there is a linear form ℓ such that $P_{\ell,A} = h_A^\vee$ the conjugate partition of h_A, see [**7**] and [**10**] for the refined version. Jordan type of a linear form for an Artinian algebra captures more information than the weak and Strong Lefschetz properties. For recent papers on Jordan types see [**9**–**11**] and their references.

The study of the Lefschetz properties of Artinian Gorenstein algebras has recently attracted the attention of several researchers in the field of commutative algebra and algebraic geometry. Artinian Gorenstein (AG) algebras are well structured and have many connections to other areas, such as combinatorics and algebraic geometry. Researchers in the field have recently become interested in studying Jordan types of such algebras. Costa and Gondim in [**3**] study Jordan types of AG algebras for a general linear form by computing the ranks of associated Hessian and mixed Hessian matrices introduced in [**13**] and [**5**].

2020 *Mathematics Subject Classification.* Primary 13E10, 13D40; Secondary 13H10, 05A17, 05E40.

The author was supported by the grant VR2021-00472. The author would like to thank the organizers of the AMS-EMS-SMF meeting at Grenoble 18-22 July 2022, special session on "Deformation of Artinian algebras and Jordan type". The computer software Macaulay2 [**6**] was important for the computations in this paper. The author thanks the referees for their useful comments and suggestions regarding the presentation of this paper.

Iarrobino, Macias Marques, and McDaniel in [**10**] study Jordan types of graded and local Artinian algebras. They introduce the *Jordan degree type* that determines the Jordan type as well as the initial degrees associated to different parts. They show for graded Artinian algebras of codimension two the Jordan degree type is equivalent to Jordan type. They also provide two Artinian algebras with the Hilbert function $(1,3,5,4,2,1)$ which have the same Jordan type, although their Jordan degree types are different. In this paper, we investigate this question for graded AG algebras. We recall the rank matrix method that has been introduced in [**1**]. Using this approach, we give a different proof than [**10**] that the Jordan degree type is determined by the Jordan type for AG algebras of codimension two. We then show that the Jordan degree type is a finer invariant than the Jordan type for AG algebras of codimension at least three. In Example 4.2 we provide two AG algebras with the Hilbert function $(1,3,6,9,9,9,6,3,1)$ which have equal Jordan type for a linear form but different Jordan degree types. It has been computationally verified that this example has the smallest socle degree among AG algebras of codimension three. We also give such example in codimension four for AG algebras with Hilbert function $(1,4,7,7,4,1)$, see Example 4.4.

2. Jordan types and Jordan degree types of AG algebras

Let $S = K[x_1,\dots,x_n]$ be a polynomial ring equipped with standard grading over a field K of characteristic zero. Let $A = S/I$ be a graded Artinian (its Krull dimension is zero) algebra where I is an homogeneous ideal. The *Hilbert function* of a graded Artinian algebra $A = S/I$ is a vector of non-negative integers and we denote it by $h_A = (1, h_1, \dots, h_d)$ where $h_A(i) = h_i = \dim_K(A_i)$. The integer d is called the *socle degree* of A, that is the largest integer i such that $h_A(i) > 0$.

Let $R = K[X_1,\dots,X_n]$ be the Macaulay dual ring of S. Given a homogeneous ideal $I \subset S$ the *inverse system* of I is defined to be a graded S-module $M \subset R$ such that S acts on R by differentiation. For more details of Macaulay's inverse system see [**4**] and [**8**]. From a result by Macaulay [**12**] it is known that an Artinian standard graded K-algebra $A = S/I$ is Gorenstein if and only if there exists $F \in R_d$ such that $I = Ann_S(F)$. In other words, the inverse system of a graded Artinian Gorenstein (AG) algebra is generated by only one form. We recall the definition of the Jordan degree type, [**10**, Definition 2.28].

DEFINITION 2.1 (Jordan degree type). Let A be a graded Artinian algebra and $\ell \in A_1$. Suppose that $P_{\ell,A} = (p_1,\dots,p_t)$ is the Jordan type for ℓ and A, then there exist elements $z_1,\dots,z_t \in A$, which depend on ℓ, such that $\{\ell^i z_k \mid 1 \le k \le t, 0 \le i \le p_k - 1\}$ is a K-basis for A. The Jordan blocks of the multiplication map by ℓ is determined by the strings $\mathsf{s}_k = \{z_k, \ell z_k, \dots, \ell^{p_k-1} z_k\}$, and A is the direct sum $A = \langle \mathsf{s}_1 \rangle \oplus \cdots \oplus \langle \mathsf{s}_t \rangle$. Denote by d_k the degree of z_k. Note that the degrees d_k of z_k are well-defined, they are uniquely determined by ℓ. Then the *Jordan degree type*, is defined to be the indexed partition $\mathcal{S}_{\ell,A} = (p_{1_{d_1}},\dots,p_{t_{d_t}})$.

We recall the definition of rank matrices for AG algebras and briefly describe its connection to Jordan degree types, see [**1**] for more details. Our convention is that the row and column indices of the matrices in this paper start from zero.

DEFINITION 2.2. Let $A = S/Ann(F)$ be an AG algebra with socle degree d. For a linear form $\ell \in A$ define the *rank matrix*, $M_{\ell,A}$, of A and ℓ to be the upper triangular square matrix of size $d+1$ with the following entries

$$(M_{\ell,A})_{i,j} = \mathrm{rk}\left(\times \ell^{j-i} : A_i \longrightarrow A_j\right),$$

for every $i \leq j$. For $i > j$ we set $(M_{\ell,A})_{i,j} = 0$.

For a given rank matrix $M_{\ell,A}$ we define the *Jordan degree type matrix*, $J_{\ell,A}$, of A and ℓ to be the upper triangular matrix of size $d+1$ with the following non-negative entries

$$(J_{\ell,A})_{i,j} :=(M_{\ell,A})_{i,j} + (M_{\ell,A})_{i-1,j+1} - (M_{\ell,A})_{i-1,j} - (M_{\ell,A})_{i,j+1},$$

where we set $(M_{\ell,A})_{i,j} = 0$ if either $i < 0$ or $j < 0$.

Matrix $J_{\ell,A}$ encodes the same information as the Jordan degree type $\mathcal{S}_{\ell,A}$ which is described in detail in (2.3).

For each AG algebra $A = S/Ann(F)$ with socle degree d and linear form ℓ define new AG algebras $A_\ell^{(i)} := S/Ann(\ell^i \circ F)$ for all $0 \leq i \leq d$.

For every $0 \leq i \leq d$ we define the i-th diagonal of $M_{\ell,A}$ to be the following vector

$$\mathrm{diag}(i, M_{\ell,A}) := \big((M_{\ell,A})_{0,i}, (M_{\ell,A})_{1,i+1}, \dots, (M_{\ell,A})_{d-i,d}\big).$$

By [**1**, Proposition 3.4], we know that $\mathrm{diag}(i, M_{\ell,A}) = h_{A_\ell^{(i)}}$ for every $0 \leq i \leq d$.

Consider the decreasing sequence

$$\mathbf{d} := (\dim_K A_\ell^{(0)}, \dim_K A_\ell^{(1)}, \dots, \dim_K A_\ell^{(d)}),$$

and recall that the second difference sequence of $\mathbf{d}$ is denoted by $\Delta^2\mathbf{d}$ and its i-th entry is given by

$$\Delta^2\mathbf{d}(i) = \dim_K A_\ell^{(i)} + \dim_K A_\ell^{(i+2)} - 2\dim_K A_\ell^{(i+1)}, \tag{2.1}$$

where we set $\dim_K A_\ell^{(i)} = 0$ for $i > d$. Then obtain the Jordan type of A and ℓ from [**1**, Proposition 3.14].

$$P_{\ell,A} = \big(\underbrace{d+1, \dots, d+1}_{n_d}, \underbrace{d, \dots, d}_{n_{d-1}}, \dots, \underbrace{2, \dots, 2}_{n_1}, \underbrace{1, \dots, 1}_{n_0}\big), \tag{2.2}$$

such that $\mathbf{n} = (n_0, n_1, \dots, n_d) = \Delta^2\mathbf{d}$.

One can see that $n_i = \sum_{k-j=i} (J_{\ell,A})_{j,k}$ for each $0 \leq i \leq d$. Moreover, to obtain the Jordan degree type, one can read off the degrees of the entries in $P_{\ell,A}$ from the row indices of non-zero elements in $J_{\ell,A}$. More precisely,

$$\begin{aligned}\mathcal{S}_{\ell,A} =\big(&\underbrace{(d+1)_0, \dots, (d+1)_0}_{n_{d,0}}, \underbrace{d_0, \dots, d_0}_{n_{d-1,0}}, \underbrace{d_1, \dots, d_1}_{n_{d-1,1}},\\ &\underbrace{(d-1)_0, \dots, (d-1)_0}_{n_{d-2,0}}, \dots, \underbrace{(d-1)_2, \dots, (d-1)_2}_{n_{(d-2),2}},\\ &\vdots\\ &\underbrace{2_0, \dots, 2_0}_{n_{1,0}}, \dots, \underbrace{2_d, \dots, 2_d}_{n_{1,d-1}}, \underbrace{1_0, \dots, 1_0}_{n_{0,0}}, \dots, \underbrace{1_{d+1}, \dots, 1_{d+1}}_{n_{0,d}}\big),\end{aligned} \tag{2.3}$$

where $n_{i,j} = (J_{\ell,A})_{j,i+j}$ and $n_i = \sum_{j=0}^{d-i} n_{i,j}$.

EXAMPLE 2.3. Let $A = S/Ann(F) = K[x,y,z]/Ann(F)$ such that $F = X^4 + XY^2Z$. Let $\ell = y$ and observe that

$$\begin{aligned} &A_y^{(0)} = A, \qquad h_A = (1,3,5,3,1),\\ &A_y^{(1)} = S/Ann(2XYZ), \qquad h_{A_y^{(1)}} = (1,3,3,1),\\ &A_y^{(2)} = S/Ann(2XZ), \qquad h_{A_y^{(1)}} = (1,2,1), \text{ and}\\ &h_{A_y^{(3)}} = h_{A_y^{(4)}} = (0). \end{aligned}$$

So we have that

$$M_{y,A} = \begin{bmatrix} 1 & 1 & 1 & 0 & 0\\ 0 & 3 & 3 & 2 & 0\\ 0 & 0 & 5 & 3 & 1\\ 0 & 0 & 0 & 3 & 1\\ 0 & 0 & 0 & 0 & 1 \end{bmatrix}, \text{ and } J_{y,A} = \begin{bmatrix} 0 & 0 & 1 & 0 & 0\\ 0 & 0 & 0 & 2 & 0\\ 0 & 0 & 1 & 0 & 1\\ 0 & 0 & 0 & 0 & 0\\ 0 & 0 & 0 & 0 & 0 \end{bmatrix}.$$

Therefore, using equations (2.2) and (2.3) we obtain that

$$P_{y,A} = (3,3,3,3,1), \text{ and } \mathcal{S}_{y,A} = (3_0, 3_1, 3_1, 3_2, 1_2).$$

REMARK 2.4. The rank matrices were further studied in [**1**] and the following three necessary conditions were provided for an upper triangular square matrix M of size d with non-negative entries that can occur as the rank matrix of some AG algebra and a linear form

(i.) For any $0 \le i \le d$ the diagonal $\operatorname{diag}(i, M)$ parallel to the main diagonal of M is a Gorenstein sequence.
(ii.) The differences between successive diagonals are O-sequences.
(iii.) Each adjacent square of entries in M satisfies

$$\begin{pmatrix} \alpha & \beta\\ \gamma & \zeta \end{pmatrix} \subset M \Rightarrow \gamma + \beta \ge \alpha + \zeta.$$

3. Artinian Gorenstein algebras of codimension two

In [**10**], Lemma 2.30, authors show that for a standard graded Artinian algebra of codimension two the Jordan types determines the Jordan degree type. Using the method of rank matrices we give a different proof for this statement for AG algebras of codimension two, which are complete intersections. The complete characterization of possible Jordan types of such algebras is provided in [**2**].

PROPOSITION 3.1. *Let $P = (P_1, \dots, P_t)$ be the Jordan type of some AG algebra of codimension two, socle degree d, and a linear form. Then the Jordan degree type associated to P is uniquely determined.*

PROOF. Note that P is of the form in equation (2.2). Using equation (2.1), we have that $\dim_K A_\ell^{(d)} = n_d$, $\dim_K A_\ell^{(d-1)} = n_{d-1} - n_d$, and recursively we obtain $\dim_K A_\ell^{(i)}$, for every $0 \le i \le d-2$ as the following

$$\dim_K A_\ell^{(i)} = n_i + 2\dim_K A_\ell^{(i+1)} - \dim_K A_\ell^{(i+2)}.$$

We claim that for each $0 \leq i \leq d$ the i-th the diagonal of the rank matrix $M_{\ell,A}$, $\operatorname{diag}(i, M_{\ell,A})$, that is equal to the Hilbert function of $A_\ell^{(i)}$ is uniquely determined by $\dim_K A_\ell^{(i)}$ and the socle degree of $A_\ell^{(i)}$ that is equal to $d-i$. To show the claim, let B be any AG algebra with socle degree s and $\dim_K B = t$. We obviously have $t \geq s+1$ and since h_B increases by at most one in each degree we have that $t = \sum_{i=0}^{r}(s+1-2i)$ where r is the largest degree i where $h_B(i) = i+1$. Therefore,

$$h_B(i) = \begin{cases} i+1, & 0 \leq i \leq r, \\ r+1, & r+1 \leq i \leq \lfloor \frac{s}{2} \rfloor, \end{cases}$$

and by symmetry we obtain the Hilbert function of B. Thus, we conclude that using equation (2.3) the Jordan degree type matrix $J_{\ell,A}$ is uniquely determined and so is the Jordan degree type $\mathcal{S}_{\ell,A}$. □

4. Artinian Gorenstein algebras of higher codimensions

In this section we show that the Jordan degree type is a finer invariant than the Jordan type for AG algebras of codimension at least three. Before that, in the following example we describe an algorithm that finds an AG algebra of codimension three and a linear form such that the associated rank matrix is equal to a given matrix satisfying the necessary conditions provided in [**1**, Corollary 3.9].
We first briefly recall the *catalecticant matrix* associated to $F \in R_d$ that is the Macaulay dual generator of $A = S/Ann(F)$. The entries of the j-th catalecticant matrix of F with respect to K-bases $\mathcal{B}_j = \{\alpha_u^{(j)}\}_u$ and $\mathcal{B}_{d-j} = \{\beta_v^{(d-j)}\}_v$ of S_j and S_{d-j} respectively is defined to be

$$\operatorname{Cat}_j(F) = \left(\alpha_u^{(j)}\beta_v^{(d-j)} \circ F\right).$$

For every j, the rank of $\operatorname{Cat}_j(F)$ is equal to $h_A(j)$, the Hilbert function of A in degree j, see [**8**].

EXAMPLE 4.1 (Constructing A and $\ell \in A_1$ having a given rank matrix). Consider

$$M = \begin{bmatrix} 1 & 1 & 1 & 1 & 0 & 0 \\ 0 & 3 & 2 & 2 & 1 & 0 \\ 0 & 0 & 4 & 3 & 2 & 1 \\ 0 & 0 & 0 & 4 & 2 & 1 \\ 0 & 0 & 0 & 0 & 3 & 1 \\ 0 & 0 & 0 & 0 & 0 & 1 \end{bmatrix}$$

and note that it satisfies the necessary conditions in [**1**, Corollary 3.9]. So we determine whether M occurs as the rank matrix of an AG algebra A and a linear form ℓ. By linear change of coordinates we may assume that $\ell = x$. Suppose $M = M_{x,A}$ for $A = S/Ann(F) = K[x,y,z]/Ann(F)$. The form F can be written as

$$F = X^5 G_0 + X^4 G_1 + \cdots + G_5,$$

such that G_i are forms of degree i in $K[Y,Z]$. We will find G_i's recursively. The last two diagonals of M are equal to $h_{A_x^{(5)}}$ and $h_{A_x^{(4)}}$ and since they are equal to zero vectors we get that $G_0 = G_1 = 0$. The dual generator of $A_x^{(3)}$ is $x^3 \circ F = G_2$. We have $h_{A_x^{(3)}} = (1,1,1)$ so G_2 is the second power of a linear form in $K[Y,Z]$, thus it can be chosen to be equal to $G_2 = Y^2$. Therefore,

$$F = X^3Y^2 + X^2G_3 + XG_4 + G_5.$$

Continuing this process we get that $h_{A_x^{(2)}} = (1,2,2,1)$ where the dual generator of $A_x^{(2)}$ is equal to $x^2 \circ F = 6XY^2 + 2G_3$. We note that for $G_3 = 0$ we get the desired Hilbert function for $A_x^{(2)}$. To obtain G_4 we compute $x \circ F = 3X^2Y^2 + G_4$ and to get $h_{A_x^{(1)}} = (1,2,3,2,1)$ we choose $G_4 = 0$. Thus we have

$$F = X^3Y^2 + G_5$$

and it remains to find for a polynomial G_5 for which we get $h_A = (1,3,4,4,3,1)$. We do so by computing the catalecticant matrices of F, we recall that the Hilbert function of A in degree 1 and 2 are equal to ranks of the first and second catalecticant matrices respectively. We denote $G_5 = \sum_{i=0}^{5} a_i Y^{5-i} Z^i$. For each j let $\mathcal{B}_j$ be the monomial basis of S_j in the lexicographic order. Then first and second catalecticant matrices of F are block matrices.

$$\mathrm{Cat}_1(F) = \left[\begin{array}{c|cc|ccc|cccc|ccccc} 0 & 0 & 0 & 12 & 0 & 0 & 0 & 0 & 0 & 0 & 0 & 0 & 0 & 0 & 0 \\ \hline 0 & 12 & 0 & 0 & 0 & 0 & 0 & 0 & 0 & 0 & a_0 & a_1 & a_2 & a_3 & a_4 \\ 0 & 0 & 0 & 0 & 0 & 0 & 0 & 0 & 0 & 0 & a_1 & a_2 & a_3 & a_4 & a_5 \end{array}\right],$$

and

$$\mathrm{Cat}_2(F) = \left[\begin{array}{c|cc|ccc|cccc} 0 & 0 & 0 & 12 & 0 & 0 & 0 & 0 & 0 & 0 \\ \hline 0 & 12 & 0 & 0 & 0 & 0 & 0 & 0 & 0 & 0 \\ 0 & 0 & 0 & 0 & 0 & 0 & 0 & 0 & 0 & 0 \\ \hline 12 & 0 & 0 & 0 & 0 & 0 & a_0 & a_1 & a_2 & a_3 \\ 0 & 0 & 0 & 0 & 0 & 0 & a_1 & a_2 & a_3 & a_4 \\ 0 & 0 & 0 & 0 & 0 & 0 & a_2 & a_3 & a_4 & a_5 \end{array}\right].$$

One can easily see that for $G_5 = Z^5$ we have $\mathrm{rkCat}_1(F) = 3$ and $\mathrm{rkCat}_2(F) = 4$. We conclude that M is the rank matrix of the AG algebra $A = S/Ann(F)$ and $\ell = x$ where

$$F = X^3Y^2 + Z^5.$$

Furthermore, using equations (2.2) and (2.3) the Jordan type and Jordan degree type of A and $\ell = x$ are equal to $P_{x,A} = (4,4,4,1,1,1,1)$ and $\mathcal{S}_{x,A} = (4_0, 4_1, 4_2, 1_1, 1_2, 1_3, 1_4)$.

Using the above method we show that Jordan degree type encodes more information than the Jordan type for AG algebras of codimension at least three.

EXAMPLE 4.2 (Two Jordan degree types for a Jordan type). The following matrices satisfy the necessary conditions given in [**1**, Corollary 3.9]. So they can occur as rank matrices for AG algebras and linear forms.

$$M_1 = \begin{bmatrix} 1 & \mathbf{1} & 1 & 1 & 1 & 1 & 1 & 0 & 0 \\ 0 & 3 & \mathbf{3} & 3 & 3 & 2 & 2 & 1 & 0 \\ 0 & 0 & 6 & \mathbf{6} & 5 & 4 & 3 & 2 & 1 \\ 0 & 0 & 0 & 9 & \mathbf{6} & 5 & 4 & 2 & 1 \\ 0 & 0 & 0 & 0 & 9 & \mathbf{6} & 5 & 3 & 1 \\ 0 & 0 & 0 & 0 & 0 & 9 & \mathbf{6} & 3 & 1 \\ 0 & 0 & 0 & 0 & 0 & 0 & 6 & \mathbf{3} & 1 \\ 0 & 0 & 0 & 0 & 0 & 0 & 0 & 3 & \mathbf{1} \\ 0 & 0 & 0 & 0 & 0 & 0 & 0 & 0 & 1 \end{bmatrix}, M_2 = \begin{bmatrix} 1 & \mathbf{1} & 1 & 1 & 1 & 1 & 1 & 0 & 0 \\ 0 & 3 & \mathbf{3} & 3 & 3 & 2 & 2 & 1 & 0 \\ 0 & 0 & 6 & \mathbf{5} & 5 & 4 & 3 & 2 & 1 \\ 0 & 0 & 0 & 9 & \mathbf{7} & 5 & 4 & 2 & 1 \\ 0 & 0 & 0 & 0 & 9 & \mathbf{7} & 5 & 3 & 1 \\ 0 & 0 & 0 & 0 & 0 & 9 & \mathbf{5} & 3 & 1 \\ 0 & 0 & 0 & 0 & 0 & 0 & 6 & \mathbf{3} & 1 \\ 0 & 0 & 0 & 0 & 0 & 0 & 0 & 3 & \mathbf{1} \\ 0 & 0 & 0 & 0 & 0 & 0 & 0 & 0 & 1 \end{bmatrix}.$$

Note that the two matrices only differ in the entries indicated in bold. Similar to the process described in Example 4.1 but with more technicality, we find that both

M_1 and M_2 occur as rank matrices. More precisely, $M_1 = M_{x,A}$ and $M_2 = M_{x,B}$ such that $A = S/Ann(F)$, $B = S/Ann(G)$,

$$F = X^6Y^2 + X^3(Y+Z)^5 + XY(Y+Z)^6 + Y^8 + Z^8,$$

and

$$G = X^6Y^2 + X^3(Y^5+Z^5) + X(Y^7+Z^7) + Y^8 + Y^7Z + YZ^7 + Z^8.$$

Because the sum of bold entries of the two matrices are equal using equation (2.2) we get that

$$P_{x,A} = P_{x,B} = (7^3, 4^4, 2^2, 1^6).$$

On the other hand, by equation (2.3) we have

$$\mathcal{S}_{x,A} = (7_0, 7_1, 7_2, 4_1, 4_2, 4_3, 4_4, 2_2, 2_5, 1_3, 1_3, 1_4, 1_4, 1_5, 1_5),$$
$$\mathcal{S}_{x,B} = (7_0, 7_1, 7_2, 4_1, 4_2, 4_3, 4_4, 2_3, 2_4, 1_2, 1_3, 1_3, 1_5, 1_5, 1_6).$$

REMARK 4.3. Using the algebra software Macaulay2, it has been computationally proven that among all AG algebras of codimension three with socle degree smaller than 8 there are no examples of AG algebras with the same Jordan type and different Jordan degree types. In other words, the above example has the smallest socle degree among all AG algebras of codimension three. On the other hand, for AG algebras of higher codimensions there exist such examples of smaller socle degrees, Example 4.4.

Using the same method as above we obtain the following example which we leave the computations to the reader. One may use the Macaulay2 codes provided bellow for further experiments.

EXAMPLE 4.4. Let $S = K[x,y,z,w]$ and $A = S/Ann(F)$ and $B = S/Ann(G)$ be AG algebras of codimension four where

$$F = X^4Y + X^2Y^2Z + XY^3W + Y^3W^2,$$

and

$$G = X^4Y + X^2Y^2Z + XZ^3Y + W^5.$$

We have that $h_A = h_B = (1,4,7,7,4,1)$. For linear form $\ell = x$ the rank matrices of A and B are equal to

$$M_{\ell,A} = \begin{bmatrix} 1 & \mathbf{1} & 1 & 1 & 1 & 0 \\ 0 & 4 & \mathbf{4} & 3 & 2 & 1 \\ 0 & 0 & 7 & \mathbf{4} & 3 & 1 \\ 0 & 0 & 0 & 7 & \mathbf{4} & 1 \\ 0 & 0 & 0 & 0 & 4 & \mathbf{1} \\ 0 & 0 & 0 & 0 & 0 & 1 \end{bmatrix}, \text{ and } M_{\ell,B} = \begin{bmatrix} 1 & \mathbf{1} & 1 & 1 & 1 & 0 \\ 0 & 4 & \mathbf{3} & 3 & 2 & 1 \\ 0 & 0 & 7 & \mathbf{6} & 3 & 1 \\ 0 & 0 & 0 & 7 & \mathbf{3} & 1 \\ 0 & 0 & 0 & 0 & 4 & \mathbf{1} \\ 0 & 0 & 0 & 0 & 0 & 1 \end{bmatrix}.$$

Using equation (2.2) and (2.3) we get that

$$P_{\ell,A} = P_{\ell,B} = (5,5,3,3,2,2,1,1,1,1),$$

and

$$\mathcal{S}_{\ell,A} = (5_0, 5_1, 3_1, 3_2, 2_1, 2_3, 1_2, 1_2, 1_3, 1_3),$$
$$\mathcal{S}_{\ell,B} = (5_0, 5_1, 3_1, 3_2, 2_2, 2_2, 1_1, 1_2, 1_3, 1_4).$$

Macaulay 2 codes: We provide Macaulay2 [**6**] codes that were primarily used for computations in this study. Let F be a form of degree d in the Macaulay dual ring, $R = K[x, y, z]$ (here, we abuse the notation and denote the variables in the dual ring by x, y, z). The following are some codes that compute the catalecticant matrix of F; using this, we compute the Hilbert function of the AG algebra A associated with F without computing A. We then for a given linear form ℓ we compute the dual generators of the AG algebras $A_\ell^{(i)}$ and then we compute their Hilbert functions which form the diagonals of the rank matrix $M_{\ell,A}$, $\operatorname{diag}(i, M_{\ell,A})$. Note that the following computations work for any number of variables, one only needs to adjust the number of variables in R and ℓ and everything else remains the same.

```
R = QQ[x,y,z]
Cat = (i,F) -> (diff(transpose basis(i,R),
diff(basis(first degree F-i,R),F)))
Hilb = (F) -> for i to first degree F list rank Cat(i,F)
l = a*x+b*y+c*z;
f = (i,l,F) -> diff(l^i,F)
diag = (i,l,F) -> Hilb(diff(l^i,F))
```

Finally, given F together with its degree, the following two functions compute the rank matrix and Jordan type of AG algebra associated with F and ℓ:

```
M = (l,F) -> (
d = first degree F;
L = for i to d list for j to d-i list
rank Cat(i,diff(l^j,F));
matrix({L_0}|(for i from 1 to d list
(for j from 1 to i list 0)|L_i))
)

P = (l,F) -> (
d = first degree F;
L1 = for i to d list sum for j to d-i list
rank Cat(j,diff(l^i,F));
L2 = L1|{0,0};
L3 = for i to d list L2_i+L2_(i+2)-2*L2_(i+1);
reverse flatten for i to d list if L3_i!= 0 then
for j to L3_i-1 list i+1 else continue
)
```

We end by posing some problems:

PROBLEM 1. Given a Jordan type occurring in a given algebra with a given Hilbert function, determine all possible compatible Jordan degree types across all algebras with that Hilbert function.

PROBLEM 2. Determine whether there is a certain family of AG algebras (for instance, complete intersections) of codimension at least three for which the Jordan degree type is equivalent to the Jordan type.

References

[1] Nasrin Altafi, *Jordan types with small parts for Artinian Gorenstein algebras of codimension three*, Linear Algebra Appl. **646** (2022), 54–83, DOI 10.1016/j.laa.2022.03.013. MR4403081

[2] Nasrin Altafi, Anthony Iarrobino, and Leila Khatami, *Complete intersection Jordan types in height two*, J. Algebra **557** (2020), 224–277, DOI 10.1016/j.jalgebra.2020.04.015. MR4092365

[3] Barbara Costa and Rodrigo Gondim, *The Jordan type of graded Artinian Gorenstein algebras*, Adv. in Appl. Math. **111** (2019), 101941, 27, DOI 10.1016/j.aam.2019.101941. MR4011109

[4] Anthony V. Geramita, *Inverse systems of fat points: Waring's problem, secant varieties of Veronese varieties and parameter spaces for Gorenstein ideals*, The Curves Seminar at Queen's, Vol. X (Kingston, ON, 1995), Queen's Papers in Pure and Appl. Math., vol. 102, Queen's Univ., Kingston, ON, 1996, pp. 2–114. MR1381732

[5] Rodrigo Gondim and Giuseppe Zappalà, *On mixed Hessians and the Lefschetz properties*, J. Pure Appl. Algebra **223** (2019), no. 10, 4268–4282, DOI 10.1016/j.jpaa.2019.01.008. MR3958091

[6] Daniel R. Grayson and Michael E. Stillman, *Macaulay2, a software system for research in algebraic geometry*, available at `http://www.math.uiuc.edu/Macaulay2/`

[7] Tadahito Harima, Toshiaki Maeno, Hideaki Morita, Yasuhide Numata, Akihito Wachi, and Junzo Watanabe, *The Lefschetz properties*, Lecture Notes in Mathematics, vol. 2080, Springer, Heidelberg, 2013, DOI 10.1007/978-3-642-38206-2. MR3112920

[8] Anthony Iarrobino and Vassil Kanev, *Power sums, Gorenstein algebras, and determinantal loci*, Lecture Notes in Mathematics, vol. 1721, Springer-Verlag, Berlin, 1999. Appendix C by Iarrobino and Steven L. Kleiman, DOI 10.1007/BFb0093426. MR1735271

[9] Anthony Iarrobino, Leila Khatami, Bart Van Steirteghem, and Rui Zhao, *Nilpotent matrices having a given Jordan type as maximum commuting nilpotent orbit*, Linear Algebra Appl. **546** (2018), 210–260, DOI 10.1016/j.laa.2018.02.007. MR3771880

[10] Anthony Iarrobino, Pedro Macias Marques, and Chris McDaniel, *Artinian algebras and Jordan type*, J. Commut. Algebra **14** (2022), no. 3, 365–414, DOI 10.1216/jca.2022.14.365. MR4492997

[11] Anthony Iarrobino, Pedro Macias Marques, and Chris McDaniel, *Free extensions and Jordan type*, J. Algebra **549** (2020), 346–364, DOI 10.1016/j.jalgebra.2020.01.003. MR4052278

[12] Francis S. Macaulay, *On a method of dealing with the intersections of plane curves*, Trans. Amer. Math. Soc. **5** (1904), no. 4, 385–410, DOI 10.2307/1986270. MR1500679

[13] Toshiaki Maeno and Junzo Watanabe, *Lefschetz elements of Artinian Gorenstein algebras and Hessians of homogeneous polynomials*, Illinois J. Math. **53** (2009), no. 2, 591–603. MR2594646

Department of Mathematics and Statistics, Queen's University, Kingston, Ontario, Canada and Department of Mathematics, KTH Royal Institute of Technology, Sweden

Email address: `nasrinar@kth.se`

Contemporary Mathematics
Volume **805**, 2024
https://doi.org/10.1090/conm/805/16127

Waring decompositions of special ternary forms with different Hilbert functions

Elena Angelini, Luca Chiantini, and Alessandro Oneto

ABSTRACT. We prove the existence of ternary forms admitting apolar sets of points of cardinality equal to the Waring rank, but having different Hilbert function and different regularity. This is done exploiting liaison theory and Cayley-Bacharach properties for sets of points in the projective plane.

1. Introduction

1.1. Waring decompositions. Let $S = \bigoplus_{d\in\mathbb{Z}} S_d = \mathbb{C}[x_0,\dots,x_n]$ be the standard graded polynomial ring in n variables with complex coefficients.

DEFINITION 1.1 (Waring rank). Given a degree-d homogeneous polynomial $f \in S_d$, a *Waring decomposition* of f is an expression as sum of d-th powers of linear forms, i.e.,

$$f = \sum_{i=1}^{r} \ell_i^d, \quad \ell_i \in S_1. \tag{1.1}$$

The minimal length r of such a decomposition is the *Waring rank*, denoted $\operatorname{rk}(f)$.

The study of Waring decompositions has its roots in the XIX century thanks to the works of Sylvester [**40, 41**], Clebsch [**18**], Richmond [**36**], Palatini [**31, 32**], Terracini [**42**], among many others. The question regained a lot of interest in the last decades. The celebrated theorem by Alexander and Hirschowitz in 1995 established the general Waring rank in any number of variables and degrees [**1**]. The attention to Waring decompositions has been motivated also by its connection to *tensor decompositions*; in this case, symmetric tensors. Indeed, tensor decompositions find a lot of uses in several applications, see for example [**26, 27**]. The computation of most tensor decompositions is known to be an NP-hard problem and the same is conjectured fo symmetric tensors [**24**]. Therefore, a theoretical study of related algebraic and geometric structures can be useful to shed lights on them.

1.2. Apolar sets of points. Waring decompositions have a geometric interpretation in terms of *Veronese varieties*. Consider the *Veronese embedding* given by

$$\nu_d : \mathbb{P}S_1 \longrightarrow \mathbb{P}S_d, \qquad [\ell] \mapsto [\ell^d].$$

2020 *Mathematics Subject Classification.* Primary 14C20, 14N07; Secondary 13A02, 13C40, 14N05, 15A69.

The authors are members of the Italian GNSAGA-INDAM.

DEFINITION 1.2 (Apolar sets of points). Given $f \in S_d$ and a set of points $A \subset \mathbb{P}S_1$, we say that A is *apolar* to f if $[f] \in \langle \nu_d(A) \rangle$. If there is no subset $A' \subset A$ such that $[f] \in \langle \nu_d(A') \rangle$ we say that A is *non-redundant*. If the cardinality of A equals the Waring rank of f we say that A *computes (the rank of) f*.

Any Waring decomposition of a polynomial $f \in S_d$ has a geometric counterpart given by a set of points $A \subset \mathbb{P}S_1$ apolar to f, i.e., the equation (1.1) is equivalent to say that $[f] \in \langle \nu_d(A) \rangle$ for $A = \{[\ell_1], \ldots, [\ell_r]\} \subset \mathbb{P}S_1$. This geometric interpretation allows us to define a notion of rank with respect to any algebraic variety $X \subset \mathbb{P}^N$. Namely, the *rank* of any $p \in \mathbb{P}^N$ with respect to X is the minimal cardinality of a set of points $A \subset X$ such that $p \in \langle A \rangle$.

A more algebraic way to relate Waring decompositions with sets of points is through *apolarity theory*. By Macaulay [**28**] (see also [**25**, Lemma 2.14]), there is a one-to-one correspondence between artinian Gorenstein graded algebras with socle degree d and homogeneous polynomials of degree d. Let $R = \bigoplus_{d \in \mathbb{Z}} R_d = \mathbb{C}[y_0, \ldots, y_n]$ be a second polynomial ring and consider the *apolar action* of R over S by partial derivatives, i.e., for any $g \in R_e, f \in S_d$, we define $g \circ f := g(\partial_0, \ldots, \partial_n) f \in S_{d-e}$. Given $f \in S_d$, the *apolar ideal* of f is the annihilator of f with respect to the apolar action, i.e.,

$$\operatorname{Ann}(f) := \{g \in R \ : \ g \circ f = 0\} \subset R.$$

Macaulay's duality affirms that an artinian algebra $A = R/J$ is Gorenstein of socle degree d if and only if $J = \operatorname{Ann}(f)$ for some $f \in R_d$. The *Apolarity Lemma*, see [**25**, Lemma 1.15], affirms that Waring decompositions of a polynomial $f \in S_d$ correspond to sets of points $A \subset \mathbb{P}S_1$ such that $I(A) \subset \operatorname{Ann}(f)$, i.e., the equation (1.1) is equivalent to say that $I(A) \subset \operatorname{Ann}(f)$ for $A = \{[\ell_1], \ldots, [\ell_r]\} \subset \mathbb{P}S_1$. In other words, the study of Waring decompositions is equivalent to study ideals of sets of points in $\mathbb{P}S_1$, i.e., radical homogeneous ideals of height one, contained in artinian Gorenstein ideals.

1.3. Hilbert function of apolar sets of points. Most of algebraic invariants of algebraic varieties are encoded by the *Hilbert function*. Given a homogeneous ideal $I \subset R$, let $I_d = I \cap R_d$ be its degree-d homogeneous part. The *Hilbert function* of R/I is the numerical function $h_{R/I} : \mathbb{Z} \to \mathbb{Z}$ with

$$h_{R/I}(d) := \dim_{\mathbb{C}}[R_d/I_d] = \dim_{\mathbb{C}} R_d - \dim_{\mathbb{C}} I_d \tag{1.2}$$

If $A \subset \mathbb{P}S_1$ is a set of points, then we write h_A for the Hilbert function of $R/I(A)$ where $I(A)$ is the defining homogeneous ideal of A. The study of Hilbert functions of reduced sets of points has been successfully used to compute Waring ranks. For example, the Hilbert function of the apolar ideal of a given polynomial provides a lower bound for the Waring rank of the polynomial. In some cases, these were shown to be optimal, see e.g. [**9, 10**].

If a polynomial admits a unique minimal Waring decomposition, then we say that the polynomial is *identifiable*. It is worth to mention that the general homogeneous polynomial with rank r strictly smaller than the general one is identifiable except for the special cases $(n, d, r) \in \{(2, 6, 9), (3, 4, 8), (5, 3, 9)\}$, see [**17**]. In [**3–5**], *Cayley-Bacharach properties* and *liaison theory* of sets of points in the projective plane were exploited to provide a criterion to certify that a form is identifiable. On the other hand, when the polynomial is not identifiable, it is interesting to study the family of apolar sets of points computing the rank of the polynomial, see [**35**].

In [**30**], a classification of low rank polynomials is given in terms of the possible Hilbert functions of apolar sets of points computing the rank of the polynomials.

It is well-known that the Hilbert function of a set of points is strictly increasing until it reaches its cardinality. We call *regularity* of A the smallest integer i for which the Hilbert function of A stabilizes, i.e., such that $h_A(i) = h_A(i+1)$.

1.4. The question. In this short note, we address the following question.

QUESTION 1.3 ([**30**, Question 2.11]). Let $f \in S_d$. Do all sets of points computing the rank of f share the same regularity? More generally, do they share the same Hilbert function?

The answer is known to be positive for binary forms ($n = 2$), because all sets of points with same cardinality on the projective line share the same Hilbert function, and for monomials $x_0^{d_0} \cdots x_n^{d_n}$ because all sets of points computing the rank are known to be complete intersections with generators of degrees $d_1 + 1, \ldots, d_n + 1$ if $d_0 \leq d_i$ for all i, see [**8**]. In the present note, we prove that the answer to Question 1.3 is negative for ternary forms ($n = 3$) if the degree is sufficiently large. We provide two cases. In Section 3.1, we consider degree-10 ternary forms having rank 22, that is the general rank. Among them, we consider the special ones which admit a unique quintic in the apolar ideal. We show in Theorem 3.1 that the general such form has a decomposition lying on the quintic and one not contained in the quintic. We recall that the study of the connection between catalecticant matrices and the Hilbert function of certain decompositions was also used in [**16**] where a complete stratification of ternary 6-ics of rank nine is given.

Along with the proof, we notice that there exists degree-10 ternary forms of rank 22 having a unique minimal (irreducible and smooth) quintic in the apolar ideal whose minimal decompositions do not lie on such quintic, see Remark 3.7. Also among the well-known case of monomials we can find examples in which generators of minimal degree of the apolar ideal cannot be used to define a minimal decomposition, however, in that case, the minimal generators are not square-free and the fact that they have to be discarted is trivial. Our example underlines that the naïve approach of looking for minimal decompositions over reduced hypersurfaces of minimal degree among the ones in the apolar ideal might fail.

In Section 3.2, we consider degree-13 ternary forms having rank 30, that is strictly subgeneric. In Theorem 3.9, we show that there exist two sets of 30 points which compute the rank of a degree-13 ternary form and have different regularity. Our proofs follow the ideas of [**3–5**], by exploiting properties of Hilbert functions of points in the projective plane and, in particular, *liaison theory*. In Section 4, we show an explicit construction with the support of the algebra software Macaulay2 [**23**]. In the present work, we do not investigate the minimality of these examples because we believe that it would require a thorough analysis of several possible cases. In particular, there might be examples in smaller degrees. Note that up to degree 6 there are no examples thanks to [**16**], but degrees between 7 and 9 might be worth to be analyzed in future works.

2. Algebraic properties of points in $\mathbb{P}^n$

2.1. Hilbert function of points in $\mathbb{P}^n$. Let $A \subset \mathbb{P}^n$ be a set of points of cardinality $\ell(A)$. We recalled in (1.2) the definition of the *Hilbert function* of A.

The *first difference* is the numerical function given by

$$Dh_A : \mathbb{Z} \to \mathbb{Z}, \quad Dh_A(i) := h_A(i) - h_A(i-1).$$

We recall some elementary properties of Hilbert functions of points in $\mathbb{P}^n$ that will be used later. We refer to [**14**] for an extensive exposition including proofs.

LEMMA 2.1. [**14**, Lemma 6.1, Proposition 6.3, Proposition 6.4] *Let $A \subset \mathbb{P}^n$ be a set of points of cardinality $\ell(A)$.*

(1) $h_A(j) \leq \ell(A)$, *for all* j;
(2) $Dh_A(j) = 0$, *for* $j < 0$;
(3) $h_A(0) = Dh_A(0) = 1$;
(4) $Dh_A(j) \geq 0$, *for all* j;
(5) $h_A(i) = \sum_{j=0}^{i} Dh_A(j)$;
(6) $h_A(j) = \ell(A)$, *for all* $j \gg 0$;
(7) $Dh_A(j) = 0$, *for* $j \gg 0$;
(8) $\sum_{j=0}^{\infty} Dh_A(j) = \ell(A)$;
(9) *for any* $A' \subset A$, $h_{A'}(j) \leq h_A(j)$ *and* $Dh_{A'}(j) \leq Dh_A(j)$ *for any* j;
(10) *if* $Dh_A(j) \leq j$, *then* $Dh_A(j) \geq Dh_A(j+1)$.

Finally, we denote

$$h^1_A(d) := \ell(A) - h_A(d) = \sum_{i=d+1}^{\infty} Dh_A(i).$$

The following proposition allows us to understand if, given two sets of points, there exists a homogeneous polynomial of given degree to which they are both apolar.

PROPOSITION 2.2. [**3**, Proposition 2.19] *Let $A, B \subset \mathbb{P}^n$ be finite sets of points and let $Z = A \cup B$. Then, for any $d \in \mathbb{N}$,*

$$\dim \left(\langle \nu_d(A) \rangle \cap \langle \nu_d(B) \rangle \right) = \ell(A \cap B) - 1 + h^1_Z(d).$$

The latter proposition, together with a result by Davis [**19**], allows to show the following result that will be used to compute the rank of the forms constructed in the examples.

PROPOSITION 2.3 ([**3**, Proposition 2.20]). *Let f be a ternary form of degree d and let A be a non-redundant finite set computing the rank of f. Then, there is no other $B \subset \mathbb{P}^2$ non-redundant finite set computing the rank of f with $A \cap B = \emptyset$, $\ell(B) \leq \ell(A)$ and such that, if $Z = A \cup B$:*

(1) $Dh_Z(j) = j + 1$ *for* $j \in \{0, \ldots, i-1\}$ *and* $Dh_Z(i) \leq i$;
(2) $Dh_Z(j_0) = Dh_Z(j_0 + 1) = e < i$ *for some* $j_0 > i - 1$.

2.2. Liaison for points in $\mathbb{P}^2$. We describe here some construction that will be used in Section 3 to construct the examples replying to Question 1.3. These are basic constructions from *liaison theory*; we refer to [**29, 34**] for a complete presentation of the theory.

DEFINITION 2.4. Let $A, B \subset \mathbb{P}^2$ be two sets of points. We say that A *and* B *are linked* if there exists a complete intersection Z such that $I_B = I_Z : I_A$. When $A \cap B = \emptyset$, this is equivalent to say that $Z = A \cup B$. We say that B is the *residue* of A with respect to Z.

A set of points in $\mathbb{P}^2$ is arithmetically Cohen-Macaulay of codimension 2, hence a minimal free resolution is provided by the *Hilbert-Burch Theorem*, see [**22**, Section 3A]. If A and B are linked, then a Hilbert-Burch matrix of B can be recovered from one of A by the *mapping cone* construction, see [**29**, Proposition 5.2.10].

Let $S(i)$ be the polynomial ring with the gradation shifted by i, i.e., $S(d)_i := S_{d+i}$. A similar notation is used for every graded S-module. If A and B are linked through a complete intersection Z of type (d_1, d_2), then we have the following free resolutions

$$\begin{array}{ccccccccc} 0 & \to & S(-d_1-d_2) & \to & S(-d_1)\oplus S(-d_2) & \to & I_Z & \to & 0 \\ & & \downarrow & & \downarrow \phi & & \downarrow & & \\ 0 & \longrightarrow & F_1 & \xrightarrow{M} & F_0 & \longrightarrow & I_A & \longrightarrow & 0 \end{array}$$

where $\phi : S(-d_1) \oplus S(-d_2) \to F_0$ is the map induced by the right-most vertical arrow which corresponds to the inclusion $I_Z \subset I_A$ and M is a Hilbert-Burch matrix of A. Then, by the mapping cone construction, a resolution of I_B is given by

$$0 \to F_0(-d_1-d_2)^\vee \xrightarrow{(\phi|M)^t} \begin{array}{c} S(-d_1)\oplus S(-d_2) \\ \oplus \\ F_1(-d_1-d_2)^\vee \end{array} \to I_B \to 0.$$

The Hilbert functions between the two linked sets of points are linked by the following formula on their first differences:

$$Dh_B(i) + Dh_A(d_1+d_2-i-2) = Dh_Z(i), \tag{2.1}$$

see [**20**, Theorem 3] or [**29**, Corollary 5.2.19].

2.3. Cayley-Bacharach properties. We recall here also some basic fact about Cayley-Bacharach properties for sets of points in projective space. They are used to understand the possible Hilbert functions of sets of points $A \cup B$ where A and B are apolar to the same homogeneous form. These properties have been crucial to provide certificates for identifiability of forms in [**3**–**5**].

DEFINITION 2.5. Let $Z \subset \mathbb{P}^n$ be a set of points. We say that it satisfies the *Cayley-Bacharach property in degree* d ($CB(d)$) if, for all $p \in Z$, it holds that every form of degree d vanishing at $Z \smallsetminus \{p\}$ also vanishes at p.

PROPOSITION 2.6. [**3**, Proposition 2.25] *Let $f \in S_d$ and let A be a non-redundant set of points apolar to f. Let B a second non-redundant set of points apolar to f such that $A \cap B = \emptyset$. Then, the set $Z = A \cup B$ satisfies $CB(d)$.*

THEOREM 2.7. [**5**, Theorem 4.9] *Let $Z \subset \mathbb{P}^n$ be a set of points satisfying $CB(d)$. Then, for any $i \in \{0, \ldots, d+1\}$,*

$$\sum_{j=0}^{i} Dh_Z(j) \le \sum_{j=0}^{i} Dh_Z(d+1-j).$$

2.4. Secant varieties and Terracini loci. The geometric framework to deal with Waring decompositions is the one of *secant varieties* of Veronese varieties, see [**7**, **11**]. Given an irreducible algebraic variety $X \subset \mathbb{P}^N$, the *r-th secant variety* is an irreducible variety defined as the Zariski closure of the union of linear spaces spanned by r-tuples of points on X, i.e.,

$$\sigma_r(X) := \overline{\bigcup_{p_1,\ldots,p_r \in X} \langle p_1, \ldots, p_r \rangle} \subset \mathbb{P}^N.$$

If X is the d-th Veronese embedding of $\mathbb{P}S_1$, then, by Definition 1.1,

$$\sigma_r(\nu_d(\mathbb{P}^n)) = \overline{\{[f] \in \mathbb{P}S_d \ : \ \mathrm{rk}(f) \leq r\}} \subset \mathbb{P}S_d.$$

Secant varieties are clearly nested into each other by definition, but, if the variety X is non-degenerate, then we have that its secant varieties eventually fill the ambient space, see [**11**, Exercise 2.6]. For example, the minimal r such that $\sigma_r(\nu_d(\mathbb{P}^n)) = \mathbb{P}S_d$ is the Waring rank of the general degree-d form in $n+1$ variables.

Understanding which secant variety fills the ambient space reduces to the computation of dimensions of secant varieties. By a simple parameter count, the *expected dimension* of the r-th secant variety of an algebraic variety $X \subset \mathbb{P}^N$ is

$$\mathrm{exp.}\dim \sigma_r(X) := \min\{r\dim(X) + r - 1, N\}.$$

The expected dimension is always an upper bound for the actual dimension and if $\dim \sigma_r(X) < \mathrm{exp.}\dim \sigma_r(X)$ then we say that X is *r-defective.* The classification of defective varieties is a very classical problem with roots in the late XIX century. By Palatini [**33**], if $\sigma_r(X)$ is not an hypersurface and does not fill the ambient space, then it is at least of codimension 2 inside $\sigma_{r+1}(X)$. As an immediate corollary, projective curves are never defective. A classification of defective surfaces is due to Severi and Terracini, see [**13**, Theorem 14 and Theorem 15], while the classification of defective threefolds is due to Scorza [**37**]. The case of fourfolds has been treated first by Scorza in [**38**] (see also [**39**]) and then completed in the recent preprint [**15**].

If we denote by $X^{(r)}$ the r-th symmetric product of the variety X, the *r-th abstract secant variety* of X is

$$ab\sigma_r(X) := \overline{\{((p_1, \ldots, p_r), q) \ : \ q \in \langle p_1, \ldots, p_r\rangle\}} \subset X^{(r)} \times \mathbb{P}^N.$$

Throughout the paper, we always denote by π_i the projection onto the i-th factor from a cartesian product. The r-th abstract secant variety is irreducible of dimension $r\dim(X) + r - 1$. Hence, if $\sigma_r(X) \subsetneq \mathbb{P}^N$, non-defectiveness is equivalent to say that the projection $\pi_2 : ab\sigma_r(X) \to \mathbb{P}^N$ has generically finite fibers. The *r-th Terracini locus* of X is

$$\mathbb{T}_r(X) := \overline{\left\{(p_1, \ldots, p_r) \ : \ {\scriptstyle p_i \in X^{\mathrm{smooth}}, \{p_1, \ldots, p_r\} \text{ are linearly independent,} \atop \scriptstyle T_{p_1}(X), \ldots, T_{p_r}(X) \text{ are linearly dependent}}\right\}} \subset X^{(r)}.$$

The main tool to study dimensions of secant varieties is the *Terracini's Lemma* which describes the general tangent space to secant varieties, see [**7**, Lemma 1]. In particular, it claims that, given general points $p_1, \ldots, p_r \in X$ and a general point $p \in \langle p_1, \ldots, p_r\rangle$, then

$$T_p\sigma_r(X) = \langle T_{p_1}(X), \ldots, T_{p_r}(X)\rangle.$$

If X is not r-defective and $\sigma_r(X)$ does not fill the space, then $\mathbb{T}_r(X)$ is a proper subvariety of $X^{(r)}$. In other words, the preimage of the Terracini's locus $\pi_1^{-1}(\mathbb{T}_r(X))$ is contained in the locus where the differential of the projection $\pi_2 : ab\sigma_r(X) \to \mathbb{P}^N$ drops rank.

The complete classification of defective Veronese varieties is known as the *Alexander-Hirschowitz Theorem* [**1**]. In the 1980s, Alexander and Hirschowitz defined a very powerful machinery to deal with problems on polynomial interpolation such as the computation of the dimension of the linear system of degree-d hypersurfaces of $\mathbb{P}^n$ having r singularities at general points. The latter problem is equivalent to compute dimensions of secant varieties of the Veronese varieties and, a posteriori, to compute the general Waring rank in S_d.

THEOREM 2.8 (Alexander-Hirschowitz Theorem, [**1**]). *The r-th secant variety of the Veronese variety $\nu_d(\mathbb{P}^n)$ is non-defective except for*

- $d = 2$, $n \geq 2$ *and* $2 \leq r \leq n$;
- $d = 3, n = 4, r = 7$;
- $d = 4, (n, r) \in \{(2, 5), (3, 9), (4, 14)\}$.

3. The examples

In this section, we give a negative answer to Question 1.3 by showing that there exist ternary forms admitting apolar sets of points computing their Waring rank with different Hilbert function as well as different regularity.

The idea is to start from a special set of points Z_1 and, by means of liaison theory, to construct a second set of points of the same cardinality Z_2, but different Hilbert function, such that $\langle\nu_d(Z_1)\rangle \cap \langle\nu_d(Z_2)\rangle$ is non-empty, i.e., there is a degree-d form f whose Z_1 and Z_2 are both apolar to. After the construction, we prove that indeed Z_1 and Z_2 compute the rank of f.

3.1. First example. Degree-10 ternary forms of rank 22.

THEOREM 3.1. *There exists a form f of degree* 10 *and of rank* 22 *having two sets of points computing its rank and presenting different Hilbert functions.*

In order to help the reader to follow the several steps of the proof, before getting into the proof, we summarize the idea of how we will proceed.

(1) Starting from a set Z_1 of 22 general points in projective plane lying on a quintic curve, we construct a new set of points Z_2 of 22 points by means of liaison theory such that $\langle\nu_{10}(Z_1)\rangle \cap \langle\nu_{10}(Z_2)\rangle \neq \emptyset$, i.e., there exists a form f which both Z_1 and Z_2 are apolar to, see Claim 3.2. Then, we have to show that the form f has indeed rank 22.
(2) We denote by $\mathcal{X}$ the variety of degree-10 forms which admit an apolar set of 22 points lying on a quintic curve and observe that the general element of $\mathcal{X}$ has rank 22, see Claim 3.3 and Claim 3.4. Our construction produces forms that are clearly in $\mathcal{X}$: we need to show that these are general enough to have rank 22.
(3) We consider the variety $\mathcal{Y}$ of degree-10 forms obtained from our construction. In Claim 3.5 and Claim 3.6, we prove that $\mathcal{Y}$ is dense in $\mathcal{X}$ through a dimension count. This is enough to conclude that the general element of $\mathcal{Y}$ has rank 22.

PROOF OF THEOREM 3.1. Let $Z_1 \subset \mathbb{P}^2$ be a set of 22 general points on a quintic curve. Then,

$$\begin{array}{rccccccccc} h_{Z_1}: & 1 & 3 & 6 & 10 & 15 & 20 & 22 & 22 & \cdots \\ Dh_{Z_1}: & 1 & 2 & 3 & 4 & 5 & 5 & 2 & - & \end{array} \tag{3.1}$$

Let $L \subset \mathbb{P}^2$ be a general line and let Y be a set of five general points on L. Consider $A = Z_1 \cup Y$. Clearly there are no quintics through A since there is a unique quintic through Z_1 and the line L is chosen generically. Moreover, Y imposes independent conditions on the space of 6-ics through Z_1: indeed, it is enough to observe that $\dim I(Z_1)_6 - \dim I(Z_1)_5 = 5 \geq \ell(Y)$, see e.g. [**12**, Lemma 1.9-(i)]. Hence,

$$\begin{array}{rccccccccc} h_A: & 1 & 3 & 6 & 10 & 15 & 21 & 27 & 27 & \cdots \\ Dh_A: & 1 & 2 & 3 & 4 & 5 & 6 & 6 & - & \end{array} \tag{3.2}$$

Let X be a complete intersection of two general 7-ics containing A. Then, we define Z_2 to be the residue of A through X. Hence, by (2.1), we have

(3.3)

		0	1	2	3	4	5	6	7	8	9	10	11	12
$Dh_X(i)$	:	1	2	3	4	5	6	7	6	5	4	3	2	1
$Dh_{Z_2}(i)$	:	1	2	3	4	5	6	1	–	–	–			
$Dh_A(12-i)$	:							6	6	5	4	3	2	1

Note that, since A is regular in degree 6, it has generators of degree at most 7. Then, by the Peskine-Szpiro Theorem, see [**29**, Theorem 6.4.4], Z_2 is smooth and $Z_1 \cap Z_2 = \emptyset$. Hence, Z_2 is a set of 22 distinct points disjoint from Z_1 and with Hilbert function different than Z_1.

Claim 3.2. Z_1 and Z_2 are apolar to a common degree-10 ternary form.

Proof. By Proposition 2.2, it is enough to compute the Hilbert function of the union $Z_1 \cup Z_2$. Note that, by construction, $Z_1 \cup Z_2$ is linked to the set Y of five collinear points. Hence, by (2.1)

		0	1	2	3	4	5	6	7	8	9	10	11	12
$Dh_X(i)$	:	1	2	3	4	5	6	7	6	5	4	3	2	1
$Dh_{Z_1\cup Z_2}(i)$	:	1	2	3	4	5	6	7	6	4	3	2	1	
$Dh_Y(12-i)$	:									1	1	1	1	1

Hence, by Proposition 2.2, $\dim\langle\nu_{10}(Z_1)\rangle \cap \langle\nu_{10}(Z_2)\rangle = 0$. I.e., there exists $f \in S_{10}$ such that $[f] \in \langle\nu_{10}(Z_1)\rangle \cap \langle\nu_{10}(Z_2)\rangle$. □

Now, we need to show that the general degree-10 form f constructed as in Claim 3.2 is indeed of rank 22. Let $\mathcal{X}$ be the variety of degree-10 forms admitting an apolar set of 22 reduced points lying on a quintic.

Claim 3.3. The variety $\mathcal{X}$ is irreducible and has codimension 2.

Proof. Note that the variety of degree-10 ternary forms $f \in S_{10}$ admitting a quintic in the apolar ideal is a hypersurface defined by the vanishing of the determinant of the central *catalecticant matrix* associated to the map

$$\mathrm{cat}_5(f) : R_5 \to S_5, g \mapsto g \circ f.$$

The fact that the variety $\{\det(\mathrm{cat}_5(f)) = 0\}$ is irreducible follows from [**21**]. Fixed an irreducible plane quintic $C \subset \mathbb{P}^2$, by Riemann-Roch Theorem, we have that $\langle\nu_{10}(C)\rangle = \mathbb{P}^{44}$. Since curves are non-defective, $\dim \sigma_{22}\nu_{10}(C) = 43$, see Section 2.4. Therefore the general form $f \in \langle\nu_{10}(C)\rangle$ has rank 23 with respect to $\nu_{10}(C)$ and $\mathcal{X}$ has codimension one in the hypersurface of forms admitting a quintic in the apolar ideal. In other words, if we consider the incidence variety of forms lying on the 22-nd secant variety of the degree-10 Veronese embedding of a plane quintic

$$\mathcal{I} = \{(C, [f]) \ : \ [f] \in \sigma_{22}(\nu_{10}(C))\} \subset \mathbb{P}S_5 \times \mathbb{P}S_{10},$$

then $\mathcal{X} = \pi_2(\mathcal{I})$. The projection π_2 is generically one-to-one because the kernel of $\mathrm{cat}_5(f)$ for a general form $[f] \in \mathcal{X}$ is one-dimensional and, generically, it is also generated by an irreducible quintic (since it is a open, this can be checked by a direct computation). Given a general irreducible quintic C, we have that $\pi_1^{-1}(C) \cong \sigma_{22}(\nu_{10}(C))$. Hence, $\dim \mathcal{X} = \dim \mathcal{I} = \dim \mathbb{P}S_5 + \dim \sigma_{22}(\nu_{10}(C)) = 20 + 43 = 63$ inside $\mathbb{P}S_{10} = \mathbb{P}^{65}$. □

Claim 3.4. The general form $[f] \in \mathcal{X}$ has rank 22.

PROOF. By Alexander-Hirschowitz Theorem (Theorem 2.8),

$$\dim \sigma_{21}(\nu_{10}(\mathbb{P}^2)) = 21 \cdot 2 + 20 = 62,$$

i.e., $\sigma_{21}(\nu_{10}(\mathbb{P}^2))$ has codimension 3. Hence, by Claim 3.3, $\mathcal{X} \cap \sigma_{21}(\nu_{10}(\mathbb{P}^2))$ is a proper subvariety of $\mathcal{X}$. □

Consider $\mathcal{Y} \subset \mathcal{X}$ the set of degree-10 forms that are constructed as in Claim 3.2. Let $\mathrm{Hilb}^{\circ}_{22}(\mathbb{P}^2)$ be the open subset of the Hilbert scheme of 22 points in $\mathbb{P}^2$ and let $\mathrm{Hilb}^{\circ}_{22,5}(\mathbb{P}^2)$ denote the set of all sets of 22 general points lying on a quintic of $\mathbb{P}^2$. Then, consider the incidence variety

$$\mathcal{J} = \{(Z_1, Z_2, [f]) \,:\, \langle \nu_{10}(Z_1)\rangle \cap \langle \nu_{10}(Z_2)\rangle = [f]\} \subset \mathrm{Hilb}^{\circ}_{22,5}(\mathbb{P}^2) \times \mathrm{Hilb}^{\circ}_{22}(\mathbb{P}^2) \times \mathbb{P}S_{10}.$$

By construction, $\mathcal{Y} = \pi_3(\mathcal{J})$. Moreover, from the construction described at the beginning of the proof, the projection π_1 is dominant, i.e., for a general choice of a set Z_1 of 22 points on a quintic in $\mathbb{P}^2$, there exists $Z_2 \in \mathrm{Hilb}^{\circ}_{22}(\mathbb{P}^2)$ and $f \in S_{10}$ such that $(Z_1, Z_2, [f]) \in \mathcal{J}$. With a similar construction, we can also show that the projection π_2 is dominant.

CLAIM 3.5. The projection $\pi_2 : \mathcal{J} \to \mathrm{Hilb}^{\circ}_{22}(\mathbb{P}^2)$ is dominant.

PROOF. Let Z_2 be a general set of 22 points in $\mathbb{P}^2$. Take a set Y of three general points collinear on a general line L and let $A = Z_2 \cup Y$. Since there are no quintics through Z_2, then the three points impose independent conditions on 6-ics through Z_2, see [**12**, Lemma 1.9]. Hence,

$$\begin{array}{rcccccccccc} h_A: & 1 & 3 & 6 & 10 & 15 & 21 & 25 & 25 & \cdots \\ Dh_A: & 1 & 2 & 3 & 4 & 5 & 6 & 4 & - & \end{array}$$

Let $\{G_1, G_2, G_3\}$ be a basis for the space of 6-ics passing through A, i.e., $[I_A]_6 = \langle G_1, G_2, G_3\rangle$. The restriction of $[I_A]_6$ on L defines a non-complete linear series g^2_3 away from the set of points Y. The linear series on L is generically base-point free, then it defines a regular map on $\mathbb{P}^2$ which maps L into a singular cubic curve C. Indeed, since being base-point free is an open condition, this can be checked directly on a specific example with the support of the algebra software Macaulay2 [**23**], see Section 4.1. For example, without loss of genericity, let $L = \{x_0 = 0\}$ and $Y = \{(0:1:0), (0:0:1), (0:1:1)\}$. Hence, for $i = 1, 2, 3$, the restriction of G_i on L is given by $G_i(0, x_1, x_2) = x_1x_2(x_1 - x_2)H_i(x_1, x_2)$ and we consider the map $\varphi : L \to \mathbb{P}^2, p \mapsto (H_1(p) : H_2(p) : H_3(p))$. Let $C = \varphi(L)$. The general cubic C constructed as above is nodal. Indeed, since being nodal is an open condition on the space of rational plane cubics, it is enough to exhibit one example in which the above construction provides a nodal curve. This is checked with the support of the algebra software Macaulay2 [**23**], see Section 4.1.

Now, let X be the set of two distinct points on L which are pre-images of the singular point of C and consider the set of five points $X \cup Y \subset L$. By construction, $X \cup Y$ does not impose independent conditions on the space of 6-ics passing through A. This implies that

$$\begin{array}{rccccccccccc} h_{X\cup A} & : & 1 & 3 & 6 & 10 & 15 & 21 & 26 & 27 & 27 & \cdots \\ Dh_{X\cup A} & : & 1 & 2 & 3 & 4 & 5 & 6 & 5 & 1 & - & \end{array}$$

Now, we consider the linkage to $X \cup A$ via a general complete intersection of type $(7, 7)$, i.e., $U = (X \cup A) \cup Z_1$. Then, Z_1 is a set of 22 points with the following

Hilbert function

(3.4)

		0	1	2	3	4	5	6	7	8	9	10	11	12
$Dh_U(i)$	:	1	2	3	4	5	6	7	6	5	4	3	2	1
$Dh_{Z_1}(i)$	:	1	2	3	4	5	5	2	–	–	–			
$Dh_{X\cup A}(12-i)$	:						1	5	6	5	4	3	2	1

Note that $X \cup A$ has regularity 7, but, by our explicit computations in Section 4.1, we also know that it is generated in degree 7. Therefore, again by Peskine-Szpiro Theorem as before, see [**29**, Theorem 6.4.4], we deduce that Z_1 is smooth and $Z_1 \cap Z_2 = \emptyset$. Hence, $Z_1 \in \mathrm{Hilb}^{\circ}_{22,5}(\mathbb{P}^2)$. Moreover, by construction, $U = (Z_1 \cup Z_2) \cup (X \cup Y)$. I.e., the union $Z_1 \cup Z_2$ is linked through U to the set $X \cup Y$ of 5 collinear points. Hence, by (2.1),

		0	1	2	3	4	5	6	7	8	9	10	11	12
$Dh_U(i)$	:	1	2	3	4	5	6	7	6	5	4	3	2	1
$Dh_{X\cup Y}(12-i)$	:									1	1	1	1	1
$Dh_{Z_1\cup Z_2}(i)$	:	1	2	3	4	5	6	7	6	4	3	2	1	

Hence, $Dh_{Z_1 \cup Z_2}(11) = 1$ and, by Proposition 2.2, there exists $f \in S_{10}$ such that $(Z_1, Z_2, [f]) \in \mathcal{J}$. □

CLAIM 3.6. $\mathcal{Y}$ is dense in $\mathcal{X}$. In particular, the general $[f] \in \mathcal{Y}$ has rank 22.

PROOF. Since the 22-nd secant variety of $\nu_{10}(\mathbb{P}^2)$ and of $\nu_{10}(C)$, for any quintic $C \subset \mathbb{P}^2$, are non-defective, we have that the Terracini loci in $\mathrm{Hilb}^{\circ}_{22,5}(\mathbb{P}^2)$ and $\mathrm{Hilb}^{\circ}_{22}(\mathbb{P}^2)$ are proper sub-varieties. In particular, we can restrict the projection π_3 to the dense subspace of $\mathcal{J}$ given by the triplets $(Z_1, Z_2, [f])$ such that Z_1 and Z_2 do not belong to the corresponding Terracini locus. This implies that, for the general $[f] \in \mathcal{Y}$, the fiber $\pi_3^{-1}([f])$ is finite, see [**6**, Proposition 6.2]. In particular, $\dim \mathcal{Y} = \dim \mathcal{J}$.

Since π_2 is dominant, for a general $Z_2 \in \mathrm{Hilb}^{\circ}_{22}(\mathbb{P}^2)$, we have that $\dim \mathcal{J} = \dim \mathrm{Hilb}^{\circ}_{22}(\mathbb{P}^2) + \dim \pi_2^{-1}(Z_2)$. Now, $\dim \mathrm{Hilb}^{\circ}_{22}(\mathbb{P}^2) = 44$. Following the construction in Claim 3.5, we see that $\dim \pi_2^{-1}(Z_2) = 19$. Indeed, it depends on the choice of three collinear points, that is five parameters (2 to fix a line and 3 to fix three points on it), and the choice of the general complete intersection of two 7-ics, that is the choice of a pencil in the 8-dimensional projective space of 7-ics through the set of 27 points A, i.e., other $14 = 2(9-2)$ parameters. In particular, $\dim \mathcal{J} = 63 = \dim \mathcal{X}$. Since $\mathcal{X}$ is irreducible, we conclude the first part of the statement. The second part of the statement follows directly from Claim 3.4. □

Therefore, the general element of $\mathcal{Y}$ evinces the statement of the theorem. □

REMARK 3.7. In the proof of Claim 3.3 we observed that, given a general quintic curve $C \subset \mathbb{P}^2$, the general degree-10 form in $\langle \nu_{10}(C) \rangle = \mathbb{P}^{44}$ has rank 23 with respect to the curve. However, since the general rank of degree-10 ternary forms is equal to 22 by Alexander-Hirschowitz Theorem (Theorem 2.8), we also deduce that the general form in $\langle \nu_{10}(C) \rangle$ has rank 22 with respect to the Veronese variety $\nu_{10}(\mathbb{P}^2)$. In other words, there exist degree-10 ternary forms whose apolar ideal has minimal generators in degree five but minimal Waring decompositions are given by sets of points whose ideal does not involve such minimal generators.

REMARK 3.8. For sake of completeness, we observe that, in the example constructed in Theorem 3.1, the space of decompositions is discrete. This has been checked employing Bertini's Theorem and numerical computations, see [**2**] for a reference about the methodology used. In particular, a degeneration argument could not be used to prove an analogous statement.

3.2. Second example. Degree-13 ternary forms of rank 30.

THEOREM 3.9. *There exists a form f of degree* 13 *and of rank* 30 *having two sets of points computing its rank and admitting different Hilbert functions and regularities.*

PROOF. Let A be a set of 12 general points lying on a cubic. I.e.,

$$\begin{array}{rcccccc} h_A: & 1 & 3 & 6 & 9 & 12 & 12 & \cdots \\ Dh_A: & 1 & 2 & 3 & 3 & 3 & - \end{array}$$

Then, we consider the set Z_1 of 30 points linked to A through a complete intersection $X = A \cup Z_1$ of 42 points defined by a general 6-ic and a 7-ic. Namely, by (2.1),

$$\begin{array}{rccccccccccccc} Dh_X: & 1 & 2 & 3 & 4 & 5 & 6 & 6 & 5 & 4 & 3 & 2 & 1 & - \\ Dh_{Z_1}: & 1 & 2 & 3 & 4 & 5 & 6 & 6 & 2 & 1 & - \end{array}$$

We link Z_1 to another set Z_2 of 30 points through a complete intersection $Y = Z_1 \cup Z_2$ of 60 points defined by a 6-ic and degree-10 form. Namely, by (2.1),

$$\begin{array}{rcccccccccccccccc} Dh_Y(i) & : & 1 & 2 & 3 & 4 & 5 & 6 & 6 & 6 & 6 & 6 & 5 & 4 & 3 & 2 & 1 & - \\ Dh_{Z_1}(14-i) & : & & & & & & & 1 & 2 & 6 & 6 & 5 & 4 & 3 & 2 & 1 \\ Dh_{Z_2}(i) & : & 1 & 2 & 3 & 4 & 5 & 6 & 5 & 4 & - \end{array} \tag{3.5}$$

We immediately observe that Z_1 and Z_2 are both apolar to a form of degree 13. Indeed, from the first difference of $Y = Z_1 \cup Z_2$ in (3.5), we have $h^1_{14}(Z_1 \cup Z_2) = 1$. By Proposition 2.2, $\langle \nu_{13}(Z_1)\rangle \cap \langle \nu_{13}(Z_2)\rangle = [f]$ for some degree 13 form f.

CLAIM 3.10. f has rank 30.

PROOF. Let Z_1 be as above and $Z' \subset \mathbb{P}^2$ be another set of points apolar to f. Our goal is to show that $\ell(Z') \geq 30$.

First, assume that $Z_1 \cap Z' = \emptyset$ and let $\overline{Z} = Z_1 \cup Z'$. By Lemma 2.1-(9) since $\sum_{i=0}^5 Dh_{Z_1}(i) = 21$ then also $\sum_{i=0}^5 Dh_{\overline{Z}}(i) = 21$, i.e., $Dh_{\overline{Z}}(i) = i+1$ for $i \in \{0,\dots 5\}$. By Lemma 2.1-(10), the function $Dh_{\overline{Z}}$ cannot increase after the first degree i for which $Dh_{\overline{Z}}(i) \leq i$. Moreover, by Proposition 2.3, we cannot have consecutive values of the first difference $Dh_{\overline{Z}}$ equal to some value strictly smaller than 6. Finally, recall that by Proposition 2.6, $\overline{Z}$ satisfies $CB(13)$ and then, by Theorem 2.7, $Dh_{\overline{Z}}(14) \geq 1$. Therefore, we deduce

$$\begin{array}{rccccccccccccc} & & \cdots & 4 & 5 & 6 & 7 & 8 & 9 & 10 & 11 & 12 & 13 & 14 \\ Dh_{\overline{Z}} & : & \cdots & 5 & 6 & \geq 6 & \geq 6 & \geq 6 & \geq 6 & \geq 5 & \geq 4 & \geq 3 & \geq 2 & \geq 1 \end{array}$$

In particular, $\ell(\overline{Z}) \geq \sum_{i=0}^{\infty} Dh_{\overline{Z}}(i) \geq 60$. Hence, $\ell(Z') \geq 30$.

Assume now that $Z_1 \cap Z' \neq \emptyset$. As observed in [**3**, Remark 3.3], if $Z'' = Z' \smallsetminus (Z' \cap Z_1)$, then there exists a form $f_0 \in S_{13}$ such that $[f_0] \in \langle \nu_{13}(Z_1)\rangle \cap \langle \nu_{13}(Z'')\rangle$. Indeed, if $Z_1 = \{\ell_1, \dots, \ell_{30}\}$ and $Z' = \{\ell_1, \dots, \ell_k, m_{k+1}, \dots, m_r\}$, then,

$$f = \sum_{i=1}^{30} a_i \ell_i^{13} = \sum_{i=1}^{k} b_i \ell_i^{13} + \sum_{i=k+1}^{r} b_i m_i^{13}$$

and

$$f_0 = \sum_{i=1}^{k}(a_i - b_i)\ell_i^{13} + \sum_{i=k+1}^{30} a_i\ell_i^{13} = \sum_{i=k+1}^{r} b_i m_i^{13}.$$

Therefore, we can proceed with the same argument as before by considering f_0 instead of f and $\overline{Z} = Z_1 \cup Z''$ to deduce that $\ell(Z_1) \geq \ell(Z'') \geq 30$. □

Therefore, we have that Z_1 and Z_2 have different Hilbert function and regularity by (3.5) and both compute the rank of f by Claim 3.10. This concludes the proof. □

4. Explicit Macaulay2 computations

In this last section, we provide a code in the language of the algebra software Macaulay2 **[23]** to construct the examples presented in the previous section.

4.1. First example. Degree-10 ternary forms of rank 22. We present here the second construction, as in the proof of Claim 3.5.

```
i1 : k = QQ;
i2 : S = k[x,y,z];
i3 : -- consider a set of 22 general points
     IZ2 = intersect for i from 1 to 22 list
                          minors(2,vars(S)||random(QQ^1,QQ^3));
i4 : degree IZ2, for i to 10 list hilbertFunction(i,IZ2)

o4 = (22, {1, 3, 6, 10, 15, 21, 22, 22, 22, 22, 22})

i5 : -- consider the set of points Y = {(0:1:0),(0:0:1),(0:1:1)} on
    the line x = 0 and the union A = Z1 u Y
     IY = intersect(ideal(x,z), ideal(x,y), ideal(x,y-z));
i6 : IA = intersect(IZ2,IY);
i7 : -- take the degree-6 part of the ideal of A
     G = ideal super basis(6,IA);
i8 : -- take the ideal generated by the restriction of G over the line
     x = 0 away from the set of points Y
     R = k[y,z];
i9 : G0 = sub(G,R);
i10 : G' = G0 : sub(IY,R);
i11 : -- G' is then defined by three cubics: let C be the cubic curve
    as image of the line L via the map defined by G'
      phi = map(R,S,first entries mingens G');
i12 : C = ker phi;
i13 : -- We compute the singular locus and we check that it is nodal
      VC = Proj (S/C);
i14 : singVC = singularLocus VC;
i15 : codim singVC, degree singVC

o15 = (2, 1)

i16 : I0 = ideal singVC;
i17 : -- We compute the pre-image X of the singular point:
      -- that is a union of two simple points
      T = QQ[a,b,c,y,z, MonomialOrder => Eliminate 3];
```

```
i18 : psi = map(T,S,{a,b,c});
i19 : J = psi(I0) + ideal(a-sub(G'_0,T),b-sub(G'_1,T),c-sub(G'_2,T));
i20 : use S;
i21 : IX=saturate(ideal(x)+sub(ideal selectInSubring(1,gens gb J),S));
i22 : dim IX, degree IX, IX == radical IX

o22 = (1, 2, true)

i23 : -- Now, the union A u X is a set of 27 points
      IA' = intersect(IA,IX);
i24 : degree IA', for i to 10 list hilbertFunction(i,IA')

o24 = (27, {1, 3, 6, 10, 15, 21, 26, 27, 27, 27, 27})

i25 : -- We can also note that IA' is generated in degree 7
      betti IA'

             0 1
o25 = total: 1 5
          0: 1 .
          1: . .
          2: . .
          3: . .
          4: . .
          5: . 2
          6: . 3

o25 : BettiTally

i26 : -- Construct a general c.i. of type (7,7)
      B7 = super basis(7,IA');
i27 : C1 = B7 * random(QQ^(numcols B7), QQ^1);
i28 : C2 = B7 * random(QQ^(numcols B7), QQ^1);
i29 : IU = ideal(C1,C2);
i30 : degree IU, for i to 12 list hilbertFunction(i,IU)

o30 = (49, {1, 3, 6, 10, 15, 21, 28, 34, 39, 43, 46, 48, 49})

i31 : -- Consider the linkage to A' through U
      IZ1 = IU : IA';
i32 : degree IZ1, for i to 10 list hilbertFunction(i,IZ1)

o32 = (22, {1, 3, 6, 10, 15, 20, 22, 22, 22, 22, 22})

i33 : -- It can be checked that IZ1 is smooth and disjoint from IZ2
      loadPackage "Divisor";
i34 : isSmooth(IZ1, IsGraded => true), dim (IZ1 + IZ2)

o34 : (true, 0)

i35 : -- we compute now the intersection of their linear spans
      perpSpace = ideal super basis(10,IZ1)+ideal super basis(10,IZ2);
i36 : B10 = basis(10,S);
```

```
i37 : f = (B10 * gens ker diff(B10, transpose mingens perpSpace))_0_0;
i38 : -- Z1 and Z2 are apolar to f and have different HFs
      fperp = inverseSystem f; -- computes the apolar ideal of f
i39 : isSubset(IZ1,fperp), isSubset(IZ2,fperp),
          netList {for i to 14 list hilbertFunction(i,IZ1), for i to
              14 list hilbertFunction(i,IZ2)}
                        +-+-+-+--+--+--+--+--+--+--+--+--+--+--+--+
o39 = (true, true,      |1|3|6|10|15|21|22|22|22|22|22|22|22|22|22|)
                        +-+-+-+--+--+--+--+--+--+--+--+--+--+--+--+
                        |1|3|6|10|15|20|22|22|22|22|22|22|22|22|22|
                        +-+-+-+--+--+--+--+--+--+--+--+--+--+--+--+
```

4.2. Second example. Degree-13 ternary forms of rank 30.

```
i1 : k = QQ;
i2 : S = k[x,y,z];

i3 : -- consider 12 random points on the rational cubic xz^2 - y^3 = 0
     a = entries random(k^12,k^2);
i4 : A = (p -> {p_0^3, p_0*p_1^2, p_1^3}) \ a;
i5 : IA = intersect ((p -> minors(2,matrix{{x,y,z},p})) \ A);
i6 : degree IA, for i to 5 list hilbertFunction(i,IA)

o6 = (12, {1, 3, 6, 9, 12, 12})

i7 : -- consider a c.i. X of type (6,7) containing A
     B6 = super basis(6,IA);    B7 = super basis(7,IA);
i9 : F1 = B6 * random(QQ^(numcols B6), QQ^1);
i10 : F2 = B7 * random(QQ^(numcols B7), QQ^1);
i11 : IX = ideal(F1,F2);
i12 : degree IX, for i to 13 list hilbertFunction(i,IX)

o12 = (42, {1, 3, 6, 10, 15, 21, 27, 32, 36, 39, 41, 42, 42, 42})

i13 : -- consider the link to A through X
      IZ1 = IX : IA;
i14 : degree IZ1, for i to 13 list hilbertFunction(i,IZ1)

o14 = (30, {1, 3, 6, 10, 15, 21, 27, 29, 30, 30, 30, 30, 30, 30})

i15 : -- consider a c.i. Y of type (6,10) containing Z1
      B6 = super basis(6,IZ1);    B10 = super basis(10,IZ1);
i17 : G1 = B6 * random(QQ^(numcols B6), QQ^1);
i18 : G2 = B10 * random(QQ^(numcols B10), QQ^1);
i19 : IY = ideal(G1,G2);
i20 : degree IY, for i to 14 list hilbertFunction(i,IY)

o20 = (60, {1, 3, 6, 10, 15, 21, 27, 33, 39, 45, 50, 54, 57, 59, 60})

i21 : -- consider the link to Z1 through Y
      IZ2 = IY : IZ1;
i22 : degree IZ2, for i to 13 list hilbertFunction(i,IZ2)
```

```
o22 = (30, {1, 3, 6, 10, 15, 21, 26, 30, 30, 30, 30, 30, 30, 30})

i23 : -- we compute now the intersection of their linear spans
      perpSpace = ideal super basis(13,IZ1)+ideal super basis(13,IZ2);
i24 : B13 = basis(13,S);
i25 : f = (B13 * gens ker diff(B13, transpose mingens perpSpace))_0_0;

i26 : -- note that Z1 and Z2 are apolar to f
      -- and have different Hilbert functions and regularity
      fperp = inverseSystem f; -- computes the apolar ideal of f
i27 : isSubset(IZ1,fperp), isSubset(IZ2,fperp),
          netList {for i to 14 list hilbertFunction(i,IZ1),
                   for i to 14 list hilbertFunction(i,IZ2)}

                         +-+-+-+--+--+--+--+--+--+--+--+--+--+--+
o27 = (true, true,       |1|3|6|10|15|21|27|29|30|30|30|30|30|30|30|)
                         +-+-+-+--+--+--+--+--+--+--+--+--+--+--+
                         |1|3|6|10|15|21|26|30|30|30|30|30|30|30|30|
                         +-+-+-+--+--+--+--+--+--+--+--+--+--+--+
```

References

[1] J. Alexander and A. Hirschowitz, *Polynomial interpolation in several variables*, J. Algebraic Geom. **4** (1995), no. 2, 201–222. MR1311347

[2] E. Angelini, *Waring decompositions and identifiability via Bertini and Macaulay2 software*, J. Symbolic Comput. **91** (2019), 200–212, DOI 10.1016/j.jsc.2018.06.021. MR3860892

[3] E. Angelini and L. Chiantini, *On the identifiability of ternary forms*, Linear Algebra Appl. **599** (2020), 36–65, DOI 10.1016/j.laa.2020.03.042. MR4083806

[4] E. Angelini and L. Chiantini, *Minimality and uniqueness for decompositions of specific ternary forms*, Math. Comp. **91** (2022), no. 334, 973–1006, DOI 10.1090/mcom/3681. MR4379984

[5] E. Angelini, L. Chiantini, and N. Vannieuwenhoven, *Identifiability beyond Kruskal's bound for symmetric tensors of degree 4*, Atti Accad. Naz. Lincei Rend. Lincei Mat. Appl. **29** (2018), no. 3, 465–485, DOI 10.4171/RLM/817. MR3819100

[6] E. Ballico and L. Chiantini, *On the Terracini locus of projective varieties*, Milan J. Math. **89** (2021), no. 1, 1–17, DOI 10.1007/s00032-020-00324-5. MR4277359

[7] A. Bernardi, E. Carlini, M. V. Catalisano, A. Gimigliano, and A. Oneto, *The hitchhiker guide to: Secant varieties and tensor decomposition*, Mathematics **6** (2018), no. 12, 314.

[8] W. Buczyńska, J. Buczyński, and Z. Teitler, *Waring decompositions of monomials*, J. Algebra **378** (2013), 45–57, DOI 10.1016/j.jalgebra.2012.12.011. MR3017012

[9] E. Carlini, M. V. Catalisano, L. Chiantini, A. V. Geramita, and Y. Woo, *Symmetric tensors: rank, Strassen's conjecture and e-computability*, Ann. Sc. Norm. Super. Pisa Cl. Sci. (5) **18** (2018), no. 1, 363–390. MR3783793

[10] E. Carlini, M. V. Catalisano, and A. V. Geramita, *The solution to the Waring problem for monomials and the sum of coprime monomials*, J. Algebra **370** (2012), 5–14, DOI 10.1016/j.jalgebra.2012.07.028. MR2966824

[11] E. Carlini, N. Grieve, and L. Oeding, *Four lectures on secant varieties*, Connections between algebra, combinatorics, and geometry, Springer Proc. Math. Stat., vol. 76, Springer, New York, 2014, pp. 101–146, DOI 10.1007/978-1-4939-0626-0_2. MR3213518

[12] M. V. Catalisano, A. V. Geramita, and A. Gimigliano, *Segre-Veronese embeddings of $\mathbb{P}^1 \times \mathbb{P}^1 \times \mathbb{P}^1$ and their secant varieties*, Collect. Math. **58** (2007), no. 1, 1–24. MR2310544

[13] L. Chiantini, *Lectures on the structure of projective embeddings*, Rend. Sem. Mat. Univ. Politec. Torino **62** (2004), no. 4, 335–388. MR2129798

[14] L. Chiantini, *Hilbert functions and tensor analysis*, Quantum physics and geometry (2019), 125–151.

[15] L. Chiantini, C. Ciliberto, and F. Russo, *On secant defective varieties, in particular of dimension 4*, Preprint, arXiv:2011.01105 (2020).

[16] L. Chiantini and G. Ottaviani, *A footnote to a footnote to a paper of B. Segre*, The art of doing algebraic geometry, Trends Math., Birkhäuser/Springer, Cham, [2023] ©2023, pp. 93–117, DOI 10.1007/978-3-031-11938-5_4. MR4592538

[17] L. Chiantini, G. Ottaviani, and N. Vannieuwenhoven, *On generic identifiability of symmetric tensors of subgeneric rank*, Trans. Amer. Math. Soc. **369** (2017), no. 6, 4021–4042, DOI 10.1090/tran/6762. MR3624400

[18] A. Clebsch, *Ueber Curven vierter Ordnung* (German), J. Reine Angew. Math. **59** (1861), 125–145, DOI 10.1515/crll.1861.59.125. MR1579172

[19] E. D. Davis, *Complete intersections of codimension* 2 *in* $\mathbf{P}^r$*: the Bezout-Jacobi-Segre theorem revisited*, Rend. Sem. Mat. Univ. Politec. Torino **43** (1985), no. 2, 333–353 (1986). MR859862

[20] E. D. Davis, A. V. Geramita, and F. Orecchia, *Gorenstein algebras and the Cayley-Bacharach theorem*, Proc. Amer. Math. Soc. **93** (1985), no. 4, 593–597, DOI 10.2307/2045527. MR776185

[21] D. Eisenbud, *Linear sections of determinantal varieties*, Amer. J. Math. **110** (1988), no. 3, 541–575, DOI 10.2307/2374622. MR944327

[22] D. Eisenbud, *The geometry of syzygies: A second course in commutative algebra and algebraic geometry*, vol. 229, Springer, 2005.

[23] D. R. Grayson and M. E. Stillman, *Macaulay2, a software system for research in algebraic geometry*, Available at http://www.math.uiuc.edu/Macaulay2/.

[24] C. J. Hillar and L.-H. Lim, *Most tensor problems are NP-hard*, J. ACM **60** (2013), no. 6, Art. 45, 39, DOI 10.1145/2512329. MR3144915

[25] A. Iarrobino and V. Kanev, *Power sums, Gorenstein algebras, and determinantal loci*, Lecture Notes in Mathematics, vol. 1721, Springer-Verlag, Berlin, 1999. Appendix C by Iarrobino and Steven L. Kleiman, DOI 10.1007/BFb0093426. MR1735271

[26] T. G. Kolda and B. W. Bader, *Tensor decompositions and applications*, SIAM Rev. **51** (2009), no. 3, 455–500, DOI 10.1137/07070111X. MR2535056

[27] J. M. Landsberg, *Tensors: geometry and applications*, Representation theory **381** (2012), no. 402, 3.

[28] F. S. Macaulay, *The algebraic theory of modular systems*, Cambridge Mathematical Library, Cambridge University Press, Cambridge, 1994. Revised reprint of the 1916 original; With an introduction by Paul Roberts. MR1281612

[29] J. C. Migliore, *Introduction to liaison theory and deficiency modules*, Progress in Mathematics, vol. 165, Birkhäuser Boston, Inc., Boston, MA, 1998, DOI 10.1007/978-1-4612-1794-7. MR1712469

[30] B. Mourrain and A. Oneto, *On minimal decompositions of low rank symmetric tensors*, Linear Algebra Appl. **607** (2020), 347–377, DOI 10.1016/j.laa.2020.06.029. MR4154967

[31] F. Palatini, *Sulla rappresentazione delle forme ed in particolare della cubica quinaria con la somma di potenze di forme lineari*, vol. 38, Atti Acad. Torino, 1902.

[32] F. Palatini, *Sulla rappresentazione delle forme ternarie mediante la somma di potenze di forme lineari*, vol. V, Tip. della R. Accademia dei Lincei, 1903.

[33] F. Palatini, *Sulle varieta algebriche per le quali sono di dimensione minore dell'ordinario senza riempire lo spazio ambiente una o alcune delle varieta formate da spazi seganti*, 1909.

[34] C. Peskine and L. Szpiro, *Liaison des variétés algébriques. I* (French), Invent. Math. **26** (1974), 271–302, DOI 10.1007/BF01425554. MR364271

[35] K. Ranestad and F.-O. Schreyer, *Varieties of sums of powers*, J. Reine Angew. Math. **525** (2000), 147–181, DOI 10.1515/crll.2000.064. MR1780430

[36] H. W. Richmond, *On canonical forms*, Quart. J. Pure Appl. Math **33** (1904), 967–984.

[37] G. Scorza, *Determinazione delle varietà tre dimensioni di* $s_r, (r \geq 7)$ *i cui* s_3 *tangenti si tangliano a due a due*, Rendiconti del Circolo Matematico di Palermo (1884-1940) **25** (1908), 193–204.

[38] G. Scorza, *Sulle varietà a quattro dimensioni di* s_r $(r > 9)$ *i cui* s_4 *tangenti si tagliano a due a due*, Rendiconti del Circolo Matematico di Palermo **27** (1909), 148–178.

[39] G. Scorza, *Opere scelte*, **1** (1960), 252–292.

[40] J. J. Sylvester, *An essay on canonical forms, supplement to a sketch of a memoir on elimination*, Collected Mathematical Paper I **34** (1851), 1851.

[41] J. J. Sylvester, *On a remarkable discovery in the theory of canonical forms and of hyperdeterminants,*, originally in Philosophical Magazine **1** (1851), 265–283.

[42] A. Terracini, *Sulla rappresentazione delle coppie di forme ternarie mediante somme di potenze di forme lineari*, Annali di Matematica Pura ed Applicata (1898-1922) **24** (1915), no. 1, 1–10.

Dip. di Ingegneria dell'Informazione e Scienze Matematiche, U. di Siena, Italy
Email address: elena.angelini@unisi.it

Dip. di Ingegneria dell'Informazione e Scienze Matematiche, U. di Siena, Italy
Email address: luca.chiantini@unisi.it

Dip. di Mathematics, U. of Trento, Via Sommarive 14, 38123 Povo (Trento), Italy
Email address: alessandro.oneto@unitn.it

Contemporary Mathematics
Volume **805**, 2024
https://doi.org/10.1090/conm/805/16128

Some remarks on varieties whose twisted normal bundle is an instanton

Vincenzo Antonelli and Gianfranco Casnati

ABSTRACT. Let $X \subseteq \mathbb{P}^N$ be a smooth variety with normal bundle $\mathcal{N}_X$. In this note we prove that if $\mathcal{N}_X \otimes \mathcal{O}_{\mathbb{P}^N}(-t)$ is an instanton with quantum number k, then $\dim(X) = N-2$, $t=1$ and, when $\dim(X) \geq 4$, also $\lfloor \dim(X)/2 \rfloor \leq \deg(X)$. Moreover, we also discuss some methods for constructing smooth varieties such that $\mathcal{N}_X \otimes \mathcal{O}_{\mathbb{P}^N}(-1)$ is an instanton with $k \neq 0$, illustrating them with explicit examples and counterexamples.

1. Introduction

In this paper a *variety* X is a closed, integral subscheme of some projective space over a field $\mathbf{k}$.

The following definition mimics the one in [**16**] extending it to each variety besides $\mathbb{P}^N$: for a more detailed discussion motivating it, see [**2**] (and [**12**] for the particular case $k=0$).

DEFINITION 1.1. Let $X \subseteq \mathbb{P}^N$ be a variety of dimension $n \geq 2$ and let $\mathcal{O}_X(h) := \mathcal{O}_X \otimes \mathcal{O}_{\mathbb{P}^N}(1)$.

A rank r sheaf $\mathcal{E}$ on X is called h–instanton with quantum number k if

$$h^0(\mathcal{E}(-h)) = h^n(\mathcal{E}(-nh)) = 0, \qquad h^1(\mathcal{E}(-h)) = h^{n-1}(\mathcal{E}(-nh)) = k,$$
$$h^i(\mathcal{E}(-(i+1)h)) = h^j(\mathcal{E}(-jh)) = 0$$

for $1 \leq i \leq n-2$ and $2 \leq j \leq n-1$.

An h–instanton sheaf $\mathcal{E}$ with quantum number 0 is called h–Ulrich.

We only notice that the vanishings in the last displayed row of the above definition contribute to it only if $n \geq 3$. When $n=2$ they are empty conditions. When $n \geq 3$ then $h^i(\mathcal{E}(-th)) = 0$ for $2 \leq t \leq n-1$ and $0 \leq i \leq n$.

On the one hand, rank two instanton bundles on $\mathbb{P}^3$ were first introduced due to their connection with the solutions of the Yang–Mills equations (see [**5**]) through the Atiyah–Penrose–Ward transformation and therefore with the physics of particles.

2020 *Mathematics Subject Classification.* Primary 14F06; Secondary 14D21, 14J45, 14J60.
Key words and phrases. instanton bundle, Steiner bundle, normal bundle.
The authors are members of GNSAGA group of INdAM and are supported by the framework of the MIUR grant Dipartimenti di Eccellenza 2018-2022 (E11G18000350001).

On the other hand, the existence of an instanton sheaf with fixed quantum number k on X is not obvious. E.g. the study of the algebraic counterpart of Ulrich sheaves, i.e. maximally generated maximal Cohen–Macaulay modules, started in the 80's of the past century and the problem of their existence on each variety is a wide open question which have been object of deep study since [**12**] appeared in 2003 (see [**6**] for more details about this case both from the algebraic viewpoint and the geometric one).

The existence of instanton and Ulrich sheaves is often also related to interesting geometric deep properties of the underlying variety X. E.g. if $X \subseteq \mathbb{P}^{n+1}$ is a hypersurface, then the existence of a locally Cohen–Macaulay h–instanton sheaf is equivalent to the existence of a representation of a power of the form defining X as the determinant of a suitable morphism of vector bundles of the same rank on $\mathbb{P}^{n+1}$ which are Steiner in the sense of [**11**] (see Definition 1.4).

Notice that every variety $X \subseteq \mathbb{P}^N$ comes equipped with certain distinctive vector bundles such as the *cotangent bundle*, i.e. the sheaf of differentials Ω^1_X, the *tangent bundle*, i.e. its dual $\mathcal{T}_X$, the *conormal bundle*, i.e. $\mathcal{C}_X := \mathcal{I}_X/\mathcal{I}_X^2$ where $\mathcal{I}_X$ is the ideal sheaf of X, and the *normal bundle*, i.e. its dual $\mathcal{N}_X$. Thus it is quite reasonable to ask if one of them is an instanton bundle with respect to $\mathcal{O}_X(h)$.

In [**8**] the second author, improving and generalizing a recent result in [**7**], shows that one can answer affirmatively to the above question only for $\mathcal{T}_X$ when either X is a twisted cubic curve in $\mathbb{P}^3$, or a Veronese surface in $\mathbb{P}^5$ and in these cases $\mathcal{T}_X$ is actually Ulrich.

Let us now focus on the normal bundle $\mathcal{N}_X$ of $X \subseteq \mathbb{P}^N$. The fact that $\mathcal{N}_X$ is never an h–instanton bundle trivially follows by combining the exact sequences

$$0 \longrightarrow \mathcal{T}_X \longrightarrow \mathcal{T}_{\mathbb{P}^N} \otimes \mathcal{O}_X \longrightarrow \mathcal{N}_X \longrightarrow 0, \tag{1.1}$$
$$0 \longrightarrow \mathcal{O}_{\mathbb{P}^N}(-1) \longrightarrow \mathcal{O}_{\mathbb{P}^N}^{\oplus N+1} \longrightarrow \mathcal{T}_{\mathbb{P}^N}(-1) \longrightarrow 0.$$

Indeed by composition we obtain a surjective morphism

$$\mathcal{O}_X^{\oplus N+1} \twoheadrightarrow \mathcal{T}_{\mathbb{P}^N} \otimes \mathcal{O}_X(-h) \twoheadrightarrow \mathcal{N}_X(-h),$$

hence $h^0(\mathcal{N}_X(-h)) \neq 0$. For this reason it is natural to raise the same question for the tensor products of $\mathcal{N}_X$ with suitable rank one coherent sheaves $\mathcal{L}$ on X.

The first result in this direction can be found in [**19**, Theorem 3.6] where the authors show that if $X \subseteq \mathbb{P}^N$ is the degeneracy locus of a matrix M of linear forms and the characteristic of $\mathbf{k}$ is zero, then there is a rank one sheaf $\mathcal{L}$ on X which is invertible outside the singular locus of X and such that $\mathcal{N}_X \otimes \mathcal{L}$ is Ulrich. In the particular case of codimension 2 varieties, then $\mathcal{L} \cong \mathcal{O}_X(-h)$.

If we restrict to smooth varieties and to sheaves $\mathcal{L} \cong \mathcal{O}_X(-th)$, there are even more precise results. Indeed, on the one hand, at least when $\mathbf{k} = \mathbb{C}$, in [**20**] the author proves that if $\mathcal{N}_X(-th)$ is Ulrich then either X is a linear subspace of $\mathbb{P}^N$, or $N = n+2$, $t = 1$ and $n = \dim(X) \leq 3$. On the other hand, for each non–degenerate complete intersection $X \subseteq \mathbb{P}^{n+2}$ the bundle $\mathcal{N}_X(-h)$ is never h–Ulrich.

In the present short note we partially extend the quoted result from [**20**] to instanton bundles proving the following result in Section 2.

THEOREM 1.2. *Let $X \subseteq \mathbb{P}^N$ be a smooth variety of dimension $n < N$ and let $\mathcal{O}_X(h) := \mathcal{O}_X \otimes \mathcal{O}_{\mathbb{P}^N}(1)$.*

Then $\mathcal{N}_X(-th)$ is an h–instanton bundle if and only if $t = 1$ and one of the following assertions holds.

(1) *X is a linear subspace in $\mathbb{P}^N$: in this case $\mathcal{N}_X(-h) \cong \mathcal{O}_{\mathbb{P}^n}^{\oplus N-n}$.*
(2) *$N = n+2$ and $h^j(\mathcal{N}_X(-(j+2)h)) = 0$ for $0 \le j \le n-2$: in this case the quantum number of $\mathcal{N}_X(-h)$ is $h^1(\mathcal{N}_X(-2h))$.*

Moreover, if the latter case occurs and $n \ge 4$, then

$$\left\lfloor \frac{n}{2} \right\rfloor \le \deg(X). \tag{1.2}$$

In view of the above result, in Section 3 we deal with the well known Serre correspondence between varieties of codimension 2 and vector bundles (see [**4**, **23**]). More precisely, we recall that for each smooth variety $X \subseteq \mathbb{P}^{n+2}$ of dimension n there exists at least one rank r vector bundle $\mathcal{F}$ on $\mathbb{P}^{n+2}$ and an injective morphism $\varphi\colon \mathcal{O}_{\mathbb{P}^{n+2}}^{\oplus r-1} \to \mathcal{F}^\vee(1)$ such that the locus $D_{r-1}(\varphi)$ of points where $\mathrm{rk}(\varphi) \le r-1$ is exactly X (see Lemma 3.1).

In Section 4 we prove the following result extending [**20**, Proposition 2.2] to instanton bundles.

THEOREM 1.3. *Let $\mathcal{F}$ be a rank r Steiner bundle with $c_1(\mathcal{F}) = f$ on $\mathbb{P}^{n+2}$, $n \ge 2$. Assume that $\varphi\colon \mathcal{O}_{\mathbb{P}^{n+2}}^{\oplus r-1} \to \mathcal{F}^\vee(1)$ is an injective morphism such that $X = D_{r-1}(\varphi)$ is smooth of dimension n and set $\mathcal{O}_X(h) := \mathcal{O}_X \otimes \mathcal{O}_{\mathbb{P}^{n+2}}(1)$.*

Then $\mathcal{N}_X(-h)$ is h–instanton with quantum number $(r-1)f$.

Trivially each linear determinantal smooth variety of codimension 2 gives examples of the above theorems. Nevertheless, we can even find non–trivial examples, i.e. varieties $X \subseteq \mathbb{P}^{n+2}$ such that $\mathcal{N}_X$ is an h–instanton with strictly positive quantum number. Several examples are exploited in Section 5.

1.1. Notation and first results. Throughout the whole paper we will work over an algebraically closed field $\mathbf{k}$ of arbitrary characteristic. The projective space of dimension N over $\mathbf{k}$ will be denoted by $\mathbb{P}^N$: $\mathcal{O}_{\mathbb{P}^N}(1)$ will denote the hyperplane line bundle.

A projective scheme X is a closed subscheme of some projective space over $\mathbf{k}$: X is a variety if it is also integral. If X is a projective scheme its structure sheaf is denoted by $\mathcal{O}_X$, its canonical sheaf by ω_X, its sheaf of p–differentials by Ω_X^p, and we set $\mathcal{T}_X := (\Omega_X^1)^\vee$.

Let X be smooth. We denote by K_X each Cartier divisor such that $\omega_X \cong \mathcal{O}_X(K_X)$. As in [**14**], $A^r(X)$ denotes the group of cycles on X of codimension r modulo rational equivalence, hence $A^1(X)$ coincides with the usual Picard group of X. The Chern classes of a coherent sheaf $\mathcal{A}$ on X are elements in $A(X)$: in particular, when $\mathcal{A}$ is locally free $c_1(\mathcal{A})$ is identified with $\det(\mathcal{A})$. When $X = \mathbb{P}^N$, then $A^r(\mathbb{P}^N)$ is generated by the class of a linear variety of codimension r, hence we make the standard identification $A^r(\mathbb{P}^N) = \mathbb{Z}$.

We recall the following definition: see [**11**].

DEFINITION 1.4. Let $N \ge 2$.

A rank r bundle $\mathcal{F}$ on $\mathbb{P}^N$ is called Steiner if it fits into an exact sequence of the form

$$0 \longrightarrow \mathcal{O}_{\mathbb{P}^N}(-1)^{\oplus f} \longrightarrow \mathcal{O}_{\mathbb{P}^N}^{\oplus r+f} \longrightarrow \mathcal{F} \longrightarrow 0 \tag{1.3}$$

for some $f \ge 0$.

Steiner bundles can be characterized cohomologically as follows.

LEMMA 1.5. *Let $N \geq 2$.*

A rank r bundle $\mathcal{F}$ on $\mathbb{P}^N$ is Steiner if and only if $h^i(\mathcal{F}(t)) = 0$ when either $0 \leq i \leq N-2$ and $t \leq -i-1$ or $1 \leq i \leq N$ and $t \geq -i$.

PROOF. In [**3**, Corollary 4.3] the authors show that the notion of Steiner bundle can be be characterized cohomologically using either [**3**, Definition 1.1] or [**3**, Proposition 4.1 (2)]. The latter is exactly the statement above. □

We will also need the following lemma.

LEMMA 1.6. *Let X be a smooth projective variety of dimension $n \geq 2$ endowed with an ample and globally generated line bundle $\mathcal{O}_X(h)$.*

A rank two bundle $\mathcal{E}$ on X with $c_1(\mathcal{E}) = (n+1)h + K_X$ is an h–instanton if and only if $h^i(\mathcal{E}(-(i+1)h)) = 0$ when $0 \leq i \leq n-2$.

PROOF. Taking into account that in this paper we only consider as instanton what is called instanton with defect $\delta = 0$ in [**2**], the statement above is exactly [**2**, Proposition 6.7]. □

For further notation and all the other results used in the paper we tacitly refer to [**15**], unless otherwise stated.

2. The proof of Theorem 1.2

In this first section we prove Theorem 1.2 stated in the introduction.

PROOF OF THEOREM 1.2. If $X \subseteq \mathbb{P}^N$ is a linear subspace, then $\mathcal{N}_X(-h) \cong \mathcal{O}_{\mathbb{P}^n}^{\oplus N-n}$ which is trivially an h–instanton.

If $N = n+2$, then (1.1) yields

$$c_1(\mathcal{N}_X(-h)) = (n+1)h + K_X. \tag{2.1}$$

In particular, thanks to Lemma 1.6, $\mathcal{N}_X(-h)$ is an h–instanton bundle if and only if $h^j(\mathcal{N}_X(-(j+2)h)) = 0$ for $0 \leq j \leq n-2$. Moreover, if these vanishings occur, then the quantum number of $\mathcal{N}_X(-h)$ is $h^1(\mathcal{N}_X(-2h))$.

The proof of the converse implication can be obtained in the same way as in [**20**, Theorem 1].

Indeed one of the two ingredients of the proof therein is that equality

$$c_1(\mathcal{E})h^{n-1} = \frac{\mathrm{rk}(\mathcal{E})}{2}((n+1)h^n + K_X h^{n-1})$$

holds for each h–Ulrich bundle $\mathcal{E}$ on a smooth variety. In [**2**, Theorem 1.6] the authors prove that the same equality also holds for each h–instanton bundle regardless of the field $\mathbf{k}$ when $\mathcal{O}_X(h)$ is very ample.

The other ingredient is that the line bundle $\mathcal{O}_X((n+1)h + K_X)$ is globally generated, hence $(n+1)h^n + K_X h^{n-1} \geq 0$ by the Nakai–Moishezon criterion. This result holds regardless of the characteristic of $\mathbf{k}$ as well, thanks to [**22**].

Thus the converse implication can be proved repeating verbatim the aforementioned proof in [**20**].

In order to prove (1.2), assume that $\mathcal{N}_X(-h)$ is an h–instanton bundle and X is not a linear subspace in $\mathbb{P}^{n+2}$. Assume $\lfloor n/2 \rfloor > \deg(X)$: thanks to [**2**, Proposition 6.3] we deduce that $\mathcal{N}_X(-h)$ is aCM, whence it is actually Ulrich. It follows from [**20**, Theorem 1] that $n \leq 3$, i.e. $\deg(X) \leq 1$ thanks to the above inequality, a contradiction. Thus the proof is complete. □

REMARK 2.1. We are unable to show that $n \le 3$ as in [**20**, Theorem 1], because the proof therein heavily depends on the fact that Ulrich bundles are aCM. We are only able prove the weaker bound (1.2). In particular, we do not know if there are non–degenerate varieties X with $n = \dim(X) \ge 4$ such that $\mathcal{N}_X(-h)$ is an h–instanton bundle.

Nevertheless, at least if $\mathbf{k} \cong \mathbb{C}$, we can look at the problem from a different perspective, suggesting the existence of a stronger bound $n \le 3$.

Indeed if $n \ge 4$, then X would satisfy $\omega_X \cong \mathcal{O}_X(e)$ for some integer, thanks to [**14**, Theorem 2.2], hence there would be an exact sequence

$$0 \longrightarrow \mathcal{O}_{\mathbb{P}^{n+2}} \longrightarrow \mathcal{E} \longrightarrow \mathcal{I}_X(c_1(\mathcal{E})) \longrightarrow 0$$

where $\mathcal{E}$ is a rank two vector bundle on $\mathbb{P}^{n+2}$ (for details see [**14**, Proposition 6.2] or the next Section 3). The above sequence is locally a Koszul complex, hence its restriction to X yields $\mathcal{O}_X \otimes \mathcal{E} \cong \mathcal{N}_X$.

Assume $\mathcal{E} \cong \mathcal{O}_{\mathbb{P}^{n+2}}(a_1) \oplus \mathcal{O}_{\mathbb{P}^{n+2}}(a_2)$. If $\mathcal{N}_X(-h)$ is an h–instanton bundle, then

$$h^0(\mathcal{O}_X((a_i-2)h)) = 0, \qquad h^n(\mathcal{O}_X((a_i-n-1)h)) = h^0(\mathcal{O}_X((n+1-a_i+e)h)) = 0,$$

thanks to Definition 1.1, hence $n+2+e \le a_i \le 1$. Since

$$\mathcal{O}_X((n+1)h + K_X) \cong \mathcal{O}_X((n+1+e)h)$$

is globally generated by [**22**], we infer that $a_i = 1$ and $e = -n-1$, i.e. $X \cong \mathbb{P}^n$ embedded linearly in $\mathbb{P}^{n+2}$ by [**17**, Theorem 1], contradicting the non–degeneracy of X. We deduce that $\mathcal{E}$ is necessarily indecomposable if $\mathcal{N}_X(-h)$ is an h–instanton bundle, contradicting the Hartshorne conjecture when $n \ge 4$.

REMARK 2.2. If $\mathcal{N}_X(-h)$ is an instanton, then X has codimension 2, hence its quantum number can be computed using the Riemann–Roch theorem (see [**15**, Theorem A.4.1]), because

$$k = h^1(\mathcal{N}_X(-2h)) = -\chi(\mathcal{N}_X(-2h)),$$

taking into account that $c_2(\mathcal{N}_X) = dh^2$ (see [**15**, A.3.C7]) and computing $c_2(\mathcal{T}_X)$ from (1.1).

E.g. if $X \subseteq \mathbb{P}^5$ is a threefold of degree d, then

$$c_2(\mathcal{T}_X) = (15-d)h^2 + 6hK_X + K_X^2.$$

Thus the Riemann–Roch theorem yields

$$k = \frac{1}{6}(7d-71)d - 4h^2K_X - \frac{1}{3}hK_X^2. \tag{2.2}$$

A similar argument for a surface $X \subseteq \mathbb{P}^4$ of degree d returns the equality

$$k = \frac{1}{2}(2d-13)d - \frac{5}{2}hK_X - 2\chi(\mathcal{O}_X). \tag{2.3}$$

3. Codimension two subvarieties

Taking into account Theorem 1.2, it is then interesting to deal with the normal bundle of codimension two smooth subvarieties $X \subseteq \mathbb{P}^{n+2}$. Since every h–instanton bundle on a curve is automatically h–Ulrich we restrict to $n \ge 2$.

The ideal of such an X has a particular resolution as the following well known result shows (see [**23**]: see also [**4**]). In order to state it, we notice that

$$\mathcal{L}(X) := \{\ \ell \in \mathbb{Z} \mid \mathcal{O}_X((n+3-\ell)h + K_X) \text{ is globally generated}\ \}$$

is always non–empty: indeed, $\mathcal{O}_X((n+1)h+K_X)$ is globally generated thanks to [**22**], hence $\mathcal{L}(X)$ certainly contains at least all the integers not greater than 2.

LEMMA 3.1. *Let $X \subseteq \mathbb{P}^{n+2}$ be a locally complete intersection of dimension $n \geq 2$ and let $\mathcal{O}_X(h) := \mathcal{O}_X \otimes \mathcal{O}_{\mathbb{P}^{n+2}}(1)$.*

For each $\ell \in \mathcal{L}(X)$ there is an integer $r_\ell \geq 2$ and a rank r_ℓ bundle $\mathcal{F}_\ell$ on $\mathbb{P}^{n+2}$ with $c_1(\mathcal{F}_\ell) = r_\ell - \ell$ which fits into an exact sequence of the form

$$0 \longrightarrow \mathcal{O}_{\mathbb{P}^{n+2}}(-\ell)^{\oplus r_\ell - 1} \xrightarrow{\varphi} \mathcal{F}_\ell^\vee(1-\ell) \longrightarrow \mathcal{I}_X \longrightarrow 0.$$

Such a sequence is uniquely determined by the choice of ℓ.

PROOF. Notice that $c_1(\mathcal{N}_X) = (n+3)h + K_X$ thanks to (1.1). By definition the line bundle $\mathcal{O}_X((n+3-\ell)h+K_X)$ is globally generated. Assume there are $r_\ell - 1$ global sections generating $\mathcal{O}_X((n+3-\ell)h+K_X)$. Thus [**4**, Theorem 1.1] yields the existence of a rank r_ℓ vector bundle $\mathcal{F}$ on $\mathbb{P}^{n+2}$ such that $c_1(\mathcal{F}^\vee(1)) = \ell$, fitting into a unique sequence as in the statement because $h^i(\mathcal{O}_{\mathbb{P}^{n+2}}(-\ell)) = 0$ for $1 \leq i \leq 2$. In particular $c_1(\mathcal{F}) = r_\ell - \ell$. $\square$

REMARK 3.2. Trivially one can take $r_\ell := h^0(\mathcal{O}_X((n+3-\ell)h+K_X)) + 1$.

The above result implies that every locally complete intersection $X \subseteq \mathbb{P}^{n+2}$ of codimension 2 satisfies the equality $X = D_{r-2}(\varphi)$ for a suitable morphism φ from a trivial bundle of rank $r-1$ to another rank r bundle $\mathcal{F}^\vee(1)$. In general X is singular along the locus $D_{r-3}(\varphi)$ whose expected dimension is $n-4$. Thus X is likely singular when $n \geq 4$.

Conversely, let $\mathcal{F}$ be a rank r bundle on $\mathbb{P}^{n+2}$, $n \geq 2$, with $c_1(\mathcal{F}) = f$ and assume that $\varphi\colon \mathcal{O}_{\mathbb{P}^{n+2}}^{\oplus r-1} \to \mathcal{F}^\vee(1)$ is injective and such that $X := D_{r-2}(\varphi) \subseteq \mathbb{P}^{n+2}$ has pure dimension n. If X is smooth, then it is certainly irreducible, because $n \geq 2$. In this case there is an exact sequence

$$0 \longrightarrow \mathcal{O}_{\mathbb{P}^{n+2}}(f-r)^{\oplus r-1} \xrightarrow{\varphi} \mathcal{F}^\vee(f-r+1) \longrightarrow \mathcal{I}_X \longrightarrow 0. \tag{3.1}$$

The following result is well known and part of folklore.

PROPOSITION 3.3. *Let P be a smooth projective variety. Assume that the characteristic of* **k** *is* 0.

If $\mathcal{A}$ and $\mathcal{B}$ are bundles of respective ranks a and b on P such that $\mathrm{Hom}_{\mathcal{O}_P}(\mathcal{A},\mathcal{B})$ *is globally generated and $\varphi\colon \mathcal{A} \to \mathcal{B}$ is general, then $D_r(\varphi)$ is either empty or has codimension $(a-r+1)(b-r+1)$ which is smooth outside $D_{r-1}(\varphi)$.*

PROOF. E.g. see [**21**, Theorem 2.6]. $\square$

In particular, on the one hand X is likely to be singular when $n \geq 4$. On the other hand, when $n \leq 3$, the variety X is smooth of pure dimension n if the sheaf

$$\mathcal{F}^\vee(1)^{\oplus r-1} \cong \mathcal{H}om_{\mathbb{P}^{n+2}}(\mathcal{O}_{\mathbb{P}^{n+2}}(-1)^{\oplus r-1}, \mathcal{F}^\vee)$$

is globally generated, φ is general and the characteristic of **k** is 0. In this case $r - f = c_1(\mathcal{F}^\vee(1)) \geq 0$ necessarily.

PROPOSITION 3.4. *Let $\mathcal{F}$ be a rank r bundle on $\mathbb{P}^{n+2}$ with $n \geq 2$ and consider a morphism $\varphi\colon \mathcal{O}_{\mathbb{P}^{n+2}}^{\oplus r-1} \to \mathcal{F}^\vee(1)$ such that $X := D_{r-2}(\varphi) \subseteq \mathbb{P}^{n+2}$ is smooth of pure dimension n.*

If $c_1(\mathcal{F}) = f$, then

$$h^0(\omega_X) = (r-1)h^0(\mathcal{O}_{\mathbb{P}^{n+2}}(r-f-n-3)) - h^0(\mathcal{F}(r-f-n-4)) + h^1(\mathcal{F}(r-f-n-4))$$

and the following assertions hold.

(1) *If $r \le f+n+2$, and $h^1(\mathcal{F}(r-f-n-4)) = 0$, then $h^0(\omega_X) = 0$.*
(2) *If $r \ge f+n+3$, then ω_X is globally generated.*
(3) *If $r \ge f+n+4$, then ω_X is very ample.*

PROOF. By applying the functor $\mathcal{H}om_{\mathbb{P}^{n+2}}(-,\omega_{\mathbb{P}^{n+2}})$ to the exact sequence obtained by glueing together (3.1) and

$$0 \longrightarrow \mathcal{I}_X \longrightarrow \mathcal{O}_{\mathbb{P}^{n+2}} \longrightarrow \mathcal{O}_X \longrightarrow 0 \tag{3.2}$$

we obtain the exact sequences

$$\begin{gathered} 0 \longrightarrow \mathcal{O}_{\mathbb{P}^{n+2}}(-n-3) \longrightarrow \mathcal{F}(r-f-n-4) \longrightarrow \mathcal{A} \longrightarrow 0, \\ 0 \longrightarrow \mathcal{A} \longrightarrow \mathcal{O}_{\mathbb{P}^{n+2}}(r-f-n-3)^{\oplus r-1} \longrightarrow \omega_X \longrightarrow 0. \end{gathered} \tag{3.3}$$

Thus ω_X is globally generated if $r \ge f+n+3$ and very ample when $r \ge f+n+4$.

Finally, we have $h^i(\mathcal{A}) = h^i(\mathcal{F}(r-f-n-4))$ for $0 \le i \le 2$ because $n \ge 2$, hence the assertion about $h^0(\omega_X)$. In particular, if $h^1(\mathcal{F}(r-f-n-4)) = 0$, then

$$0 \le h^0(\omega_X) \le (r-1)h^0(\mathcal{O}_{\mathbb{P}^{n+2}}(r-f-n-3)) - h^0(\mathcal{F}(r-f-n-4)):$$

the first summand on the right vanishes if $r \le f+n+2$, hence $h^0(\omega_X) = 0$ necessarily. □

REMARK 3.5. The cohomology of (3.1) yields

$$H^i_*(\mathcal{F}) \cong H^{n+2-i}_*(\mathcal{F}^\vee(-n-3)) \cong H^{n+2-i}_*(\mathcal{I}_X(r-f-n-4)),$$

by the Serre duality. Thus X is aCM if and only if the latter direct sum is zero in the range $2 \le i \le n+1$, i.e. if and only if $H^i_*(\mathcal{F}) = 0$ in the same range.

EXAMPLE 3.6. Assume that the characteristic of $\mathbf{k}$ is 0 and $2 \le n \le 3$.

Let $\mathcal{G}$ be any Steiner bundle on $\mathbb{P}^{n+2}$ of rank r and $c_1(\mathcal{G}) = g$. In what follows we consider $\mathcal{F} := \mathcal{G}^\vee(1-t)$ where $t \ge 0$. The bundle $\mathcal{F}$ satisfies $H^i_*(\mathcal{F}) = 0$ for $2 \le i \le n+1$ and $\mathcal{F}^\vee(1) \cong \mathcal{G}(t)$ is globally generated because $\mathcal{G}$ is Steiner (see (1.3)), hence we can construct smooth varieties $X \subseteq \mathbb{P}^N$ whose sheaf of ideals fits into the exact sequence

$$0 \longrightarrow \mathcal{O}_{\mathbb{P}^{n+2}}(-rt-g)^{\oplus r-1} \longrightarrow \mathcal{G}(t-rt-g) \longrightarrow \mathcal{I}_X \longrightarrow 0.$$

Thanks to Remark 3.5 all such varieties are aCM.

Indeed we can actually recover a determinantal resolution of $\mathcal{I}_X$ simply by taking the mapping cone of the above sequence with (1.3). In this case we obtain

$$\begin{aligned} 0 \longrightarrow \mathcal{O}_{\mathbb{P}^{n+2}}(-rt-g)^{\oplus r-1} &\oplus \mathcal{O}_{\mathbb{P}^{n+2}}(t-rt-g-1)^{\oplus g} \\ &\longrightarrow \mathcal{O}_{\mathbb{P}^{n+2}}(t-rt-g)^{\oplus g+r} \longrightarrow \mathcal{I}_X \longrightarrow 0. \end{aligned}$$

There are many varieties of codimension 2 which are not aCM. E.g. in **[9]** the author classifies the arithmetically Buchsbaum ones in $\mathbb{P}^{n+2}$ which are not of general type, listing certain locally free resolutions of their sheaves of ideals. Some of them are not in the shape described in Lemma 3.1. In the following example we show how to recover a locally free resolution with that shape.

EXAMPLE 3.7. Assume that the characteristic of $\mathbf{k}$ is 0 and consider the arithmetically Buchsbaum $K3$ surfaces $X \subseteq \mathbb{P}^4$ of degree $d = 8$ whose sheaf of ideals fits into the exact sequence

$$0 \longrightarrow \mathcal{O}_{\mathbb{P}^4}(-4)^{\oplus 2} \oplus \mathcal{O}_{\mathbb{P}^4}(-5) \longrightarrow \Omega^1_{\mathbb{P}^4}(-2) \longrightarrow \mathcal{I}_X \longrightarrow 0. \tag{3.4}$$

In order to compute an integer ℓ and the corresponding sheaf $\mathcal{F}_\ell$ defined in Lemma 3.1 we apply the functor $\mathcal{H}om_{\mathbb{P}^4}(-, \omega_{\mathbb{P}^4})$ to the sequence above glued with (3.2). Thus we obtain the exact sequence

$$0 \longrightarrow \mathcal{O}_{\mathbb{P}^4}(-5) \longrightarrow \mathcal{T}_{\mathbb{P}^4}(-3) \longrightarrow \mathcal{O}_{\mathbb{P}^4}(-1)^{\oplus 2} \oplus \mathcal{O}_{\mathbb{P}^4} \longrightarrow \omega_X \longrightarrow 0.$$

In particular we can take $\ell = 4$ in Lemma 3.1 and, consequently, $r_\ell = 7$. Thus there is a vector bundle $\mathcal{F}_4$ of rank 8 fitting into an exact sequence of the form

$$0 \longrightarrow \mathcal{O}_{\mathbb{P}^4}(-4)^{\oplus 7} \longrightarrow \mathcal{F}_4^\vee(-3) \longrightarrow \mathcal{I}_X \longrightarrow 0. \tag{3.5}$$

The cohomologies of (3.4) and (3.5) yield that $h^q(\mathcal{F}_4^\vee(p+1))$ in the range $-4 \le p \le 0$ is as follows

0	0	0	0	0	$q=4$
1	0	0	0	0	$q=3$
0	0	0	0	0	$q=2$
0	0	1	0	0	$q=1$
0	0	0	0	15	$q=0$
$p=-4$	$p=-3$	$p=-2$	$p=-1$	$p=0$	

Table 1: Values of $h^q(\mathcal{F}_4^\vee(p+1))$ in the range $-4 \le p \le 0$

Thanks to [**1**, Beilinson's theorem (strong form)], we know that $\mathcal{F}_4^\vee(1)$ fits into an exact sequence of the following form

$$0 \longrightarrow \mathcal{O}_{\mathbb{P}^4}(-1) \oplus \Omega^2_{\mathbb{P}^4}(2) \longrightarrow \mathcal{O}_{\mathbb{P}^4}^{\oplus 15} \longrightarrow \mathcal{F}_4^\vee(1) \longrightarrow 0.$$

4. The proof of Theorem 1.3.

In this section we prove Theorem 1.3 stated in the introduction. Let $X \subseteq \mathbb{P}^{n+2}$ be a smooth variety of dimension n and set $\mathcal{O}_X(h) := \mathcal{O}_X \otimes \mathcal{O}_{\mathbb{P}^N}(1)$. Thanks to (2.1) and Lemma 1.6, in order to show that $\mathcal{N}_X(-h)$ is an h–instanton bundle it suffices to check that $H^j(\mathcal{N}_X(-(j+2)h)) = 0$ for all $0 \le j \le n-2$.

In order to compute such vanishings we will use the isomorphism

$$H^j(\mathcal{N}_X(th)) \cong \operatorname{Ext}^{j+1}\big(\mathcal{I}_X, \mathcal{I}_X(t)\big) \tag{4.1}$$

for $0 \le j \le n$ and $t \in \mathbb{Z}$ (see [**18**, Remark 2.2.6]).

Proof of Theorem 1.3. By applying the contravariant functor $\operatorname{Hom}(-, \mathcal{I}_X(-j-2))$ to (3.1) we obtain

$$\begin{aligned} &H^j(\mathcal{I}_X(r-f-j-2))^{\oplus r-1} \longrightarrow H^j(\mathcal{N}(-(j+2)h)) \\ &\quad\longrightarrow H^{j+1}(\mathcal{F} \otimes \mathcal{I}_X(r-f-j-3)) \longrightarrow H^{j+1}(\mathcal{I}_X(r-f-j-2))^{\oplus r-1}. \end{aligned} \tag{4.2}$$

Thus

$$\begin{aligned} h^j(\mathcal{N}(-(j+2)h)) \le (r-1)&h^j(\mathcal{I}_X(r-f-j-2)) \\ &+ h^{j+1}(\mathcal{F} \otimes \mathcal{I}_X(r-f-j-3)). \end{aligned}$$

First we compute $h^j(\mathcal{I}_X(r-f-j-2))$. Tensoring (3.1) by $\mathcal{O}_{\mathbb{P}^{n+2}}(r-f-j-2)$ we obtain

$$h^j(\mathcal{I}_X(r-f-j-2)) \le h^j(\mathcal{F}^\vee(-j-1)) + (r-1)h^{j+1}(\mathcal{O}_{\mathbb{P}^{n+2}}(-j-2)).$$

The right–hand side of the above inequality vanishes in the range $0 \le j \le n$, because trivially $h^{j+1}(\mathcal{O}_{\mathbb{P}^{n+2}}(-j-2)) = 0$ and the first summand vanishes by Lemma 1.5. In particular we also deduce that

$$h^j(\mathcal{I}_X(r-f-j-2-t)) = 0$$

for $0 \le j \le n$ and $t \ge 0$, thanks to [**2**, Proposition 2.1]. Thus, tensoring (1.3) by $\mathcal{I}_X(r-f-j-3)$, the above equality implies

$$\begin{aligned} h^{j+1}(\mathcal{F}\otimes\mathcal{I}_X(r-f-j-3)) &\le (r+f)h^{j+1}(\mathcal{I}_X(r-f-j-3)) \\ &\quad + fh^{j+2}(\mathcal{I}_X(r-f-j-4)) = 0 \end{aligned}$$

for $0 \le j \le n-2$. Thus the proof that $\mathcal{N}_X(-h)$ is an instanton bundle is complete.

It remains to compute $k := h^1(\mathcal{N}_X(-2h)) = h^{n-1}(\mathcal{N}_X(-(n+1)h))$. Arguing as above, (1.3) tensored by $\mathcal{O}_{\mathbb{P}^{n+2}}(r-f-n-1)$ yields $h^n(\mathcal{I}_X(r-f-n-1)) = 0$, hence (4.2) for $j = n-1$ returns

$$h^{n-1}(\mathcal{N}_X(-(n+1)h)) = h^n(\mathcal{F}\otimes\mathcal{I}_X(r-f-n-2)).$$

By combining the cohomologies of (1.3) and (3.1) tensored by $\mathcal{I}_X(r-f-n-2)$ and $\mathcal{I}_X(r-f-n-3)$ respectively, the same argument used above yields

$$\begin{aligned} h^n(\mathcal{F}\otimes\mathcal{I}_X(r-f-n-2)) &= fh^{n+1}(\mathcal{I}_X(r-f-n-3)) \\ &= f(r-1)h^{n+2}(\mathcal{O}_{\mathbb{P}^{n+2}}(-n-3)) = f(r-1). \end{aligned}$$

The statement is then completely proved. □

In what follows we will construct examples of smooth varieties $X \subseteq \mathbb{P}^N$ of dimension n such that $\mathcal{N}_X(-h)$ is an instanton.

In view of Theorems 1.2, 1.3 and Proposition 3.3 the first possible approach is to construct them starting from a Steiner bundle $\mathcal{F}$ such that $\mathcal{F}^\vee(1)$ is globally generated. We do not know the classification of such interesting bundles on $\mathbb{P}^N$. Nevertheless, we are able to classify below Steiner bundles $\mathcal{F}$ such that $\mathcal{F}^\vee(1)$ is regular, i.e. such that $h^i(\mathcal{F}(-i)) = 0$ for $1 \le i \le N$. Recall that regular sheaves are globally generated.

LEMMA 4.1. *Let $\mathcal{F}$ be a Steiner bundle on $\mathbb{P}^N$. Then $\mathcal{F}^\vee(1)$ is regular if and only if $\mathcal{F} \cong \Omega^{N-1}_{\mathbb{P}^N}(N)^{\oplus a} \oplus \mathcal{O}^b_{\mathbb{P}^N}$ for some non–negative integers a and b.*

PROOF. Let $\mathcal{F}$ be a Steiner bundle such that $\mathcal{F}^\vee(1)$ is regular. We set

$$b = h^{N-1}(\mathcal{F}(-N-1)), \qquad c = h^{N-1}(\mathcal{F}(-N)), \qquad a = h^{N-1}(\mathcal{F}(-N)).$$

Thanks to Lemma 1.5 and [**1**, Beilinson's theorem (weak form)] applied to the vector bundle $\mathcal{F}(-1)$ we obtain a spectral sequence whose $E_1^{p,q}$ sheaves are as in Table 2.

$\mathcal{O}_{\mathbb{P}^N}^{\oplus b}(-1)$	0	0	$q = N$
$\mathcal{O}_{\mathbb{P}^N}^{\oplus c}(-1)$	$\Omega_{\mathbb{P}^N}^{N-1}(N-1)^{\oplus a}$	0	$q = N-1$
0	0	0	$0 \le q \le N-2$
$p = -N$	$p = -N+1$	$-N+2 \le p \le 0$	

Table 2: Values of $E_1^{p,q}$ in the range $-N \le p \le 0$

The only possibly non–zero differential in this first page is

$$d_1^{-N,N-1}\colon \mathcal{O}_{\mathbb{P}^N}^{\oplus c}(-1) \longrightarrow \Omega_{\mathbb{P}^N}^{N-1}(N-1)^{\oplus a}.$$

Let us continue by considering the second page E_2^{pq} of the spectral sequence in Table 3.

$\mathcal{O}_{\mathbb{P}^N}^{\oplus b}(-1)$	0	0	$q = N$
$\ker(d_1^{-N,N-1})$	$\operatorname{coker}(d_1^{-N,N-1})$	0	$q = N-1$
0	0	0	$0 \le q \le N-2$
$p = -N$	$p = -N+1$	$-N+2 \le p \le 0$	

Table 3: Values of $E_2^{pq} = E_\infty^{pq}$ in the range $-N \le p \le 0$

Notice that the spectral sequence degenerates after the second page, thus $E_2^{pq} = E_\infty^{pq}$ and in particular $\ker(d_1^{-N,N-1}) = 0$. Again[**1**, Beilinson's theorem (weak form)] implies that

$$\bigoplus_{p=0}^{N} E_\infty^{-p,p} \cong \operatorname{coker}(\varphi) \oplus \mathcal{O}_{\mathbb{P}^N}^{\oplus b}(-1)$$

is the graded sheaf associated to the filtration $0 \subseteq \mathcal{G}(-1) \subseteq \mathcal{F}(-1)$, where $\mathcal{G}(-1) := \operatorname{coker}(d_1^{-N,N-1})$. Thus we obtain the short exact sequences

$$0 \longrightarrow \mathcal{G}(-1) \longrightarrow \mathcal{F}(-1) \longrightarrow \mathcal{O}_{\mathbb{P}^N}^{\oplus b}(-1) \longrightarrow 0,$$
$$0 \longrightarrow \mathcal{O}_{\mathbb{P}^N}^{\oplus c}(-1) \longrightarrow \Omega_{\mathbb{P}^N}^{N-1}(N-1)^{\oplus a} \longrightarrow \mathcal{G}(-1) \longrightarrow 0.$$

Notice that $\mathcal{G}$ is Steiner thanks to [**3**, Example 4.5], hence

$$\operatorname{Ext}^1\big(\mathcal{O}_{\mathbb{P}^N}^{\oplus b}(-1), \mathcal{G}(-1)\big) \cong H^1(\mathcal{G})^{\oplus b} = 0$$

by Lemma 1.5. It follows that the first sequence above splits, whence $\mathcal{F} \cong \mathcal{O}_{\mathbb{P}^N}^{\oplus b} \oplus \mathcal{G}$. If $\mathcal{F}^\vee(1)$ is regular then

$$c = h^{N-1}(\mathcal{G}(-N-1)) = h^{N-1}(\mathcal{F}(-N-1)) = h^1(\mathcal{F}^\vee) = 0,$$

hence

$$\mathcal{F} \cong \Omega_{\mathbb{P}^N}^{N-1}(N)^{\oplus a} \oplus \mathcal{O}_{\mathbb{P}^N}^{\oplus b}.$$

Conversely, if $\mathcal{F} \cong \Omega_{\mathbb{P}^N}^{N-1}(N)^{\oplus a} \oplus \mathcal{O}_{\mathbb{P}^N}^{b}$, then it is easy to check that $\mathcal{F}$ is Steiner and $\mathcal{F}^\vee(1)$ is regular. □

The above lemma motivates the following example.

EXAMPLE 4.2. Let $2 \le n \le 3$ and assume that the characteristic of $\mathbf{k}$ is 0. If we set

$$\mathcal{F}_{n;a,b} := \Omega^{n+1}_{\mathbb{P}^N}(n+2)^{\oplus a} \oplus \mathcal{O}^{\oplus b}_{\mathbb{P}^{n+2}}$$

thanks to Lemma 4.1 and Proposition 3.3 we can construct a family of smooth varieties $X_{n;a,b} \subseteq \mathbb{P}^{n+2}$ whose sheaf of ideals fits into the exact sequence

$$0 \longrightarrow \mathcal{O}_{\mathbb{P}^{n+2}}(-a-an-b)^{\oplus an+2a+b-1} \longrightarrow \mathcal{F}^\vee_{n;a,b}(1-a-an-b) \longrightarrow \mathcal{I}_{X_{n;a,b}} \longrightarrow 0.$$

For each $a \ge 1$ the bundle $\mathcal{N}_{X_{n;a,b}}(-h)$ is h–instanton with quantum number

$$h^1(\mathcal{N}_{X_{n;a,b}}(-2h)) = (an+2a+b-1)a,$$

thanks to Theorem 1.3.

Thanks to Proposition 3.4 we now that $\omega_{X_{n;a,b}}$ is ample and globally generated when $a \ge 2$. It follows that $X_{n;a,b}$ is of general type for $a \ge 2$.

The surface $X_{2;1,0}$ is the isomorphic projection in $\mathbb{P}^4$ of the Veronese surface of $\mathbb{P}^5$. It is well known that $X_{2;1,0}$ supports h–Ulrich bundles of minimal rank two (see [**12**, Proposition 5.9]): notice that $\mathcal{T}_{X_{2;1,0}}$ is one of them (see [**8**]).

The threefold $X_{3;1,0}$ is the Palatini scroll (see [**9**]), which supports h–Ulrich bundles of ranks both one and two (see [**13**, Theorem 0.1 (2)]). Its general hyperplane section is $X_{2;1,1}$: trivially it still supports Ulrich bundles of ranks one and two.

The threefold $X_{3;1,1}$ is such that the linear system $|h+K_{X_{3;1,1}}|$ gives a birational morphism φ onto a cubic threefold $\overline{X} \subseteq \mathbb{P}^4$. Moreover, φ fails to be injective along a curve $C \subseteq \overline{X}$ which has degree 14 and genus 15. We do not know if such threefold $X_{3;1,1}$ actually supports h–Ulrich bundles.

5. Further examples

Theorem 1.3 gives a sufficient condition for $\mathcal{N}_X(-h)$ to be an instanton. We show in the following examples that such a condition is actually not necessary.

EXAMPLE 5.1. Let us consider the smooth threefold whose ideal sheaf fits into

$$0 \longrightarrow \mathcal{O}_{\mathbb{P}^5}(-4)^{\oplus 9} \longrightarrow \Omega^3_{\mathbb{P}^5} \longrightarrow \mathcal{I}_X \longrightarrow 0. \tag{5.1}$$

In particular X is a $K3$ scroll of degree 9 over the surface $B = \{\mathrm{Gr}(\mathbb{P}^1, \mathbb{P}^5) \cap \mathbb{P}^8\}$. Thanks to Lemma 1.6, in order to show that $\mathcal{N}_X(-h)$ is an instanton bundle it is enough to check that $H^j(\mathcal{N}_X(-(j+2)h)) \cong 0$ for $j = 0, 1$. Arguing in the same way as in the proof of Theorem 1.3 one obtains the exact sequence

$$\begin{aligned} H^j(\mathcal{I}_X(2-j))^{\oplus 9} &\longrightarrow H^j(\mathcal{N}_X(-(j+2)h)) \\ &\longrightarrow H^{j+1}(\Omega^{3\vee}_{\mathbb{P}^5} \otimes \mathcal{I}_X(-j-2)) \longrightarrow H^{j+1}(\mathcal{I}_X(2-j))^{\oplus 9}. \end{aligned}$$

Let us start by computing the cohomology of $\mathcal{I}_X(2-j)$. Twisting (5.1) by $\mathcal{O}_{\mathbb{P}^5}(2-j)$ we get $H^i(\mathcal{I}_X(2-j)) \cong 0$ for all i and for $j = 0, 1$ since both $\mathcal{O}_{\mathbb{P}^5}(-j-2)^{\oplus 9}$ and $\Omega^3_{\mathbb{P}^5}(2-j)$ are acyclic in that range. Now we compute $H^{j+1}(\Omega^{3\vee}_{\mathbb{P}^5} \otimes \mathcal{I}_X(-j-2))$. In order to do so, recall that $\Omega^{3\vee}_{\mathbb{P}^5} \cong \Omega^2_{\mathbb{P}^5}(6)$ and let us consider the Euler sequences

$$0 \longrightarrow \Omega^p_{\mathbb{P}^5}(p) \longrightarrow \mathcal{O}^{\oplus \binom{6}{p}}_{\mathbb{P}^5} \longrightarrow \Omega^{p-1}_{\mathbb{P}^5}(p) \longrightarrow 0 \tag{5.2}$$

for $p = 1, 2$. If we tensor (5.2) by $\mathcal{I}_X(4-p-j)$ we obtain

$$H^{j+1}(\Omega^p_{\mathbb{P}^5} \otimes \mathcal{I}_X(4-j)) \cong H^j(\Omega^{p-1}_{\mathbb{P}^5} \otimes \mathcal{I}_X(4-j)) \cong 0$$

for $1 \le p \le 2$ and $0 \le j \le 1$ since $\mathcal{I}_X(4-p-j)$ is acyclic in this range, thus $\mathcal{N}_X(-h)$ is an instanton bundle. Its quantum number is $k = 6$ and can be computed by (2.2).

It is possible to repeat the same argument to deal with other varieties. E.g. taking into account [**10**, Theorem] and Lemma 4.1, it is quite natural to inspect the case of codimension two arithmetically Buchsbaum threefolds and surfaces. Such varieties have been classified in [**9**], when not of general type.

We summarise what we obtain in the tables 4 and 5, the quantum numbers being computed by (2.2) and (2.3). Notice that cases (a), (e), (2) and (3) can be directly obtained using Theorem 1.3.

	Resolution of $\mathcal{I}_X$	$\mathcal{N}_X(-h)$ **instanton**	**Charge**
(a)	$0 \to \mathcal{O}_{\mathbb{P}^5}(-4)^{\oplus 4} \to \Omega^1_{\mathbb{P}^5}(-2) \to \mathcal{I}_X \to 0$	yes	4
(b)	$0 \to \mathcal{O}_{\mathbb{P}^5}(-4)^{\oplus 9} \to \Omega^3_{\mathbb{P}^5} \to \mathcal{I}_X \to 0$	yes	6
(c)	$0 \to \mathcal{O}_{\mathbb{P}^5}(-5)^{\oplus 8} \oplus \mathcal{O}_{\mathbb{P}^5}(-6) \to \Omega^3_{\mathbb{P}^5}(-1) \to \mathcal{I}_X \to 0$	yes	7
(d)	$0 \to \mathcal{O}_{\mathbb{P}^5}(-5)^{\oplus 10} \to \Omega^3_{\mathbb{P}^5}(-1) \oplus \mathcal{O}_{\mathbb{P}^5}(-4) \to \mathcal{I}_X \to 0$	yes	5
(e)	$0 \to \mathcal{O}_{\mathbb{P}^5}(-5)^{\oplus 5} \to \Omega^1_{\mathbb{P}^5}(-3) \oplus \mathcal{O}_{\mathbb{P}^5}(-4) \to \mathcal{I}_X \to 0$	yes	5
(f)	$0 \to \mathcal{O}_{\mathbb{P}^5}(-5)^{\oplus 3} \oplus \mathcal{O}_{\mathbb{P}^5}(-6) \to \Omega^1_{\mathbb{P}^5}(-3) \to \mathcal{I}_X \to 0$	??	??

Table 4: Arithmetically Buchsbaum threefolds of non-general type

	Resolution of $\mathcal{I}_X$	$\mathcal{N}_X(-h)$ **instanton**	**Charge**
(1)	$0 \to \mathcal{O}_{\mathbb{P}^4}(-3)^{\oplus 5} \to \Omega^2_{\mathbb{P}^4} \to \mathcal{I}_X \to 0$	yes	5
(2)	$0 \to \mathcal{O}_{\mathbb{P}^4}(-3)^{\oplus 3} \to \Omega^1_{\mathbb{P}^4}(-1) \to \mathcal{I}_X \to 0$	yes	3
(3)	$0 \to \mathcal{O}_{\mathbb{P}^4}(-4)^{\oplus 4} \to \Omega^1_{\mathbb{P}^4}(-2) \oplus \mathcal{O}_{\mathbb{P}^4}(-3) \to \mathcal{I}_X \to 0$	yes	4
(4)	$0 \to \mathcal{O}_{\mathbb{P}^4}(-4)^{\oplus 6} \to \Omega^2_{\mathbb{P}^4}(-1) \oplus \mathcal{O}_{\mathbb{P}^4}(-3) \to \mathcal{I}_X \to 0$	??	??
(5)	$0 \to \mathcal{O}_{\mathbb{P}^4}(-4)^{\oplus 4} \oplus \mathcal{O}_{\mathbb{P}^4}(-5) \to \Omega^2_{\mathbb{P}^4}(-1) \to \mathcal{I}_X \to 0$	yes	6
(6)	$0 \to \mathcal{O}_{\mathbb{P}^4}(-4)^{\oplus 2} \oplus \mathcal{O}_{\mathbb{P}^4}(-5) \to \Omega^1_{\mathbb{P}^4}(-2) \to \mathcal{I}_X \to 0$	yes	3

Table 5: Arithmetically Buchsbaum surfaces of non-general type

In view of the cases (f) and (4) in the table 4 and 5 above, it might be perhaps natural to ask the following question

QUESTION 5.2. *Let X be a codimension two smooth, arithmetically Buchsbaum variety not of general type. Is $\mathcal{N}_X(-h)$ an instanton bundle?*

In the following examples we show that the Question 5.2 has no general answer for varieties of general type.

EXAMPLE 5.3. Consider any non–degenerate complete intersection $X \subseteq \mathbb{P}^{n+2}$ of two hypersurfaces the bundle $\mathcal{N}_X(-h)$. It is easy to check that $h^0(\mathcal{N}_X(-2h)) \neq 0$, hence $\mathcal{N}_X(-h)$ cannot be an h–instanton.

EXAMPLE 5.4. Let the characteristic of $\mathbf{k}$ be 0. Thanks to Proposition 3.3 we can consider the smooth surface in $\mathbb{P}^4$ whose ideal sheaf admits the resolution

$$0 \longrightarrow \mathcal{O}_{\mathbb{P}^4}(-5)^{\oplus 2} \oplus \mathcal{O}_{\mathbb{P}^4}(-7) \longrightarrow \Omega^1_{\mathbb{P}^4}(-3) \longrightarrow \mathcal{I}_X \longrightarrow 0. \tag{5.3}$$

If we apply the functor $\mathcal{H}om_{\mathbb{P}^4}(-,\omega_{\mathbb{P}^4})$ to the sequence above glued with (3.2) as in Example 3.7, we deduce that X is of general type. Let us verify that $\mathcal{N}_X(-h)$ is not an instanton, by showing that $h^0(\mathcal{N}_X(-2h)) > 0$. Let us apply the contravariant functor $\mathrm{Hom}(-,\mathcal{I}_X(-2))$ to (5.3) obtaining the exact sequence

$$\begin{aligned} H^0(\Omega^3_{\mathbb{P}^4}(6)\otimes\mathcal{I}_X) \longrightarrow & H^0(\mathcal{I}_X(3))^{\oplus 2} \oplus H^0(\mathcal{I}_X(5)) \longrightarrow H^0(\mathcal{N}_X(-2h)) \\ & \longrightarrow H^1(\Omega^3_{\mathbb{P}^4}(6)\otimes\mathcal{I}_X) \longrightarrow H^1(\mathcal{I}_X(3))^{\oplus 2} \oplus H^1(\mathcal{I}_X(5)). \end{aligned}$$

We start by computing $h^0(\Omega^3_{\mathbb{P}^4}(6)\otimes\mathcal{I}_X)$. If we tensor (5.2) by $\mathcal{I}_X(3)$ we get

$$h^0(\Omega^3_{\mathbb{P}^4}(6)\otimes\mathcal{I}_X) \leq 10h^0(\mathcal{I}_X(3)).$$

Thus, taking the cohomology of (5.3) twisted by $\mathcal{O}_{\mathbb{P}^4}(3)$ we obtain $h^0(\mathcal{I}_X(3)) = 0$, hence $h^0(\Omega^3_{\mathbb{P}^4}(6)\otimes\mathcal{I}_X) = 0$. Tensoring sequence (5.3) by $\mathcal{O}_{\mathbb{P}^4}(5)$ one can directly obtain $h^0(\mathcal{I}_X(5)) = 8$, thus $h^0(\mathcal{N}_X(-2h)) > 8$ and $\mathcal{N}_X(-h)$ is not an instanton bundle.

6. Acknowledgments

The authors would like to thank the reviewer for her/his corrections, remarks and suggestions.

Moreover, the second author heartily thanks Tony Iarrobino, Pedro Macias Marques, Maria Evelina Rossi and Jean Vallès, for their organization of the beautiful conference AMS–EMS–SMF special session on Deformation of Artinian algebras and Jordan Type, held in Grenoble on July 18–22, 2022.

References

[1] V. Ancona and G. Ottaviani, *Some applications of Beilinson's theorem to projective spaces and quadrics*, Forum Math. **3** (1991), no. 2, 157–176, DOI 10.1515/form.1991.3.157. MR1092580

[2] V. Antonelli and G. Casnati, *Instanton sheaves on projective schemes*, J. Pure Appl. Algebra **227** (2023), no. 4, Paper No. 107246, 40, DOI 10.1016/j.jpaa.2022.107246. MR4510797

[3] V. Antonelli and G. Casnati, *Steiner representations of hypersurfaces*. Available at arXiv:2210.03408.

[4] E. Arrondo, *A home-made Hartshorne-Serre correspondence*, Rev. Mat. Complut. **20** (2007), no. 2, 423–443, DOI 10.5209/rev_REMA.2007.v20.n2.16502. MR2351117

[5] M. F. Atiyah and R. S. Ward, *Instantons and algebraic geometry*, Comm. Math. Phys. **55** (1977), no. 2, 117–124. MR494098

[6] A. Beauville, *An introduction to Ulrich bundles*, Eur. J. Math. **4** (2018), no. 1, 26–36, DOI 10.1007/s40879-017-0154-4. MR3782216

[7] V. Benedetti, P. Montero, Y. Prieto-Montañez, and S. Troncoso, *Projective manifolds whose tangent bundle is Ulrich*, J. Algebra **630** (2023), 248–273, DOI 10.1016/j.jalgebra.2023.03.046. MR4586748

[8] G. Casnati, *Tangent, cotangent, normal and conormal bundles are almost never instanton bundles*, Comm. Algebra, **52** (2024), no. 2, 572–581, https://doi.org/10.1080/00927872.2023.2245911. MR4703541

[9] M.-C. Chang, *Classification of Buchsbaum subvarieties of codimension* 2 *in projective space*, J. Reine Angew. Math. **401** (1989), 101–112, DOI 10.1515/crll.1989.401.101. MR1018055

[10] M.-C. Chang, *Characterization of arithmetically Buchsbaum subschemes of codimension* 2 *in* $\mathbf{P}^n$, J. Differential Geom. **31** (1990), no. 2, 323–341. MR1037403

[11] I. Dolgachev and M. Kapranov, *Arrangements of hyperplanes and vector bundles on* $\mathbf{P}^n$, Duke Math. J. **71** (1993), no. 3, 633–664, DOI 10.1215/S0012-7094-93-07125-6. MR1240599

[12] D. Eisenbud and F.-O. Schreyer, *Resultants and Chow forms via exterior syzygies*, J. Amer. Math. Soc. **16** (2003), no. 3, 537–579, DOI 10.1090/S0894-0347-03-00423-5. With an appendix by Jerzy Weyman. MR1969204

[13] M. L. Fania, M. Lelli-Chiesa, and J. Pons-Llopis, *Ulrich bundles on three-dimensional scrolls*, Int. Math. Res. Not. IMRN **17** (2021), 13478–13507, DOI 10.1093/imrn/rnz288. MR4307693

[14] R. Hartshorne, *Varieties of small codimension in projective space*, Bull. Amer. Math. Soc. **80** (1974), 1017–1032, DOI 10.1090/S0002-9904-1974-13612-8. MR384816

[15] R. Hartshorne, *Algebraic geometry*, Graduate Texts in Mathematics, No. 52, Springer-Verlag, New York-Heidelberg, 1977. MR463157

[16] M. Jardim, *Instanton sheaves on complex projective spaces*, Collect. Math. **57** (2006), no. 1, 69–91. MR2206181

[17] Y. Kachi and J. Kollár, *Characterizations of* $\mathbf{P}^n$ *in arbitrary characteristic*, Asian J. Math. **4** (2000), no. 1, 115–121, DOI 10.4310/AJM.2000.v4.n1.a8. Kodaira's issue. MR1802915

[18] J. O. Kleppe, *The Hilbert-flag scheme, its properties and its connection with the Hilbert scheme. Applications to curves in the* 3*–space.* Ph.D. Thesis, University of Oslo (1981).

[19] J. O. Kleppe and R. M. Miró-Roig, *On the normal sheaf of determinantal varieties*, J. Reine Angew. Math. **719** (2016), 173–209, DOI 10.1515/crelle-2014-0041. MR3552495

[20] A. F. Lopez, *On varieties with Ulrich twisted normal bundles.* Available at `arXiv:2205.06602`. To appear in Perspectives on four decades: Algebraic Geometry 1980-2020. In memory of Alberto Collino'. Trends in Mathematics, Birkhäuser.

[21] G. Ottaviani, *Some constructions of projective varieties.* Barcelona, April 28, 2005. Available at `https://web.math.unifi.it/users/ottavian/bcn.pdf`.

[22] K. E. Smith, *Fujita's freeness conjecture in terms of local cohomology*, J. Algebraic Geom. **6** (1997), no. 3, 417–429. MR1487221

[23] J. A. Vogelaar, *Constructing Vector Bundles from Codimension–two Subvarieties.* Ph.D. Thesis, Leiden (1978).

Dipartimento di Scienze Matematiche, Politecnico di Torino – c.so Duca degli Abruzzi 24 – 10129 Torino, Italy

Email address: `vincenzo.antonelli@polito.it`

Dipartimento di Scienze Matematiche, Politecnico di Torino – c.so Duca degli Abruzzi 24 – 10129 Torino, Italy

Email address: `gianfranco.casnati@polito.it`

Contemporary Mathematics
Volume **805**, 2024
https://doi.org/10.1090/conm/805/16129

On the Koiran-Skomra's question about Hessians

Edoardo Ballico and Emanuele Ventura

Abstract. Extending a construction due to B. Segre, we give a negative answer to a question of Koiran and Skomra about Hessians, motivated by Kayal's algorithm for the equivalence problem to the Fermat polynomial. We conjecture that our counterexamples are the only ones. We also study a local version of their question.

1. Introduction

Let $g \in \mathbb{C}[x_1, \ldots, x_n]_d$ be a homogeneous polynomial of degree d and let $\mathrm{Hess}(g)$ be its Hessian matrix. The *Hessian map* is the polynomial map

$$H : \mathbb{C}[x_1, \ldots, x_n]_d \longrightarrow \mathbb{C}[x_1, \ldots, x_n]_{n(d-2)},$$

defined by $H(g) = \det(\mathrm{Hess}(g))$. This determinant is often simply called the *Hessian* of g. *Kayal's algorithm* takes as input a homogeneous polynomial $g \in \mathbb{C}[x_1, \ldots, x_n]_d$ and determines whether g is in the $\mathrm{GL}(\mathbb{C}^n)$-orbit of (or *it is equivalent to*) the Fermat polynomial $f = x_1^d + \cdots + x_n^d$; if it is so, the algorithm outputs linearly independent linear forms $\ell_i \in \mathbb{C}[x_1, \ldots, x_n]_1$ such that $g = \sum_{i=1}^n \ell_i^d$. See [**11**] and [**13**, §3.1] for a detailed account. This algorithm is based on three steps:

(1) Check that the Hessian $H(g)$ is nonzero and can be factored as $H(g) = \alpha \prod_{i=1}^n h_i^{d-2}$ where $h_i \in \mathbb{C}[x_1, \ldots, x_n]_1$ are linear forms. If it not possible, reject.
(2) Try to find complex constants $\alpha_i \in \mathbb{C}$ such that $g = \sum_{i=1}^n \alpha_i \ell_i^d$. If this is not possible, reject.
(3) Declare g to be equivalent to f and output $\ell_i = \beta_i h_i$ where $\beta_i^d = \alpha_i$.

Motivated by Kayal's algorithm, Koiran and Skomra [**13**, Question 1] asked the following question:

Question 1.1 (Koiran-Skomra). Let $n \geq 2$ and $d \geq 3$. Let $f \in \mathbb{C}[x_1, \ldots, x_n]_d$ be such that $H(f) = \alpha(x_1 \cdots x_n)^{d-2}$, for some $\alpha \neq 0$.

Does it follow that $f = \alpha_1 x_1^d + \cdots + \alpha_n x_n^d$, for some constants $\alpha_j \in \mathbb{C}$?

2020 *Mathematics Subject Classification.* Primary 05E40, 05E16, 05E14; Secondary 14N05, 14L35.

Key words and phrases. Hessians, Kayal's algorithm, Fermat polynomial, stabilizer group.
The authors are members of GNSAGA of INdAM (Italy).

This question is interesting also because it is so simple to state and has a delightful invariant theory flavour. Hessians of forms have been the subject of many classical works in algebraic geometry and commutative algebra (for instance, Hessians are related to Lefschetz properties of artinian Gorenstein algebras); see e.g. [**17**, Chapter 7] and [**3**], along with the several references therein. Hessians are also naturally related to ranks of homogeneous polynomials [**4, 8**]. Beniamino Segre [**18**, p. 174] found a family of counterexamples in four variables of even degree to the Koiran-Skomra question; his aim was to construct polynomials whose Hessians equal to a square in the field of fractions of the corresponding hypersurfaces. We extend Segre's counterexamples to an arbitrary number of variables and composite degrees in Theorem 2.1.
Employing the algorithm in [**14**, §4.1] to decide whether the factorization in step (1) of Kayal's algorithm *exists* (and not to compute the explicit factorization), a positive answer to Question 1.1 would provide a polynomial time algorithm for the equivalence problem over $\mathbb{C}$. In their article [**13**], Koiran and Skomra did provide a polynomial time algorithm for the equivalence problem over $\mathbb{C}$ in degree $d = 3$, which was later extended by Koiran and Saha [**12**] to $d > 3$. However, their algorithm has to work in a very different way than Kayal's algorithm because of our Theorem 2.1: this shows that over any field $K \subset \mathbb{C}$, the answer is negative for infinite pairs of integers (n, d). An instance of interest in [**13**] is $d = 3$: in this case, the answer is positive if and only if $n = 2$ (Corollary 2.5).

Our monomial counterexamples are *homaloidal polynomials*, i.e. their first partial derivatives define a Cremona transformation. Theorem 2.1 also shows that these monomials are *totally Hessian* [**3**, Remark 3.5]. Therefore they give examples of totally Hessian polynomials (although reducible) of arbitrarily large degree in any number of variables $n \geq 2$.

In Theorem 2.8 and Theorem 2.9, we deal with the case of singular binary homogeneous polynomials, providing the only ones whose Hessians have the desired form. Our approach is combinatorial and we wonder whether an invariant theoretic strategy could be put in place. We conjecture that for any $n \geq 2$ the only smooth homogeneous polynomials satisfying the assumptions of Question 1.1 are the Fermat polynomials; see Conjecture 2.16.

Computing the differential of the Hessian map, we formulate a local version of Koiran-Skomra's question, which has a positive answer when $d \geq n+1$ (Proposition 3.2).

Acknowledgments. We thank Enrico Carlini, Pascal Koiran, Francesco Russo, and Mateusz Skomra for useful discussions. The second author would like to thank Tony Iarrobino, Pedro Marques, Maria Evelina Rossi, and Jean Vallès for organizing the nice workshop *Deformation of Artin algebras and Jordan type* within the AMS-EMS-SMF meeting in Grenoble on 18-22 July 2022. We are very grateful to an anonymous referee for several precious remarks: especially for providing a precise reference to the contribution of Beniamino Segre [**18**] and for an alternative proof of Theorem 2.1 described in Remark 2.2.

2. Koiran-Skomra's question

2.1. Negative result. The first counterexamples to Question 1.1 were constructed by Beniamino Segre [**18**, p. 174], where he considers polynomials in four variables of degree $d = 2k$ of the form $g(x_1, x_2, x_3, x_4) = x_1^{2k} + (1-2k)x_2^{2k} - (x_3 \cdot x_4)^k$, whose Hessian matrix has determinant

$$H(g) = \left(2k^2(2k-1)^2(x_1 \cdot x_2 \cdot x_3 \cdot x_4)^{k-1}\right)^2.$$

We extend this family of counterexamples as follows.

THEOREM 2.1. *Let $K \subset \mathbb{C}$ be a field. The answer to Question 1.1 over K is negative for all $n \geq 2$ and $d \geq 3$ such that there exists $2 \leq q \leq n$ with $q \mid d$.*

PROOF. Let $d = kq$ with $k \geq 1$ (or $k \geq 2$ if $q = 2$). Then define

$$g = (x_1 \cdots x_q)^k + x_{q+1}^d + \cdots + x_n^d \in K[x_1, \ldots, x_n].$$

We show that

$$H((x_1 \cdots x_q)^k) = (1-d)(-k)^q (x_1 \cdots x_q)^{d-2} \neq 0. \tag{2.1}$$

This is enough to conclude that $H(g)$ has the desired form. We first prove that every monomial in the determinant expansion of the Hessian above is $(x_1 \cdots x_q)^{d-2}$.

To see this, fix an index $\ell \in \{1, \ldots, q\}$ and look at the total degree in x_ℓ of an arbitrary monomial in the determinant expansion. By symmetry, we may assume $x_\ell = x_1$.

Let $f_{i,j} = \left(\mathrm{Hess}((x_1 \cdots x_q)^k)\right)_{i,j}$ be the (i,j)-entry of the Hessian matrix of $(x_1 \cdots x_q)^k$. Let m be a monomial in the expansion of $H((x_1 \cdots x_q)^k)$. This is a product of entries $f_{i,j}$, where the product ranges over the indices of all rows and columns, by definition of determinant. Then exactly one of the following cases can occur:

(I) m is a multiple of the $(1,1)$-entry (this choice gives no contribution for $k = 1$);

(II) m is a multiple of two distinct entries in the first row and in the first column.

In case (I), the total degree of m with respect to x_1 is $(k-2) + (q-1)k = qk - 2 = d - 2$. In case (II), the total degree of m with respect to x_1 is $2(k-1) + (q-2)k = qk - 2 = d - 2$. In other words, for any monomial in the expansion of $H((x_1 \cdots x_q)^k)$, the total degree of every variable is $d-2$. Let C be the $q \times q$ matrix whose (i,j)-entry is the coefficient of the monomial $f_{i,j}$. One has

$$C = \begin{bmatrix} k(k-1) & k^2 & \cdots & k^2 \\ k^2 & k(k-1) & \cdots & k^2 \\ \vdots & \vdots & \vdots & \vdots \\ k^2 & k^2 & \cdots & k(k-1) \end{bmatrix}. \tag{2.2}$$

The argument above shows that

$$H((x_1 \cdots x_q)^k) = \det(C)(x_1 \cdots x_q)^{d-2}.$$

Write $C = -kI_q + k^2 J_q$, where I_q is the $q \times q$ identity matrix and J_q is the $q \times q$ matrix whose entries are all ones. Note that $J_q = \mathbf{1}_q \cdot (\mathbf{1}_q)^T$, where $\mathbf{1}_q$ is the column vector with q ones.

One may write $C = -kI_q \cdot C'$, where $C' = I_q - \mathbf{1}_q \cdot (k\mathbf{1}_q)^T$. Note that

$$\begin{bmatrix} I_q & 0 \\ (k\mathbf{1}_q)^T & 1 \end{bmatrix} \cdot \begin{bmatrix} I_q - (\mathbf{1}_q) \cdot (k\mathbf{1_q})^T & -\mathbf{1}_q \\ 0 & 1 \end{bmatrix} \cdot \begin{bmatrix} I_q & 0 \\ -(k\mathbf{1}_q)^T & 1 \end{bmatrix} = \begin{bmatrix} I_q & -\mathbf{1}_q \\ 0 & 1-(k\mathbf{1}_q)^T \cdot \mathbf{1}_q \end{bmatrix}. \tag{2.3}$$

Then $\det(C')$ is the determinant of the left-hand side of (2.3), thus the determinant of the right-hand side too. So $\det(C') = 1 - (k\mathbf{1}_q)^T \cdot \mathbf{1}_q = 1 - qk$ and $\det(C) = (1-d)(-k)^q \neq 0$, which shows (2.1). Finally, the polynomial g has Hessian $H(g) = \alpha(x_1 \cdots x_n)^{d-2} \neq 0$ and is not a smooth homogeneous polynomial over K. Hence g cannot be equivalent to a Fermat polynomial over K. □

REMARK 2.2. A homogeneous polynomial $f \in \mathbb{C}[x_1, \ldots, x_n]$ is *homaloidal* when its first partial derivatives define a birational gradient map $\operatorname{grad}(f) : \mathbb{P}^{n-1} \dashrightarrow \mathbb{P}^{n-1}$ (i.e. a Cremona transformation). Equivalently, their *polar degree* is one. Let $g = f^k$ with $k \geq 2$. By the chain rule, we have

$$\operatorname{grad}(g) = kf^{k-1}\operatorname{grad}(f).$$

Hence the two rational maps $\operatorname{grad}(g)$ and $\operatorname{grad}(f)$ coincide outside the locus $\mathcal{V}(f) = \{f = 0\}$. In particular, they have the same polar degree; see also [**5**, Corollary 2] for a more general deep result. The polynomial $f = x_1 \cdots x_n$ is homaloidal, i.e. the rational map $\operatorname{grad}(f) : \mathbb{P}^{n-1} \dashrightarrow \mathbb{P}^{n-1}$ is an isomorphism onto its image outside $\mathcal{V}(f)$, and so is $g = (x_1 \cdots x_n)^k$. Theorem 2.1 also shows that these polynomials are *totally Hessian* [**3**, Remark 3.5] (although reducible) of arbitrarily large degree in any number of variables $n \geq 2$. These observations lead to an alternative and more conceptual proof of Theorem 2.1, kindly suggested to us by an anonymous referee. Since $\operatorname{grad}(f)$ and $\operatorname{grad}(g)$ coincide outside $\mathcal{V}(f)$, the ramification divisors of the birational maps $\operatorname{grad}(f)$ and $\operatorname{grad}(g)$ have the same support. Hence $H(g) = \alpha \cdot x_1^{b_1} \cdots x_n^{b_n}$, with $b_1 + \cdots + b_n = n(nk-2)$. By symmetry, all the b_i's must be equal and so $b_i = nk - 2$. Thus $H((x_1 \cdots x_n)^k) = \alpha \cdot (x_1 \cdots x_n)^{nk-2}$.

REMARK 2.3. The homogeneous polynomial g in the proof of Theorem 2.1 is a sum of pairwise coprime monomials. Then [**1**, Theorem 3.2] shows that the complex Waring rank of g satisfies the equality $\operatorname{rk}_K(g) \geq \operatorname{rk}_{\mathbb{C}}(g) = (k+1)^{q-1} + (n-q) > n = \operatorname{rk}_K(f)$, where $f = x_1^d + \cdots + x_n^d$.

REMARK 2.4. There exist homogeneous polynomials whose Hessians are monomials distinct from $(x_1 \cdots x_n)^{d-2}$. Three examples:

(i) for $k \geq 2$, let $f = x_2^{k-1}x_1^2 - x_3^{k+1} \in \mathbb{C}[x_1, x_2, x_3]_{k+1}$. It has singularities at $(0:1:0)$ and $(1:0:0)$, and $H(f) = -2k^2(k+1)(k-1)x_1^2x_2^{2(k-2)}x_3^{k-1}$.

(ii) $g = x_1(x_2^2 + x_1x_3 + \sum_{i=4}^n x_i^2) \in \mathbb{C}[x_1, x_2, x_3, \ldots, x_n]_3$ is a singular cubic form which splits as a product between a smooth quadric and a tangent hyperplane to it. Then $H(f) = -2^n x_1^n$.

(iii) The determinant of a generic sub-Hankel matrix [**3**, 4.1.1] is a homaloidal polynomial whose Hessian is a power of a variable [**3**, Theorem 4.4]. Since these are homaloidal of degree $d \geq 3$, they are singular.

A case of interest in [**13**] is when $d = 3$.

COROLLARY 2.5. *Let* $d = 3$. *Then the answer to Question* 1.1 *is positive if and only if* $n = 2$.

PROOF. If $n \geq 3 = d$, by Theorem 2.1 the answer is negative. Suppose $n = 2$ and let $f = a_3x_1^3 + a_2x_1^2x_2 + a_1x_1x_2^2 + a_0x_2^3$. So

$$H(f) = (-4a_2^2 + 12a_3a_1)x_1^2 + (-4a_2a_1 + 36a_3a_0)x_1x_2 + (-4a_1^2 + 12a_2a_0)x_2^2,$$

where $H(f) = \alpha x_1x_2 \neq 0$. If $a_2 = 0$, then $a_3 \neq 0$ and so $a_1 = 0$. But then $f = a_3x_1^3 + a_0x_2^3$. If not, by symmetry, we may assume $a_1, a_2 \neq 0$. Therefore $a_0 = a_1^2/3a_2$ and $a_3 = a_2^2/3a_1$, which implies that the coefficient of the monomial x_1x_2 in $H(f)$ is zero. □

For $n = d = 3$, the following statement is based on a well-known classification and we record it from our perspective.

PROPOSITION 2.6. *For $n = d = 3$, the only cubic homogenous polynomials satisfying the assumptions in Question* 1.1 *are $f = a_1x_1^3 + a_2x_2^3 + a_3x_3^3$ and $f = a_{123}x_1x_2x_3$ (with $a_1a_2a_3 \neq 0$ and $a_{123} \neq 0$ resp.).*

PROOF. Let $f \in \mathbb{C}[x_1, x_2, x_3]$ be a cubic form. The Hessian map $H : \mathbb{C}[x_1, x_2, x_3]_3 \to \mathbb{C}[x_1, x_2, x_3]_3$, $H : g \mapsto H(g)$ is an $\mathrm{SL}(\mathbb{C}^3)$-coinvariant [**7**, §5.1]. Thus, it is enough to check the $\mathrm{SL}(\mathbb{C}^3)$-orbits of ternary cubic homogenous polynomials, which are well-known. Among these, we see that the only cubic homogenous polynomials whose Hessian splits into three linearly independent linear forms are the Fermat polynomials $g = \ell_1^3 + \ell_2^3 + \ell_3^3$ and the triangles $g = \ell_1\ell_2\ell_3$. Since by assumption $H(g) = \alpha x_1x_2x_3 \neq 0$, by Lemma 2.10 proved below, we see that $\ell_i = x_i$ (up to relabeling and scaling). This shows the statement. □

2.2. Singular binary homogeneous polynomials. We deal with the case of singular binary homogeneous polynomials, providing the only ones whose Hessians have the desired form.

REMARK 2.7. For $n \geq 2$, let $f \in \mathbb{C}[x_1, \ldots, x_n]_d$ and let $X_f \subset \mathbb{P}^{n-1}$ be the reduced projective hypersurface defined by f. Then the reduced singular locus $\mathrm{Sing}(X_f)$ sits inside $X_{H(f)}$, where $H(f) = \det(\mathrm{Hess}(f))$. This may be seen by interpreting the Hessian as the Jacobian of the $\frac{\partial f}{\partial x_i}$'s.

THEOREM 2.8. *For $n = 2$ and $d \geq 3$ odd, there are no singular homogeneous polynomials satisfying the assumptions in Question* 1.1.

PROOF. Let $d = 2p + 1$. Let $f = f(x_1, x_2) = \sum_{i=0}^{d} a_ix_1^ix_2^{d-i} \in \mathbb{C}[x_1, x_2]_d$ be such that $H(f) = \alpha(x_1x_2)^{d-2} \neq 0$. Up to the automorphism of $\mathbb{C}[x_1, x_2]$ swapping x_1 and x_2 and by Remark 2.7, we may assume that $f = \sum_{i\geq 2}^{d} a_ix_1^ix_2^{d-i}$, i.e. $a_0 = a_1 = 0$. The coefficient of the monomial $x_1^jx_2^{2d-j-4}$ in $H(f)$ is of the form

$$\sum_{k+\ell=j+2} \gamma_{(k,\ell)}a_ka_\ell.$$

For all $2 \leq s \leq p$ and $j = 2(s-1)$, the coefficient $\gamma_{(s,s)}$ in front of a_s^2 appearing in the monomial $x_1^{2s-2}x_2^{2d-2s-2}$ in $H(f)$ is

$$\gamma_{(s,s)} = -s(d-1)(d-s) \neq 0.$$

Now, suppose that for some $1 \leq j' < p$ we have that $a_{j''} = 0$ for all $0 \leq j'' \leq j'$. We claim that $a_{j'+1} = 0$.

To show the claim, look at the coefficient of the monomial $x_1^{2j'}x_2^{2d-2j'-4}$ given by

$$\sum_{k+\ell=2j'+2} \gamma_{(k,\ell)} a_k a_\ell. \tag{2.4}$$

Since $k+\ell = 2(j'+1)$, if $\ell \geq j'+2$ then $k \leq j'$. By assumption, in this case, we have $a_k a_\ell = 0$ and so this summand does not contribute to (2.4). Then the only summand left in (2.4) is $\gamma_{(j'+1,j'+1)}a_{j'+1}^2$. Since $j' < p$, $2j' \leq 2(p-1) = 2p-2 = d-3$. Thus the monomial $x_1^{2j'}x_2^{2d-2j'-2}$ does not appear in $H(f)$, by assumption. Therefore $\gamma_{(j'+1,j'+1)}a_{j'+1}^2 = 0$ and since $\gamma_{(j'+1,j'+1)} \neq 0$, we find $a_{j'+1} = 0$.

Since $a_0 = a_1 = 0$, then the argument above proves that $a_j = 0$ for all $j \leq p$. Now, look at the coefficient of $x^{d-2}y^{d-2}$. This is of the form

$$\sum_{k+\ell=d} \gamma_{(k,\ell)} a_k a_\ell.$$

If $\ell \geq p+1$, then $k \leq p$. So $a_k a_\ell = 0$ by what we have shown. Hence the coefficient above vanishes. This is in contradiction with the assumption on $H(f)$. □

THEOREM 2.9. *For $n = 2$ and $d \geq 4$ even, the only singular binary homogeneous polynomials satisfying the assumptions in Question* 1.1 *are of the form* $f = a_{d/2}x_1^{d/2}x_2^{d/2}$.

PROOF. Let $d = 2p$. Let $f = f(x_1,x_2) = \sum_{i=0}^d a_i x_1^i x_2^{d-i} \in \mathbb{C}[x_1,x_2]_d$ be such that $H(f) = \alpha(x_1x_2)^{d-2} \neq 0$. As in the proof of Theorem 2.8, we may assume that $f = \sum_{i\geq 2}^d a_i x_1^i x_2^{d-i}$, i.e. $a_0 = a_1 = 0$. Moreover, again looking at $\gamma_{j,j}$ for $j \leq p-1$, we find that $a_j = 0$ for all $j \leq p-1$ and $a_p = a_{d/2} \neq 0$.

Let $1 \leq t \leq p$ and consider the coefficient a_{p+t} of the monomial $x_1^{p+t}x_2^{d-p-t}$ in f. Look at the coefficient of the monomial $x_1^{2p+t-2}x_2^{2p-t-2}$ in $H(f)$. This is of the form

$$\sum_{k+\ell=2p+t} \gamma_{(k,\ell)} a_k a_\ell \tag{2.5}$$

and, since $2p+t-2 > 2p-2 = d-2$, it must vanish by the assumption on $H(f)$. Since $k+\ell = 2p+t$, if $\ell \geq p+t+1$ then $k \leq p-1$. In this case, $a_k a_\ell = 0$ by what we have shown above. So the only left summand in (2.5) is $\gamma_{(p,p+t)}a_p a_{p+t}$, where $a_p \neq 0$ but $\gamma_{(p,p+t)}a_p a_{p+t} = 0$.

If $\gamma_{(p,p+t)} \neq 0$, for any given p and every $1 \leq t \leq p$, then $a_{p+t} = 0$ for all $1 \leq t \leq p$ and we would be done.

The coefficient $\gamma_{(p,p+t)}$ is given by

$$\gamma_{(p,p+t)} = 2p(2p-1)(t^2-p). \tag{2.6}$$

In other words, given $p \geq 2$, $\gamma_{(p,p+t)} = 0$ if and only if $t = \sqrt{p}$ (i.e. when p is a perfect square and t is its positive square root).

From now on, we shall then assume that p is a perfect square, otherwise we are done. Since $\gamma_{(p,p+t')} \neq 0$ for all $1 \leq t' \leq \sqrt{p}-1$, then the coefficient (2.5) for the monomial $x_1^{2p+t'-2}x_2^{2p-t'-2}$ is

$$\gamma_{(p,p+t')}a_p a_{p+t'} = 0.$$

Hence $a_{p+t'} = 0$ for all $1 \leq t' \leq \sqrt{p}-1$.

CLAIM 1. Let $1 \le t \le p$ be an integer such that $m\sqrt{p} < t < (m+1)\sqrt{p}$, for some integer $0 \le m \le \sqrt{p} - 1$. Suppose that $a_{p+t'} = 0$ for all t' such that $m'\sqrt{p} < t' < (m'+1)\sqrt{p}$ for all $0 \le m' < m$. Then $a_{p+t} = 0$.

PROOF. Write $t = m\sqrt{p} + n$, where $1 \le n < \sqrt{p}$. We look at the coefficient of the monomial $x_1^{2p+t-2} x_2^{2p-t-2} = x_1^{2p+m\sqrt{p}+n-2} x_2^{2p-m\sqrt{p}-n}$. This is of the form

$$\sum_{k+\ell=2p+m\sqrt{p}+n} \gamma_{(k,\ell)} a_k a_\ell. \tag{2.7}$$

Note that $2p + m\sqrt{p} + n - 2 > 2p - 2 = d - 2$ and so the coefficient (2.7) must vanish. If ℓ or k are of the form $p + q\sqrt{p} + q'$ with $0 \le q \le m-1$ and $1 \le q' < \sqrt{p}$, then the product $a_k a_\ell = 0$, by assumption. If $\ell = p + z\sqrt{p}$ with $0 \le z \le m-1$, then $k = p + z'\sqrt{p} + n$ with $1 \le z' \le m-1$. Thus the product $a_k a_\ell = 0$, by assumption. In conclusion, the only left summand in the coefficient (2.7) is $\gamma_{p,p+m\sqrt{p}+n} a_p a_{p+m\sqrt{p}+n} = 0$. Since $a_p \neq 0$ and $\gamma_{p,p+m\sqrt{p}+n} \neq 0$ (because $m\sqrt{p} + n \neq \sqrt{p}$ for all choices of m, n defined above), we find $a_{p+m\sqrt{p}+n} = a_{p+t} = 0$. $\square$

Note that we have already proven that $a_{p+t'} = 0$ for all $1 \le t' \le \sqrt{p} - 1$. This is the case $m = 1$ in the assumption of Claim 1. Then applying the conclusion of Claim 1 iteratively, we find that all the coefficients of f vanish, *unless* they are the following $(\sqrt{p}+1)$ coefficients: $a_p, a_{p+\sqrt{p}}, a_{p+2\sqrt{p}}, \ldots, a_{2p}$. To conclude, we have to show that, except a_p, they all must vanish.

To this aim, for $0 \le s, r \le \sqrt{p}$ and $(s,r) \neq (0,0)$, we look at the coefficient of $a_{p+s\sqrt{p}} a_{p+r\sqrt{p}}$ in the coefficient of the monomial $x_1^{2p+(s+r)\sqrt{p}-2} x_2^{2p-(s+r)\sqrt{p}-2}$ in $H(f)$. This has the form $\gamma_{p+s\sqrt{p},p+r\sqrt{p}} = 2p(2p-1)(ps^2 - 2srp + pr^2 + sr - p)$. We study the vanishing (and the sign) of $F(p,s,r) = ps^2 - 2srp + pr^2 + sr - p$. Write $F(p,s,r) = p(s-r)^2 + sr - p$. If $s = r$, then $s = r < \sqrt{p}$, because the exponent of x_1 must satisfy $2p + (s+r)\sqrt{p} - 2 \le 4p - 4 = 2d - 4$. Thus, if $s = r$ then $F(p,s,s) < 0$. If $s, r \ge 1$ and $s \neq r$, then $p(s-r)^2 - p \ge 0$ and the product $sr \ge 1$. So $F(p,s,r) > 0$. If $s = 0$ (or $r = 0$), then $F(p,0,r) = pr^2 - p = 0$ if and only $r = 1$ (or $s = 1$). Otherwise, if $s = 0$, then $F(p,0,r) > 0$ for any $r \ge 2$. In conclusion, $\gamma_{p+s\sqrt{p},p+r\sqrt{p}} \neq 0$ unless $s = 0$ and $r = 1$ (or the way around). This is exactly the case of (2.6).

CLAIM 2. Suppose that for all $1 \le i \le u - 1 < \sqrt{p} - 1$ we know that $a_{p+(\sqrt{p}-i)\sqrt{p}} = a_{2p-i\sqrt{p}} = 0$. Then $a_{p+(\sqrt{p}-u)\sqrt{p}} = 0$.

PROOF. The remaining coefficients are $a_{p+\sqrt{p}}, a_{p+2\sqrt{p}}, \ldots, a_{2p}$ (besides a_p) and all the others vanish.

Note that $2p - i\sqrt{p} > p + (\sqrt{p} - u)\sqrt{p}$. In other words, we are assuming that all the a_h with $h > p + (\sqrt{p} - u)\sqrt{p}$ vanish.

We look at the coefficient of the monomial $x_1^{4p-2u\sqrt{p}-2} x_2^{2u\sqrt{p}-2}$ in $H(f)$, which must vanish by assumption. This coefficient has the form

$$\sum_{k+\ell=4p-2u\sqrt{p}} \gamma_{(k,\ell)} a_k a_\ell = 0. \tag{2.8}$$

Write $k = p + h'\sqrt{p}$ and $\ell = p + h''\sqrt{p}$. Then $k + \ell = 4p - 2u\sqrt{p} = 2p + (h' + h'')\sqrt{p}$. Therefore $h' + h'' = 2(\sqrt{p} - u)$. If $h' \ge \sqrt{p} - u + 1$, then $k = p + h'\sqrt{p} > p + (\sqrt{p} - u)\sqrt{p}$ and hence $a_k = 0$, by assumption. Thus the coefficient (2.8) becomes

$\gamma_{(p+(\sqrt{p}-u)\sqrt{p},p+(\sqrt{p}-u)\sqrt{p})}a^2_{p+(\sqrt{p}-u)\sqrt{p}} = 0$. Since $\gamma_{(p+(\sqrt{p}-u)\sqrt{p},p+(\sqrt{p}-u)\sqrt{p})} \neq 0$ by the part before this claim, we find $a_{p+(\sqrt{p}-u)\sqrt{p}} = 0$. □

Now, look at the coefficient of the monomial $x_1^{4p-\sqrt{p}-2}x_2^{\sqrt{p}-2}$. The only coefficient appearing here is the product $\gamma_{(2p,p+(\sqrt{p}-1)\sqrt{p})}a_{2p}a_{p+(\sqrt{p}-1)\sqrt{p}}$, which must vanish. Since $\gamma_{(2p,p+(\sqrt{p}-1)\sqrt{p})} \neq 0$ by the part of the proof before Claim 2, either a_{2p} or $a_{p+(\sqrt{p}-1)\sqrt{p}}$ vanishes.

If $a_{2p} = 0$, applying Claim 2, we are done. If $a_{p+(\sqrt{p}-1)\sqrt{p}} = 0$, then applying Claim 2, all coefficients $a_{p+\sqrt{p}}, a_{p+2\sqrt{p}}, \ldots, a_{2p}$ are zero, except possibly a_{2p}. However, assuming $a_{2p} \neq 0$ (with all the other coefficients being zero, except a_p) leads to an immediate contradiction with the assumption on $H(f)$. In conclusion, $f = a_{d/2}x_1^{d/2}x_2^{d/2}$ and this establishes the statement. □

2.3. Stabilizers and Jacobian rings.

LEMMA 2.10. *One has the following descriptions for the stabilizers:*

(i) $\mathrm{Stab}_{\mathrm{GL}(\mathbb{C}^n)}((x_1\cdots x_n)^{d-2}) \cong (\mathbb{C}^*)^{n-1} \rtimes \mathfrak{S}_n$.

(ii) *Let* $h = (x_1\cdots x_n)^{d-2}$ *and let* $\mathcal{V}(h)$ *be the corresponding projective hypersurface in the projective space* $\mathbb{P}^{n-1}_{\mathbb{C}}$ *equipped with the natural action of* $\mathrm{PGL}(\mathbb{C}^n)$. *Then* $\mathrm{Stab}_{\mathrm{PGL}(\mathbb{C}^n)}(\mathcal{V}(h)) \cong (\mathbb{C}^*)^n \rtimes \mathfrak{S}_n/\mathbb{C}^*$.

PROOF. (i). Let $A \in \mathrm{Stab}_{\mathrm{GL}(\mathbb{C}^n)}((x_1\cdots x_n)^{d-2})$ be an element of the stabilizer in $\mathrm{GL}(\mathbb{C}^n)$ of the indicated polynomial. Since a polynomial ring is a unique factorization domain and the x_i's are irreducible in this ring, for each i we find $A \circ x_i = \gamma_i x_{\sigma(i)}$, for some $\sigma \in \mathfrak{S}_n$ and $\gamma_i \in \mathbb{C}^*$ with $\prod_{i=1}^n \gamma_i^{d-2} = 1$. The last condition provides an isomorphism with the $(n-1)$-dimensional algebraic torus. Statement (ii) is proven similarly (here we have to further quotient out by the $\mathbb{C}^*$ given by the scalar multiples of the identity). □

LEMMA 2.11. *Let* $f \in \mathbb{C}[x_1,\ldots,x_n]_d$ *and* $h = H(f) \in \mathbb{C}[x_1,\ldots,x_n]_{n(d-2)}$ *be its Hessian.*

(i) *Their stabilizers in* $\mathrm{SL}(\mathbb{C}^n)$ *satisfy* $\mathrm{Stab}_{\mathrm{SL}(\mathbb{C}^n)}(f) \subset \mathrm{Stab}_{\mathrm{SL}(\mathbb{C}^n)}(h)$.

(ii) *Suppose* $h = \alpha(x_1\cdots x_n)^{d-2} \neq 0$ *and let* $s = \dim \mathrm{Stab}_{\mathrm{SL}(\mathbb{C}^n)}(f)$. *Then every irreducible component of the fiber* $H^{-1}(h)$ *containing* f *has dimension* $\geq (n-1) - s$.

PROOF.

(i) Since the Hessian map is an $\mathrm{SL}(\mathbb{C}^n)$-coinvariant [**7**, §5.1], for any $A \in \mathrm{Stab}_{\mathrm{SL}(\mathbb{C}^n)}(f)$, we have $h = H(A\circ f) = A \circ h$, which proves the containment.

(ii) Let $X \subset H^{-1}(h)$ be an irreducible component containing f and let G be the connected component of the identity in $\mathrm{Stab}_{\mathrm{SL}(\mathbb{C}^n)}(h)$, which has dimension $\dim(G) = n-1$. Note that the orbit $G\cdot f \subset X$ and so $\dim X \geq \dim(G\cdot f) \geq n-1-s$. □

The two statements in Lemma 2.11 may be used as necessary conditions for the equality $H(f) = \alpha(x_1\cdots x_n)^{d-2} \neq 0$ to be satisfied by a given $f \in \mathbb{C}[x_1,\ldots,x_n]_d$. An application of that comes next.

PROPOSITION 2.12. *Let* $n \geq 2$ *and* $d \geq 3$. *There is no* $g \in \mathbb{C}[x_1,\ldots,x_n]_d$ *such that* $H(g) = \alpha(x_1\cdots x_n)^{d-2} \neq 0$ *with complex Waring rank* $\mathrm{rk}_{\mathbb{C}}(g) = n+1$.

PROOF. Up to the action of $\mathrm{GL}(\mathbb{C}^n)$, we may assume that $g = x_1^d + \cdots + x_n^d + \ell^d$, where $\ell = \alpha_1 x_1 + \cdots + \alpha_n x_n$, for some $\alpha_i \in \mathbb{C}$. If $\ell \in \langle x_{i_1}, \ldots, x_{i_s} \rangle$ for some $s \leq n-1$, then setting $h = g - \sum_{j \neq i_1, \ldots, i_s} x_j^d \in \mathbb{C}[x_{i_1}, \ldots, x_{i_s}]$, we find that $H(g) = \beta H(h) \cdot \prod_{j \neq i_1, \ldots, i_s} x_j^{d-2}$. Thus we reduce to the case where $\alpha_i \neq 0$ for all i. It is a direct computation to see that $\dim \mathrm{Stab}_{\mathrm{GL}(\mathbb{C}^n)}(g) = 0$. By the assumption $H(g) = \alpha(x_1 \cdots x_n)^{d-2} \neq 0$ and Lemma 2.11, $\dim \mathrm{Ker}(dH_g) \geq n-1$. (This is because if the dimension of the fiber at $H(g)$ satisfies $\dim H^{-1}(H(g)) \geq q$ then $\dim \mathrm{Ker}(dH_g) \geq q$, for any $q \in \mathbb{N}$.) Now, following the argument in the proof of [**2**, Lemma 7.2], we find that $\mathrm{Ker}(dH_g) = 0$, which is a contradiction. □

One of the tools at our disposal is a finite-dimensional algebra attached to a smooth polynomial called the *Jacobian ring*. Although in Proposition 2.15 we will establish a sufficient condition on a smooth homogeneous polynomial with a monomial Hessian to be a Fermat polynomial, this approach seems to be weak towards answering Question 1.1 in the smooth case.

DEFINITION 2.13. Let $g \in \mathbb{C}[x_1, \ldots, x_n]_d$. The Jacobian ring of g is the quotient ring $R(g) = \mathbb{C}[x_1, \ldots, x_n]/J(g)$, where $J(g) = (\partial g/\partial x_1, \ldots, \partial g/\partial x_n)$. If g is a smooth homogeneous polynomial, then $R(g)$ is a zero-dimensional local ring and a finite-dimensional graded Gorenstein $\mathbb{C}$-algebra [**9**, Corollary 4.2]. Moreover, the highest non-zero degree summand (the *socle*) of $R(g)$ is $R(g)_{n(d-2)} \cong \mathbb{C}$ and generated by the class of its Hessian, i.e., $H(g) \notin J(g)$.

PROPOSITION 2.14. *Let f be a Fermat polynomial. Suppose a smooth homogenous polynomial $g \in \mathbb{C}[x_1, \ldots, x_n]$ of degree $d \geq 3$ in $n \geq 2$ variables is such that its Jacobian ring satisfies $R(g) = R(f)$ and its Hessian $H(g) = \alpha(x_1 \cdots x_n)^{d-2}$. Then g is a Fermat polynomial.*

PROOF. Since $R(g) = R(f)$, a result of Donagi [**9**, Proposition 4.9] shows that there exists $A \in \mathrm{PGL}(\mathbb{C}^n)$ such that $A \circ \mathcal{V}(f) = \mathcal{V}(g)$, where these are the corresponding degree d smooth projective hypersurfaces. Note that A must fix $\mathcal{V}(H(f))$, as $H(g)$ is a multiple of $H(f)$. Therefore $A \in \mathrm{Stab}_{\mathrm{PGL}(\mathbb{C}^n)}(\mathcal{V}(H(f)))$. By Lemma 2.10(ii), one then finds $g = \beta_1 x_1^d + \cdots + \beta_n x_n^d$. □

PROPOSITION 2.15. *Suppose a smooth homogenous polynomial $g \in S = \mathbb{C}[x_1, \ldots, x_n]$ of degree $d \geq 3$ in $n \geq 2$ variables is such that $H(g) = x_1^{b_1} \cdots x_n^{b_n}$. Suppose its Jacobian ideal $J(g)$ is a monomial ideal. Then $H(g) = (x_1 \cdots x_n)^{d-2}$ and g is a Fermat polynomial.*

PROOF. The dimension of the Jacobian ring $R(g)$ as a complex vector space depends only on n and d; see [**9**, Proposition 4.3]. Hence it coincides with the one of $R(f)$, where f is a Fermat polynomial. One then has $\dim_{\mathbb{C}} R(g) = \dim_{\mathbb{C}} R(f) = (d-1)^n$. To see the last equality, let $f = \alpha_1 x_1^d + \cdots + \alpha_n x_n^d$ be a Fermat polynomial. Its Jacobian ring is the quotient $R(f) = S/(x_1^{d-1}, \ldots, x_n^{d-1})$. A monomial $\mathbb{C}$-basis for this algebra is formed by all monomial divisors of $H(f) = \alpha(x_1 \cdots x_n)^{d-2}$, whose cardinality is $(d-1)^n$. Since $R(g)$ is a finite-dimensional graded artinian Gorenstein $\mathbb{C}$-algebra, Macaulay's theorem [**10**, Lemma 2.12] gives a bijection between these Gorenstein algebras, whose socle is in degree $n(d-2)$, and homogenous polynomials of degree $n(d-2)$ up to $\mathbb{C}^*$. This implies that there exists $h \in T = \mathbb{C}[y_1, \ldots, y_n]_{n(d-2)}$ such that $J(g) = \mathrm{Ann}(h)$, where S acts by differentiation on T. Recall that $H(g) = x_1^{b_1} \cdots x_n^{b_n} \notin J(g) = \mathrm{Ann}(h)$. Since $\mathrm{Ann}(h)$ is a monomial ideal and $\dim_{\mathbb{C}} R(g)_{n(d-2)} = 1$, every monomial of degree $n(d-2)$ different from $H(g)$ is in $\mathrm{Ann}(h)$. Thus, writing a monomial expansion of h, we find that $h = y_1^{b_1} \cdots y_n^{b_n}$, up to scaling. So $\mathrm{Ann}(h) = (x_1^{b_1+1}, \ldots, x_n^{b_n+1})$ and then $(d-1)^n = \dim_{\mathbb{C}} R(g) = \dim_{\mathbb{C}} T/\mathrm{Ann}(h) = \prod_{i=1}^n (b_i + 1)$. Moreover, by definition of the Hessian, we have $\sum_{i=1}^n (b_i+1) = (\sum_{i=1}^n b_i) + n = n(d-2) + n = n(d-1)$.

For any finite collection of nonnegative real numbers $\{b_i + 1\}_{i \in [n]}$, we have the inequality between their arithmetic and geometric means (the AM-GM inequality):

$$d - 1 = \frac{\sum_{i=1}(b_i+1)}{n} \geq \sqrt[n]{\prod_{i=1}^{n}(b_i+1)} = d-1, \tag{2.9}$$

where the first and last equalities are consequences of the two identities on the b_i's. Equality in the AM-GM inequality (2.9) is verified if and only if $b_1 + 1 = \cdots = b_n + 1$. This is the case, and so $b_i = d-2$ for all $1 \leq i \leq n$. Therefore we find $H(g) = (x_1 \cdots x_n)^{d-2}$ and $J(g) = \mathrm{Ann}(h) = (x_1^{d-1}, \ldots, x_n^{d-1}) = J(f)$. By Proposition 2.14, g is a Fermat polynomial. □

We do not know whether after removing the assumption on $H(g)$ from Proposition 2.15 the conclusion still holds true. Note that the homogeneous polynomials in Remark 2.4 are all singular.

CONJECTURE 2.16. *Suppose a homogeneous polynomial $g \in S = \mathbb{C}[x_1, \ldots, x_n]$ of degree $d \geq 3$ in $n \geq 2$ variables is such that its Hessian $H(g)$ is a monomial. Then g is smooth if and only if $H(g) = (x_1 \cdots x_n)^{d-2}$ and g is a Fermat polynomial.*

REMARK 2.17. Conjecture 2.16 is true also for $n = 3$ and $d = 4$. This follows from a result of Kuribayashi and Komiya [**15**] stating that any smooth quartic curve with 12 hyperflexes and with monomial Hessian is projectively equivalent to a Fermat quartic.

DEFINITION 2.18. Let $A \in \mathrm{GL}(\mathbb{C}^n)$ and let I_n be the $n \times n$ identity matrix. The linear transformation A is a *unitary reflection* if it has finite order and $\mathrm{Ker}(A - I_n)$ is a codimension-one subspace, i.e. A has finite order and fixes pointwise a hyperplane in $\mathbb{C}^n$. A finite group generated by unitary reflections is called a *unitary group generated by reflections (u.g.g.r.)*.

REMARK 2.19. A complete classification of u.g.g.r. was found by Shephard and Todd [**20**]. These include as particular cases the finite Euclidean reflection groups (called *Coxeter groups*). An interesting subfamily of u.g.g.r. is the one of *Shephard groups* arising as a symmetry group of a *regular complex polytope*, defined and classified by Shephard [**19**]. The equivalence stated in [**16**, Theorem 5.10] characterises Shephard and Coxeter groups among all the u.g.g.r. groups. Item (v) in *loc. cit.* states that the Hessian of a minimal degree invariant form under a Shephard group G is a suitable product of powers of linear functionals (each functional corresponds to a hyperplane fixed by an element in G). This very last statement and its similarity with Question 1.1 was our motivation to look at these finite groups.

EXAMPLE 2.20. The group $G(d,1,n) \cong \mathbb{Z}_d^n \rtimes \mathfrak{S}_n \subset (\mathbb{C}^*)^n \rtimes \mathfrak{S}_n$ is a Shephard group. As a matrix group, it is the group of permutation matrices whose nonzero entries are d-th roots of unity.

EXAMPLE 2.21. Let $n, d \geq 2$ and $n|d$. As a matrix group, the subgroup $G(d,n,n) \subset G(d,1,n)$ consists of permutation matrices of the form $\mathrm{Diag}(\theta^{a_1}, \ldots, \theta^{a_n}) \circ \sigma$, where $\sigma \in \mathfrak{S}_n$ acts by permutation and Diag is a diagonal matrix with the indicated entries, θ is a primitive d-th root of unity and $\sum_{i=1}^{n} a_i \equiv 0 \mod n$. The group $G(d,n,n)$ is a u.g.g.r. but not a Shephard group. In particular, [**16**, Theorem 5.10](v) fails.

REMARK 2.22. The groups described above are contained in the stabilizers of homogeneous polynomials that are related to Question 1.1.

(i) Let $f = x_1^d + \cdots + x_n^d$. Then $\mathrm{Stab}_{\mathrm{GL}(\mathbb{C}^n)}(f) = G(d,1,n)$.
(ii) Let $n, d \geq 2$ and $n|d$. If $f = (x_1 \cdots x_n)^{d/n}$ then $G(d,n,n) \subset \mathrm{Stab}_{\mathrm{GL}(\mathbb{C}^n)}(f)$.

PROPOSITION 2.23. *Let $n \geq 2$ and $d \geq 3$ with $n|d$. Let $f \in \mathbb{C}[x_1, \ldots, x_n]_d$ be such that its Hessian $H(f) = \alpha(x_1 \cdots x_n)^{d-2} \neq 0$. Then $f = \beta(x_1 \cdots x_n)^{d/n}$ or $f = \beta(x_1^d + \cdots + x_n^d)$ if and only if $G(d,n,n) \subset \mathrm{Stab}_{\mathrm{GL}(\mathbb{C}^n)}(f)$.*

PROOF. One implication is Remark 2.22. Suppose $G(d,n,n) \subset \mathrm{Stab}_{\mathrm{GL}(\mathbb{C}^n)}(f)$, i.e. f is an invariant under $G(d,n,n)$. Then $f = \alpha_1(x_1^d + \cdots + x_n^d) + \alpha_2(x_1 \cdots x_n)^{d/n}$ as the invariant ring $\mathbb{C}[x_1, \ldots, x_n]^{G(d,n,n)}$ is the polynomial ring generated by the elementary symmetric functions of degrees $1, \ldots, n-1$ evaluated at $x_1^d, \ldots, x_n^d$ and by $(x_1 \cdots x_n)^{d/n}$ [**20**, Part II, §6]. (Indeed, by the Jacobian criterion these n invariant polynomials are algebraically independent and the product of their degrees is equal to $d^n \cdot (n-1)!$, the cardinality of $G(d,n,n)$; see [**20**, Theorem 5.1].) Suppose $\alpha_1, \alpha_2 \neq 0$. In the monomial expansion of $H(f)$ we look for monomials divisible by $(x_1 \cdots x_{n-2})^{d-2+2\frac{d}{n}}$. Equivalently, we search for monomials m in the expansion of $H(f)$ such that the x_i-degree of m, for $i \in \{1, \cdots, n-2\}$, is at least $d-2+2\frac{d}{n}$. One may write $\deg_{x_i}(m) = h_1(d-2) + h_2\left(\frac{d}{n}-2\right) + h_3\left(\frac{d}{n}-1\right) + h_4\frac{d}{n} \geq d-2+2\frac{d}{n}$, where h_1, h_2, h_3 gives a choice of a monomial in the i-th row of $\mathrm{Hess}(f)$, i.e. the row where the entries are of the form $\frac{\partial^2 f}{\partial x_i \partial x_j}$, and h_4 represents a choice of a monomial outside the i-th row. Thus $h_1 + h_2 + h_3 = 1$ and $h_1 + h_2 + h_3 + h_4 \leq n$. Suppose $h_1 = 0$. If $h_2 = 0$, then $h_3 = 1$ and $h_4 \leq n-1$. So $h_3\left(\frac{d}{n}-1\right) + h_4\frac{d}{n} \leq \frac{d}{n} - 1 + (n-1)\frac{d}{n} = d - 1 < d - 2 + 2\frac{d}{n}$ and hence this case is not possible. Similarly, if $h_3 = 0$, then $h_2 = 1$ and this case is not possible. Therefore $h_1 = 1$ and $h_2 = h_3 = 0$. Hence $h_1(d-2) + h_4\frac{d}{n} = d-2+h_4\frac{d}{n} \geq d-2+2\frac{d}{n}$, which implies $h_4 \geq 2$. Since this holds for every $i \in \{1, \ldots, n-2\}$, the monomials we are looking for can only be monomials in the expansion of the product between $\gamma(x_1 \cdots x_{n-2})^{d-2}$ ($\gamma \neq 0$ because $\alpha_1 \neq 0$) and the determinant of the lower-right 2×2 minor of $\mathrm{Hess}(f)$ given by

(2.10)
$$\begin{vmatrix} \alpha_1 d(d-1)x_{n-1}^{d-2} + \alpha_2\frac{d}{n}\left(\frac{d}{n}-1\right)(x_1 \ldots x_{n-2})^{\frac{d}{n}} x_{n-1}^{\frac{d}{n}-2} x_n^{\frac{d}{n}} & \alpha_2\left(\frac{d}{n}\right)^2 (x_1 \ldots x_{n-2})^{\frac{d}{n}} x_{n-1}^{\frac{d}{n}-1} x_n^{\frac{d}{n}-1} \\ \alpha_2\left(\frac{d}{n}\right)^2 (x_1 \ldots x_{n-2})^{\frac{d}{n}} x_{n-1}^{\frac{d}{n}-1} x_n^{\frac{d}{n}-1} & \alpha_1 d(d-1)x_n^{d-2} + \alpha_2\frac{d}{n}\left(\frac{d}{n}-1\right)(x_1 \ldots x_{n-2})^{\frac{d}{n}} x_{n-1}^{\frac{d}{n}} x_n^{\frac{d}{n}-2} \end{vmatrix}.$$

From (2.10), we see that the monomial $(x_1 \cdots x_{n-2})^{d-2+2\frac{d}{n}} x_{n-1}^{2\frac{d}{n}-2} x_n^{2\frac{d}{n}-2}$ appearing in $H(f)$ has coefficient

$$\gamma\alpha_2^2 \cdot \left[\left(\frac{d}{n}\right)^2\left(\frac{d}{n}-1\right)^2 - \left(\frac{d}{n}\right)^4\right] = \gamma\alpha_2^2\left(\frac{d}{n}\right)^2\left(1 - 2\frac{d}{n}\right) \neq 0.$$

This is a contradiction and hence either $\alpha_1 = 0$ or $\alpha_2 = 0$. $\square$

CONJECTURE 2.24. *Let $n \geq 2$ and $d \geq 3$. Let $f \in \mathbb{C}[x_1, \dots, x_n]_d$ be such that its Hessian is $H(f) = \alpha(x_1 \cdots x_n)^{d-2} \neq 0$. Then, up to conjugation,* $\mathrm{Stab}_{\mathrm{GL}(\mathbb{C}^n)}(f)$ *contains a product of groups of the form $G(d, 1, n-q) \times G(d, q, q)$ (with $G(d, 1, 0)$ and $G(d, 0, 0)$ being trivial groups by convention), and f is an invariant of minimal degree for such a group.*

3. Local Koiran-Skomra's question

We formulate a local version of the Koiran-Skomra's Question 1.1: whether there exist homogeneous polynomial solutions g to the equality $H(g) = \alpha(x_1 \cdots x_n)^{d-2} \neq 0$, that are close to a Fermat polynomial $f = \alpha_1 x_1^d + \cdots + \alpha_n x_n^d$. To deal with this local version, we first compute the differential of the Hessian map.

LEMMA 3.1 ([**2**, Lemma 7.2]). *Let $f, g \in \mathbb{C}[x_1, \dots, x_n]_d$ be two homogeneous polynomials of degree d. The linear function $dH_f : \mathbb{C}[x_1, \dots, x_n]_d \to \mathbb{C}[x_1, \dots, x_n]_{n(d-2)}$, defined as $g \mapsto \frac{d}{dt} H(f + tg)_{|t=0}$, is the differential at f of the Hessian map $H : \mathbb{C}[x_1, \dots, x_n]_d \to \mathbb{C}[x_1, \dots, x_n]_{n(d-2)}$. The image $dH_f(g)$ is the sum of the determinants of n matrices H^i whose i-th row is $H^i_{ij} = \frac{\partial^2 g}{\partial x_i \partial x_j}$, and whose k-th row for $k \neq i$ is $H^i_{kj} = \frac{\partial^2 f}{\partial x_k \partial x_j}$.*

PROOF. Viewing $p(t) = H(f + tg) = \det(\mathrm{Hess}(f + tg))$ as a polynomial in t, the image of g under dH_f is by definition the coefficient of t in $p(t)$. By definition of determinant, every contribution to this coefficient must involve a unique row of the matrix $\mathrm{Hess}(f + tg)$: the i-th such contribution is the determinant of H^i in the statement. □

PROPOSITION 3.2. *Let $f = \sum_{i=1}^n x_i^d \in \mathbb{C}[x_1, \dots, x_n]_d$.*

(i) *If $d \geq n + 1$, then* $\dim(\mathrm{Ker}(dH_f)) = n - 1$.
(ii) *If $d \leq n$, then* $\dim(\mathrm{Ker}(dH_f)) = n - 1 + \binom{n}{d}$.

PROOF. (i). Let $g \in \mathbb{C}[x_1, \dots, x_n]_d$. Then, using the formula in Lemma 3.1, we find $dH_f(g) = (d(d-1))^{n-1} \cdot \sum_{i=1}^n \prod_{j \neq i} x_j^{d-2} \frac{\partial^2 g}{\partial x_i^2}$. Denote $g_i = \prod_{j \neq i} x_j^{d-2} \frac{\partial^2 g}{\partial x_i^2}$. For two distinct indices $0 \leq h, k \leq n$, consider the monomial expansion of g_h and g_k. Note that $\deg(g_h) = \deg(g_k) = n(d-2)$. A monomial shared by g_h and g_k must be divisible by both $\prod_{j \neq h} x_j^{d-2}$ and $\prod_{j \neq k} x_j^{d-2}$. Thus such a monomial is divisible by $\prod_{i=1}^n x_i^{d-2}$. Since any monomial of the g_i has degree $n(d-2)$, the only shared monomial by any two of the g_i's is $\prod_{i=1}^n x_i^{d-2}$.

Suppose $g \in \mathrm{Ker}(dH_f)$. Then the condition $dH_f(g) = 0$ implies that the coefficients of all monomials appearing only in a single g_i vanish. Since $d \geq n+1$, all the coefficients of g appear in a unique g_i, except the coefficients β_i of the powers x_i^d appearing in the monomial expansion of g. The fact that the only shared monomial by all the g_i's is $\prod_{i=1}^n x_i^{d-2}$ implies the linear equation $\sum_{i=1}^n \beta_i = 0$. (Note that this is the tangent space of the algebraic torus in the fiber at f.) Thus $\dim(\mathrm{Ker}(dH_f)) = n - 1$.
(ii). Keep the notation from above. Since $d \leq n$, the coefficients of monomials in g annihilated by all the quadratic differentials, i.e. the square-free monomials $x_{i_1} \cdots x_{i_d}$ for some $1 \leq i_1, \dots, i_d \leq n$, do not appear in any of the g_i's summands.

There are $\binom{n}{d}$ of those. This shows the lower bound $\dim(\mathrm{Ker}(dH_f)) \geq n-1+\binom{n}{d}$. The equality comes from the direct verification that any other coefficient in g appears as coefficient of a unique monomial in a unique g_i and thus must vanish. □

REMARK 3.3. Proposition 3.2 implies that, when $d \geq n+1$, the local version of Koiran-Skomra's Question 1.1 has a positive answer.

For binary homogeneous polynomials, we offer a description of the differential at all monomials. We believe that the differential is important to understand better the Hessian map. For this part we were inspired by [**2**], where the authors provide ample evidence on how useful the differential is.

PROPOSITION 3.4. *Let $d \geq 2$ and let $f = x_1^k x_2^{d-k} \in \mathbb{C}[x_1, x_2]_d$, with $0 < k \leq d$.*

(i) *If $k = d$, then $\dim(\mathrm{Ker}(dH_f)) = 2$.*
(ii) *If $k = d-1$, then $\dim(\mathrm{Ker}(dH_f)) = 1$, unless $d = 2$. In the last case, $\dim(\mathrm{Ker}(dH_f)) = 2$.*
(iii) *If $k \leq d-2$, then $\dim(\mathrm{Ker}(dH_f))$ is the number of indices $0 \leq j \leq d$ such that $d(k-j)^2 - d(k+j) + 2kj = 0$.*
(iv) *Let $d \geq 3$ and $k \leq d-1$. Then $\dim(\mathrm{Ker}(dH_f)) \leq 1$, unless $d = 2k$ and k is a square. In this case $\mathrm{Ker}(dH_f)$ is the span of $x_1^{k+\sqrt{k}} x_2^{k-\sqrt{k}}$ and $x_1^{k-\sqrt{k}} x_2^{k+\sqrt{k}}$.*

PROOF. Statement (i) is a direct computation using the formula in Lemma 3.1. By Lemma 3.1, for $g \in \mathbb{C}[x_1, x_2]_d$, we have

$$dH_f(g) = \begin{vmatrix} \partial^2 g/\partial x_1^2 & \partial^2 g/\partial x_1 \partial x_2 \\ (d-k)k x_1^{k-1} x_2^{d-k-1} & (d-k)(d-k-1) x_1^k x_2^{d-k-2} \end{vmatrix} + \begin{vmatrix} k(k-1) x_1^{k-2} x_2^{d-k} & (d-k)k x_1^{k-1} x_2^{d-k-1} \\ \partial^2 g/\partial x_1 \partial x_2 & \partial^2 g/\partial x_2^2 \end{vmatrix}.$$

Let $g = \sum_{i=0}^d a_i x_1^i x_2^{d-i}$. Thus

$$dH_f(g) = (d-k)(d-k-1)\left[\sum_{i=0}^d a_i i(i-1) x_1^{k+i-2} x_2^{2d-k-i-2}\right] +$$

$$-2(d-k)k\left[\sum_{i=0}^d a_i (d-i) i x_1^{k+i-2} x_2^{2d-k-i-2}\right] +$$

$$+k(k-1)\left[\sum_{i=0}^d a_i (d-i)(d-i-1) x_1^{k+i-2} x_2^{2d-k-i-2}\right].$$

The coefficient of the monomial $a_j x_1^{k+j-2} x_2^{2d-k-j-2}$ is

(3.1)
$$g(d,k,j) = (d-k)(d-k-1)j(j-1) - 2(d-k)k(d-j)j + k(k-1)(d-j)(d-j-1) =$$
$$= (d-1)\left[d(k-j)^2 - d(k+j) + 2kj\right].$$

(ii). Suppose $k = d-1$. Then $g(d, d-1, j) = (d-1)(d-j)(d^2 - dj - 3d + 2)$. Thus we have $g(d, d-1, d) = 0$. For $j = d-s \leq d-1$, we find that $g(d, d-1, d-s) = (d-1)(d-j)(sd - 3d + 2)$. So $g(d, d-1, d-s) = 0$ if and only if $d = -\frac{2}{s-3}$, which is only possible if $s = 1$ or $s = 2$. If $s = 1$, then $d = 1$, which is excluded from the assumptions as it is a trivial case. For $s = 2$, we have $d = 2$ and $j = 0$. In this case, in the expansion of $dH_f(g)$ the only vanishing coefficients are those of a_0 and a_2

and whence $\dim(\mathrm{Ker}(dH_f)) = 2$. Otherwise, when $k = d-1$, in the expansion of $dH_f(g)$ the only vanishing coefficient is the one of a_d and so $\dim(\mathrm{Ker}(dH_f)) = 1$.
(iii). Using the approach in (ii) above, when $k \le d-2$ the only vanishing coefficients of the monomials $a_j x_1^{k+j-2} x_2^{2d-k-j-2}$ are for those indices j satisfying $g(d,k,j) = 0$. Then it is immediate to see that $\dim(\mathrm{Ker}(dH_f))$ is the number of indices j such that $g(d,k,j) = 0$. (Clearly, the equation is quadratic in j and hence $\dim(\mathrm{Ker}(dH_f)) \le 2$.)
(iv). By part (iii), $\dim \mathrm{Ker}(H_f) > 1$ if and only if the following polynomial in j

$$j^2 + j\left(-2k - 1 + \frac{2p}{d}\right) + k^2 - k = 0 \tag{3.2}$$

has two integral roots $0 \le j_1 < j_2 \le d$. Assume the existence of such j_1, j_2. Thus

$$j_1 j_2 = k(k-1) \text{ and } j_1 + j_2 = 2k + 1 - \frac{2k}{d}. \tag{3.3}$$

Since $0 < k < d$, (3.3) gives $d = 2k$ and $j_1 + j_2 = 2k$. So $j_1 = k - \sqrt{k}$ and $j_2 = k + \sqrt{k}$. Assuming the existence of these integral solutions, $d = 2k$ where k is a square. □

In some cases, one finds a large subset of binary homogeneous polynomials whose differential is injective:

COROLLARY 3.5. *Let $d \ge 3$ be a prime and $2 \le k \le d-2$. Let $W_k \subset \mathbb{C}[x_1, x_2]_d$ be the set of homogenous polynomials of the form $\sum_{i\ge k}^{d} a_i x_1^i x_2^{d-i}$ such that $a_k \ne 0$. Then, for any $f \in W_k$, $\dim(\mathrm{Ker}(dH_f)) = 0$.*

PROOF. For $\lambda \in \mathbb{C}^*$ and $f \in W_k$, define $f_\lambda(x_1, x_2) = f(\lambda x_1, x_2)/\lambda^k$. One has $\dim(\mathrm{Ker}(dH(f_\lambda))) = \dim(\mathrm{Ker}(dH_f))$. Note that $g = \lim_{\lambda\to 0} f_\lambda(x_1, x_2) = a_k x_1^k x_2^{d-k} \ne 0$. By semicontinuity, we have $\dim(\mathrm{Ker}(dH_f)) \le \dim(\mathrm{Ker}(dH(g)))$. By Proposition 3.4(iii), the latter is the number of indices $0 \le j \le d$ satisfying the equation $d(k-j)^2 - d(k+j) + 2kj = 0$. If $j = 0$, then $k = 0$ or $k = 1$, which are excluded. For $j \ne 0$, since d is a prime dividing the nonzero integer $2kj$ and $k \le d-2$, we must have $j = d$. For $j = d$, one finds $d(k-j)^2 - d(k+j) + 2kj = d(k-d)(k-d+1)$. So, given $2 \le k \le d-2$, there is no j satisfying the equation above. Thus $\dim(\mathrm{Ker}(dH_f)) = 0$ for all $f \in W_k$. □

In Theorem 2.9, the exceptional cases appear when $d = 2m^2$. These are also somewhat exceptional instances from the perspective of the differential of the Hessian map, as witnessed by Proposition 3.4(iv) and, as a consequence, by the next result.

THEOREM 3.6. *Let $d \ge 3$ and let $f \in \mathbb{C}[x_1, x_2]_d$ with $H(f) \ne 0$. Then $\dim H^{-1}(H(f)) \le 1$ and $\dim(\mathrm{Ker}(dH_f)) \le 1$, unless $d = 2m^2$ and $f = \ell_1^{m^2}\ell_2^{m^2}$ (where ℓ_i are linearly independent linear forms). In the latter case, $\dim(\mathrm{Ker}(dH_f)) = 2$.*

PROOF. First, suppose $d \ne 2m^2$. It is sufficient to prove that $\dim \mathrm{Ker}(dH_f) \le 1$. (This is because if the dimension of the fiber at $H(f)$ satisfies $\dim H^{-1}(H(f)) \ge q$ then $\dim(\mathrm{Ker}(dH_f)) \ge q$, for any $q \in \mathbb{N}$.) For any $k \ge 1$, let W_k be the set of all $g \in \mathbb{C}[x_1, x_2]_d$ such that $g = \sum_{i=k}^{d} a_i x_1^i x_2^{d-i}$ and $a_k \ne 0$. Up to the action of $\mathrm{GL}(\mathbb{C}^2)$, we may assume $f(0,1) = 0$, i.e. we may assume $f = \sum_{i\ge 1}^{d} a_i x_1^i x_2^{d-i}$. Let

$i_{\min}$ be the smallest positive integer such that $a_{i_{\min}} \neq 0$ and hence $f \in W_{i_{\min}}$. We have $i_{\min} < d$, because $H(f) \neq 0$ and so $1 \leq i_{\min} \leq d-1$. As in the proof of Corollary 3.5, for any $\lambda \in \mathbb{C}^*$, define $g_\lambda(x_1, x_2) = f(\lambda x_1, x_2)/\lambda^{i_{\min}}$. Note that $g = a_{i_{\min}} x_1^{i_{\min}} x_2^{d-i_{\min}} = \lim_{\lambda \to 0} g_\lambda$. One then has $\dim(\operatorname{Ker}(dH_{g_\lambda})) = \dim(\operatorname{Ker}(dH_f))$ for all $\lambda \in \mathbb{C}^*$. By semicontinuity, we have $\dim(\operatorname{Ker}(dH_f)) \leq \dim(\operatorname{Ker}(dH_g)) \leq 1$, where the last inequality is Proposition 3.4(iv).
Now, assume $d = 2m^2$. Since $H(f) \neq 0$ and $d \geq 3$, we may assume that f has at least two distinct zeros. Up to the action of $\mathrm{GL}(\mathbb{C}^2)$, we may assume $f(0,1) = f(1,0) = 0$, i.e. we may assume $a_0 = a_d = 0$. The first part of this proof works if $i_{\min} \neq m^2$. Thus, we may assume $i_{\min} = m^2$. Let $j_{\min}$ be the minimal positive integer such that $a_{d-j_{\min}} \neq 0$ by Proposition 3.4(iv). Since $a_d = 0$, $j_{\min} > 0$. Define $h_\lambda(x_1, x_2) = f(x_1, \lambda x_2)/\lambda^{j_{\min}}$. Since $h = a_{d-j_{\min}} x_1^{d-j_{\min}} x_2^{j_{\min}} = \lim_{\lambda \to 0} h_\lambda$, by semicontinuity we obtain $\dim(\operatorname{Ker}(dH_f)) \leq \dim(\operatorname{Ker}(dH_h))$. If $j_{\min} < m^2$, then Proposition 3.4(iv) gives $\dim(\operatorname{Ker}(dH_f)) \leq 1$. Otherwise, $i_{\min} = j_{\min} = m^2$ and so $f = x_1^{m^2} x_2^{m^2}$. In this case, Proposition 3.4(iv) shows that $\dim(\operatorname{Ker}(H_f)) = 2$. □

REMARK 3.7. Theorem 3.6 also shows that the Hessian map at the monomial counterexamples is not necessarily a local embedding.

References

[1] E. Carlini, M. V. Catalisano, and A. V. Geramita, *The solution to the Waring problem for monomials and the sum of coprime monomials*, J. Algebra **370** (2012), 5–14, DOI 10.1016/j.jalgebra.2012.07.028. MR2966824

[2] C. Ciliberto and G. Ottaviani, *The Hessian map*, Int. Math. Res. Not. IMRN **8** (2022), 5781–5817, DOI 10.1093/imrn/rnaa288. MR4406119

[3] C. Ciliberto, F. Russo, and A. Simis, *Homaloidal hypersurfaces and hypersurfaces with vanishing Hessian*, Adv. Math. **218** (2008), no. 6, 1759–1805, DOI 10.1016/j.aim.2008.03.025. MR2431661

[4] T. Dias and R. Gondim, *Waring problems and the Lefschetz properties*, Selecta Math. (N.S.) **29** (2023), no. 3, Paper No. 32, 21, DOI 10.1007/s00029-023-00840-3. MR4578470

[5] A. Dimca and S. Papadima, *Hypersurface complements, Milnor fibers and higher homotopy groups of arrangments*, Ann. of Math. (2) **158** (2003), no. 2, 473–507, DOI 10.4007/annals.2003.158.473. MR2018927

[6] I. V. Dolgachev, *Classical algebraic geometry*, Cambridge University Press, Cambridge, 2012. A modern view, DOI 10.1017/CBO9781139084437. MR2964027

[7] I. Dolgachev, *Lectures on invariant theory*, London Mathematical Society Lecture Note Series, vol. 296, Cambridge University Press, Cambridge, 2003, DOI 10.1017/CBO9780511615436. MR2004511

[8] H. Huang, M. Michałek, and E. Ventura, *Vanishing Hessian, wild forms and their border VSP*, Math. Ann. **378** (2020), no. 3-4, 1505–1532, DOI 10.1007/s00208-020-02080-8. MR4163534

[9] D. Huybrechts, *The geometry of cubic hypersurfaces*, notes available on the webpage of the author, `https://www.math.uni-bonn.de/ huybrech/Notes.pdf`.

[10] A. Iarrobino and V. Kanev, *Power sums, Gorenstein algebras, and determinantal loci*, Lecture Notes in Mathematics, vol. 1721, Springer-Verlag, Berlin, 1999. Appendix C by Iarrobino and Steven L. Kleiman, DOI 10.1007/BFb0093426. MR1735271

[11] N. Kayal, *Efficient algorithms for some special cases of the polynomial equivalence problem*, Proceedings of the Twenty-Second Annual ACM-SIAM Symposium on Discrete Algorithms, SIAM, Philadelphia, PA, 2011, pp. 1409–1421. MR2858410

[12] P. Koiran and S. Saha, *Black box absolute reconstruction for sums of powers of linear forms*, 42nd IARCS Annual Conference on Foundations of Software Technology and Theoretical Computer Science, LIPIcs. Leibniz Int. Proc. Inform., vol. 250, Schloss Dagstuhl. Leibniz-Zent. Inform., Wadern, 2022, pp. Paper No. 24, 17, DOI 10.4230/lipics.fsttcs.2022.24. MR4526778

[13] P. Koiran and M. Skomra, *Derandomization and absolute reconstruction for sums of powers of linear forms*, Theoret. Comput. Sci. **887** (2021), 63–84, DOI 10.1016/j.tcs.2021.07.005. MR4312932

[14] P. Koiran and N. Ressayre, *Orbits of monomials and factorization into products of linear forms*, preprint at arXiv:1807.03663, 2018.

[15] A. Kuribayashi and K. Komiya, *On Weierstrass points of non-hyperelliptic compact Riemann surfaces of genus three*, Hiroshima Math. J. **7** (1977), no. 3, 743–768. MR472831

[16] P. Orlik and L. Solomon, *The Hessian map in the invariant theory of reflection groups*, Nagoya Math. J. **109** (1988), 1–21, DOI 10.1017/S0027763000002737. MR931948

[17] F. Russo, *On the geometry of some special projective varieties*, Lecture Notes of the Unione Matematica Italiana, vol. 18, Springer, Cham; Unione Matematica Italiana, Bologna, 2016, DOI 10.1007/978-3-319-26765-4. MR3445582

[18] B. Segre, *Variazione continua ed omotopia in geometria algebrica* (Italian), Ann. Mat. Pura Appl. (4) **50** (1960), 149–186, DOI 10.1007/BF02414510. MR121698

[19] G. C. Shephard, *Regular complex polytopes*, Proc. London Math. Soc. (3) **2** (1952), 82–97, DOI 10.1112/plms/s3-2.1.82. MR48041

[20] G. C. Shephard and J. A. Todd, *Finite unitary reflection groups*, Canad. J. Math. **6** (1954), 274–304, DOI 10.4153/cjm-1954-028-3. MR59914

Università di Trento, Via Sommarive 14, 38123 Povo (Trento), Italy
Email address: edoardo.ballico@unitn.it

Politecnico di Torino, Dipartimento di Scienze Matematiche "G. L. Lagrange", Corso Duca degli Abruzzi 24, 10129 Torino, Italy
Email address: emanuele.ventura@polito.it

Contemporary Mathematics
Volume **805**, 2024
https://doi.org/10.1090/conm/805/16130

Standard Artinian algebras and Lefschetz properties: A geometric approach

Davide Bricalli and Filippo Francesco Favale

Abstract. In this paper we deal with Lefschetz properties for standard Artinian Gorenstein algebras (SAGAs). We will briefly present some geometric methods already used by the authors in collaboration with G.P. Pirola which allowed to prove the validity of some weak and strong Lefschetz properties for complete intersection SAGAs presented by quadrics of codimension 5 and 6. We focus on complete intersection SAGAs presented by quadrics where "special" elements appear. In particular, we are interested in elements of degree 1 and 2 whose associated multiplication map from R^1 has not maximal rank and in the information they give about the structure of the starting SAGA or, when the SAGA is a Jacobian ring of a smooth cubic X, about the geometry of X itself and its hessian.

Introduction

A *Standard Artinian Gorenstein Algebra* over a field $\mathbb{K}$ (which is supposed to be algebraically closed and of characteristic 0), also called *SAGA* for brevity, is an Artinian graded algebra $R = \bigoplus_{i=0}^{N} R^i$ which is standard (i.e. it is generated in degree 1) and which satisfies the Poincaré-Gorenstein duality (namely the multiplication map $R^a \times R^{N-a} \to R^N \simeq \mathbb{K}$ is a perfect pairing). One of the algebraic properties of central importance for SAGAs are Lefschetz properties. One says that a SAGA R satisfies the *weak (strong) Lefschetz property* in degree k if the multiplication map from R^k induced by a general element $x \in R^1$ (by x^{N-2k}, respectively, for $k \leq N/2$) is of maximal rank (see Definition 1.2 for details). One of the main problems concerning SAGAs is to identify classes of SAGAs which satisfy weak and/or strong Lefschetz properties (WLP and SLP, for brevity).

2020 *Mathematics Subject Classification.* Primary 13E10; Secondary 14J70, 14N05, 14M10, 13F20.

Key words and phrases. Artinian Gorenstein algebras, Lefschetz properties, Cubic hypersurfaces, Hessian varieties, Jacobian rings.

The authors are thankful to Gian Pietro Pirola for his priceless suggestions and his constant support and to the anonymous referee for the comments. The authors are partially supported by MIUR: Dipartimenti di Eccellenza Program (2018-2022) - Dept. of Math. Univ. of Pavia and by the INdAM – GNSAGA Project, "Classification Problems in Algebraic Geometry: Lefschetz Properties and Moduli Spaces" (CUP_E55F22000270001).

Several authors have tried to answer to this kind of questions (also in a more general framework) with an algebraic approach (see, for example, [**HMNW03**], [**MMRN11**], [**MMRO13**], [**MN13**], [**BMMRN18**], [**GZ18**], [**AAI$^+$23**]...). Nevertheless, an exhaustive answer seems to be out of reach at the moment, even if a precise conjecture is well-known in this field (see Conjecture 3.46 [**HMM$^+$13**]): it is expected that complete intersection SAGAs, i.e. SAGAs which can be written as $\mathbb{K}[x_0, \ldots, x_n]/I$ where I is an ideal generated by a regular sequence, should always satisfy SLP. This conjecture is widely open also if one restricts to Jacobian rings, a class of SAGAs which plays a fundamental role in algebraic geometry, in particular in the study of smooth complex hypersurfaces of projective spaces. Indeed, if $X = V(f) \subseteq \mathbb{P}^n$ is a hypersurface, one can construct its Jacobian ring $R_X = \mathbb{C}[x_0, \ldots, x_n]/J_f$ where J_f is the Jacobian ideal of f, i.e. the ideal generated by the partial derivatives of f. When X is smooth, R_X is a SAGA which codifies a portion of the primitive part of the Dolbeault cohomology of X and it is a crucial ingredient in the study of the infinitesimal variation of the Hodge structure of X itself; see for example the seminal works [**Gri68**], [**CG80**] and [**CGGH83**].

By taking into account only Jacobian rings of smooth hypersurfaces, it is known that SLP holds for smooth curves in $\mathbb{P}^2$ up to degree $d = 4$ (the case $d \leq 3$ follows from a classical theorem of Gordan and Noether, whereas the case $d = 4$ is studied in [**DGI20**]). In $\mathbb{P}^3$ and $\mathbb{P}^4$ the result is known up to degree $d = 3$: in $\mathbb{P}^3$ this is again a consequence of Gordan-Noether's theorem, whereas in $\mathbb{P}^4$ it has been proved in [**AR19**] and [**BFP23a**]. Finally, by exploiting the geometric methods developed in the aforementioned work, the authors proved in [**BF22**] the SLP in degree 1 for Jacobian rings of cubic fourfolds. We also stress that the results in [**BFP23a**] and [**BF22**] hold more generally for complete intersection SAGAs presented in degree 2. In particular, for $d \geq 4$ and $n \geq 3$ the conjecture is still widely open.

Regarding WLP, other results are known, especially for SAGAs R with low codimensions (i.e. the dimension of R^1). In particular, in [**HMNW03**] it is proved the validity of WLP for complete intersection SAGAs of codimension 3 and in [**MN13**] the WLP in degree 1 for complete intersection SAGAs presented by forms of degree $d \geq 2$. In [**BF22**], the authors proved a "lifting criterion" for WLP in degree 2 in the case of complete intersection SAGAs presented by quadrics (see Theorem 2.1), under the assumption of the existence of some non-Lefschetz elements in degree 1 (see Section 1 for the definition). One of the aims of this paper is to generalize a part of this result by considering, instead of non-Lefschetz elements in degree 1, other "special" elements in degree 2. More precisely, we will deal with elements in $R^2 \setminus \{0\}$ inducing multiplication maps $R^1 \to R^3$ whose kernels have maximal dimension (i.e. dimension 2 by Proposition 1.5). This is done in Theorem 2.4 and Corollary 2.5.

Non-Lefschetz loci, which are clearly related to the validity of Lefschetz properties, have been studied from different perspectives (and also in more general contexts) by several authors (as, for example, [**BMMRN18**] and [**AR19**]). Particular non-Lefschetz elements are the elements in $\mathrm{Rad}(R) = \sqrt{(0)}$, i.e. nilpotents of R. Nilpotent elements of a given degree d define subvarieties of $\mathbb{P}(R^d)$.

For a complete intersection SAGA R presented by forms of degree 2, in [**BF22**], the authors started to analyse the rich geometry of the *nilpotent loci*

$$\mathcal{N}_s = \{[x] \in \mathbb{P}(R^1) \,|\, x^s = 0\}$$

and their connections with the algebraic structure of the SAGA itself. In this paper, we keep on studying these objects, in the case of Jacobian rings of smooth cubic hypersurfaces. In particular, (see Propositions 3.1 and 3.2), we derive some interesting consequences at the level of the Hessian hypersurface of a smooth cubic hypersurface $X = V(f)$, coming from the existence of points in the nilpotent locus $\mathcal{N}_2$ in $R_X = S/J_f$.

The interest of hessians of polynomials is not new for those studying Lefschetz properties. Indeed, if one writes a SAGA as apolar ring associated to a polynomial g (see Section 1), one can characterise the failure of SLP_k for R by the vanishing of the *kth higher hessian* of g, by the seminal result in [**MW09**].

The structure of the paper is as follows. In Section 1, we recall some basic definitions and most of the results necessary for the following sections. In Section 2, we study non-Lefschetz loci for complete intersection SAGAs presented by quadrics and prove Theorem 2.4 and Corollary 2.5. Finally, in Section 3, we analyze the relations between nilpotent elements of order 2 in the Jacobian ring of a smooth cubic hypersurface $X = V(f)$ and the structure of f and the decomposition of its hessian, by proving Propositions 3.1 and 3.2.

1. Preliminaries

We start by introducing all the main objects that we will use through the article. The interested reader can refer to [**HMM+13**] for a comprehensive treatment of the subject.

We denote by $\mathbb{K}$ an algebraically closed field of characteristic 0 and we set $S = \mathbb{K}[x_0, \cdots, x_n] = \bigoplus_{k \geq 0} S^k$, where $S^k = H^0(\mathcal{O}_{\mathbb{P}^n}(k))$. Moreover, we denote by $\mathcal{D}$ the ring of linear differential operators in the variables $x_0, \cdots, x_n$, i.e. $\mathcal{D} = \mathbb{K}[y_0, \cdots, y_n]$ where we denote for brevity $y_i = \partial_{x_i}$. We have a natural map $S \times \mathcal{D} \to S$ given by derivation.

First of all, let us define *standard Artinian Gorenstein algebras*, which will be briefly called SAGAs:

DEFINITION 1.1. A graded Artinian $\mathbb{K}$-algebra $R = \bigoplus_{i=0}^{N} R^i$ is a SAGA if it is standard (i.e. if it is generated in degree 1) and if it satisfies the Gorenstein-Poincaré duality, i.e. if $R^N \simeq R^0 = \mathbb{K}$ and the multiplication map $R^s \times R^{N-s} \to R^N$ is a perfect pairing for all $0 \leq s \leq N$. The codimension of a SAGA R is the dimension of R^1 as $\mathbb{K}$-vector space. Moreover, R^N is said to be the socle of R.

Examples of SAGAs can be given by considering smooth hypersurfaces in projective spaces: indeed, if $f \in S^d$ is such that $X = V(f) \subset \mathbb{P}^n$ is smooth, the Jacobian $J_f = (\partial_{x_i}(f))$ is spanned by a regular sequence of polynomials of degree $d-1$ and it is well known that the Jacobian ring $R_X = S/J_f$ is a SAGA (whose codimension is $n+1$ and socle degree is $N = (n+1)(d-2)$). More generally, one can consider *complete intersection SAGAs* (or *CI SAGA*, for brevity) by taking the quotient of S by an ideal spanned by a regular sequence $\{g_0, \ldots, g_n\}$. If $\deg(g_i) = e$ for all $i = 0, \ldots, n$ one talks about CI SAGA presented in degree e.

If $g \in S^n$ one can define the annihilator ideal $\operatorname{Ann}_{\mathcal{D}}(f) = \{D \in \mathcal{D} \mid D(g) = 0\}$. One can show that the so-called *apolar ring* associated to g, namely $\mathcal{D}/\operatorname{Ann}_{\mathcal{D}}(g)$, is always a SAGA and a very important and interesting fact is that every SAGA can be written in this form. This follows by the theory of Inverse Systems of Macaulay (see [**Mac94**] for a revisited reprint of the original work by Macaulay of 1916).

Let us now define the Lefschetz properties (the weak and the strong ones) for a SAGA R.

DEFINITION 1.2. A SAGA R with socle degree N satisfies

- the *weak Lefschetz property* in degree k (WLP_k in short), if there exists $x \in R^1$ such that the multiplication map $x\cdot : R^k \to R^{k+1}$ has maximal rank;
- the *strong Lefschetz property* in degree $k \leq N/2$ (SLP_k in short), if there exists $x \in R^1$ such that the multiplication map $x^{N-2k}\cdot : R^k \to R^{N-k}$ is an isomorphism.

We also say that R satisfies *weak (strong) Lefschetz property* - WLP (respectively SLP) in short - if it satisfies WLP_k (respectively SLP_k) for all k.

We stress that the above definition for SLP is not the usual one that is given for more general $\mathbb{K}$-algebras, but for SAGAs the two definitions coincide (see Definition 3.18 and subsequent discussion in [**HMM**$^+$**13**]).

These properties are expected to hold for "a lot" of SAGAs. For example, it is conjectured the following (see [**HMM**$^+$**13**, Conjecture 3.46]):

CONJECTURE 1.3. *Any CI SAGA satisfies SLP.*

Unfortunately, the proof of this conjecture seems out of reach at the moment, despite the effort of several authors through the years. The validity of both weak and strong Lefschetz properties has been proved for a few "classes" of SAGAs, as discussed in the Introduction.

We briefly present here the framework used by the authors in [**BFP23a**] and [**BF22**], also in collaboration with Gian Pietro Pirola, to study Lefschetz properties on SAGAs (a similar approach was used also in [**AR19**]).

For brevity and simplicity, given a SAGA R with codimension $\dim R^1 = n+1$ with $n \geq 1$ and socle in degree N, we focus on the study of the strong Lefschetz property in degree 1, i.e. the one involving the multiplication map $x^{N-2}\cdot : R^1 \to R^{N-1}$. With the same methods, as done in the above-mentioned articles, one can approach the study of Lefschetz properties in higher degrees.

First of all, let us define

$$\Gamma = \{([x],[y]) \in \mathbb{P}(R^1) \times \mathbb{P}(R^1) \,|\, x^{N-2}y = 0\} \subset \mathbb{P}(R^1) \times \mathbb{P}(R^1)$$

and the two projections p_1 and p_2 from Γ on the factors. We assume then that R does not satisfy SLP_1. One can show that the failure of the validity of SLP_1 is equivalent to the surjectivity of the first projection $p_1 : \Gamma \to \mathbb{P}(R^1)$. From the surjectivity of p_1 and since its fibers are projective spaces, we have that there exists a unique irreducible component of Γ which dominates $\mathbb{P}(R^1)$ via the first projection and which will be denoted by Θ. Let us also set π_i as the restriction of p_i to Θ for $i = 1, 2$, $Y := \pi_2(\Theta)$ and, for any $[y] \in Y$, $F_y := \pi_1(\pi_2^{-1}([y]))$ so that we have the following diagram:

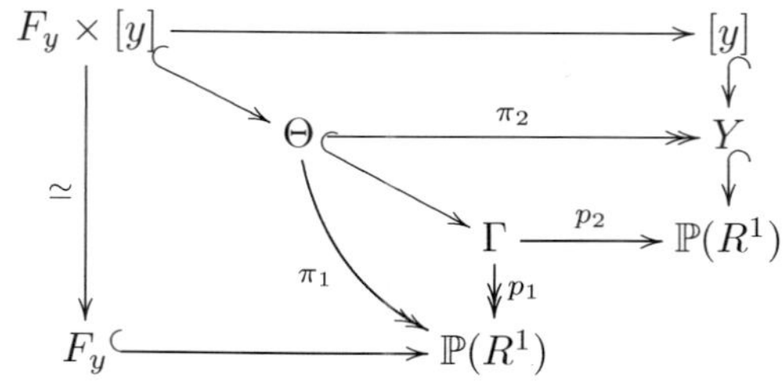

Very roughly, the aim is to obtain relations involving the dimensions of the objects in the diagram and then to rule out all the possibilities. In such a way, we deny the existence of the component Θ and, equivalently, the surjectivity of the projection p_1 from Γ, by proving the validity of the strong Lefschetz property in degree 1. With this approach, two of the main results proved in the papers cited above are the following :

THEOREM 1.4 ([**BFP23a**, Theorem C], [**BF22**, Theorem 3.1]). *Let R be a CI SAGA presented by quadrics.*

- *If the codimension of R is 5, then R satisfies SLP;*
- *If the codimension of R is 6, then R satisfies SLP_1.*

In particular, Jacobian rings of smooth cubic threefolds and of smooth cubic fourfolds satisfy the SLP and the SLP_1 respectively.

For the topics of this paper, a central role is played by vector spaces which are the kernels of multiplication maps defined by an element in R: for $\eta \in R^h$ we set

$$K^i_\eta := \ker\left(\eta\cdot : R^i \to R^{i+h}\right).$$

Notice that, in the above framework, the fiber of p_1 over $[x] \in \mathbb{P}(R^1)$ is given by $\mathbb{P}(K^1_{x^{N-2}})$. We will use extensively the following result.

PROPOSITION 1.5 ([**BFP23a**, Prop.4.1]). *Let R be a CI SAGA presented by forms of degree $d-1$ and with socle in degree N. Assume that $1 \le h \le N-1$. The following hold:*

(a) *If $\eta \in R^h \setminus \{0\}$, then $(d-2)\dim(K^1_\eta) \le h$;*
(b) *Let $\eta, \zeta \in R^h \setminus \{0\}$ and assume $h = (d-2)\dim(K^1_\eta) = (d-2)\dim(K^1_\zeta)$. Then $K^1_\eta = K^1_\zeta$ if and only if $[\eta] = [\zeta]$ in $\mathbb{P}(R^h)$.*

By focusing on elements in degree 1, we say that x is a *Lefschetz element (in degree 1)* when $K^1_x = \{0\}$. We denote by

$$\mathcal{L} := \{[x] \in \mathbb{P}(R^1) \mid K^1_x = \{0\}\}$$

the *Lefschetz locus (in degree 1)*, i.e. the set of Lefschetz elements (in degree 1). Its complement $\mathcal{L}^c$ is the *non-Lefschetz locus (in degree 1)*. One easily sees that $\mathcal{L}^c$ is an algebraic variety whose dimension is at most 1 (see [**AR19**]). Other special loci associated to SAGAs are *nilpotent loci*, namely

$$\mathcal{N}_s = \{[x] \in \mathbb{P}(R^1) \,|\, x^s = 0\}.$$

In [**BFP23a**] the authors gave a bound for the dimension of these loci for CI SAGAs presented in degree e: one has

$$\dim(\mathcal{N}_s) \le \frac{s-1}{e-1} - 1.$$

One easily sees that $\mathcal{N}_2$ is a subvariety of $\mathcal{L}^c$. Moreover, it is empty for $e \geq 3$ and, by the above result, of dimension at most 0 for $e = 2$. Here we will be interested mainly in studying $\mathcal{N}_2$ for CI SAGA presented by quadrics. We recall now an interesting geometrical description of $\mathcal{N}_2$, reflecting also, in some cases, some properties of the smooth cubic hypersurface $X = V(f)$, in the case where $R_X = S/J_f$ is its Jacobian ring.

PROPOSITION 1.6 ([**BF22**, Prop. 2.1, Cor. 2.2]). *Let R be a CI SAGA presented by quadrics. If $[x_1], \cdots, [x_k] \in \mathcal{N}_2$ are distinct points, then $\prod_{i=1}^k x_i \neq 0$ in R and $[x_1], \cdots, [x_k]$ are in general position in $\mathbb{P}(R^1)$. In particular, $\#\mathcal{N}_2 \leq n+1 = N$ and in the case where this is actually an equality, then R is the Jacobian ring of a Fermat cubic in $\mathbb{P}^n$.*

In [**BF22**], other geometrical properties of the loci $\mathcal{N}_s$ have been investigated in particular when $\mathcal{N}_s$ contains linear spaces of suitable dimension.

We recall now a result regarding Hessian varieties associated to smooth cubic hypersurfaces that will be used in Section 3. If $f \in S^3$ and $v \in \mathbb{K}^{n+1}$, we set $\partial_v = \sum_{i=0}^n v_i y_i$, where $y_i = \partial_{x_i}$. The Hessian matrix associated to f is then $\operatorname{Hess}(f) = (y_i y_j f)_{i,j=0,\dots,n}$ and one defines the *hessian of f* $\operatorname{hess}(f)$ as the determinant of $\operatorname{Hess}(f)$. The *Hessian hypersurface* $\mathcal{H}$ associated to $X = V(f)$ is $V(\operatorname{hess}(f))$. We have the following easy but important lemma (see [**BFP23b**]):

LEMMA 1.7. *Let f be an element in $S^3 \setminus \{0\}$ and consider $v, w \in \mathbb{K}^{n+1}$. Then*

$$\operatorname{Hess}(f)|_v \cdot w = \nabla(\partial_v \partial_w(f)),$$

where $\nabla(g)$ denotes the gradient of g. Moreover, $[v] \in \mathbb{P}^n$ is singular for $V(f)$ if and only if $\operatorname{Hess}(f)|_v \cdot v = 0$, or equivalently, ${\partial_v}^2(f) = 0$.

We conclude with an interesting connection between Hessian hypersurfaces and non-Lefschetz loci.

PROPOSITION 1.8. *Let $f \in S^3 \setminus \{0\}$ for which $X = V(f)$ is not a cone and consider the apolar ring $A = \mathcal{D}/\operatorname{Ann}_{\mathcal{D}}(f)$. The non-Lefschetz locus (in degree 1) of A is isomorphic to the Hessian hypersurface $\mathcal{H}$ associated to $X = V(f)$.*

PROOF. The non-Lefschetz locus $\mathcal{L}^c$ of A is given by the elements $[\partial_v] \in \mathbb{P}(A^1)$ for which there exists an element $[\partial_w] \in \mathbb{P}(A^1)$ such that $\partial_v \partial_w(f) = 0$. On the other hand, if $\partial_v \partial_w(f) = 0$, then by Lemma 1.7 we also have $\operatorname{Hess}(f)|_v \cdot w = 0$, which is equivalent to say that $[v] \in \mathcal{H}$. $\square$

With the notation as above, it is then clear from the Proposition that for the SAGA A the WLP_1 is satisfied and so the same holds for the SLP, since the two properties coincide for a SAGA with socle in degree 3. This also follows from one of the main results of [**MW09**].

2. WLP: Lifting and nilpotent elements

In this section, we deal with CI SAGAs presented by quadrics. In [**BF22**], by considering quotients of SAGAs over ideals generated by a non-Lefschetz element, the authors proved a lifting criterion for the weak Lefschetz property in degree 2.

More precisely:

THEOREM 2.1 ([**BF22**]). *Let R be a CI SAGA presented by quadrics, with codimension $n+1$, and let $[z],[w] \in \mathcal{L}^c$ such that $zw = 0$. Then:*

(a) *$\bar{R} = R/(z)$ is a CI SAGA of codimension n presented by quadrics and one has $(z) = (0 : w)$;*

(b) *If $\bar{R}$ satisfies WLP_2, then the same holds for R.*

The validity of the WLP_2 for CI SAGAs presented by quadrics and with codimension at most 5 is known: the case of codimension smaller than or equal to 4 follows from the famous theorem of Gordan and Noether (see [**GN76**]) by using the above-mentioned result of Maeno and Watanabe ([**MW09**]), whereas for the case of codimension 5 one can refer for example to [**AR19**] and [**BFP23a**]). When the codimension is 6, the question whether the WLP_2 is satisfied or not is still open only in the case where the Lefschetz locus is the whole $\mathbb{P}(R^1)$. By using Proposition 1.6 we can actually prove something more:

COROLLARY 2.2. *If a CI SAGA R presented by quadrics has codimension $5+k$ and $\#\mathcal{N}_2 \geq k$, then R satisfies the WLP_2.*

PROOF. If $[z_1], \ldots, [z_k] \in \mathcal{N}_2$ are distinct, by Proposition 1.6 they are in general position. Moreover, $\bar{z}_i^2 = 0$ and $\bar{z}_i \neq 0$ in $\bar{R} = R/(z_1)$ for $i = 2, \ldots, k$. By Theorem 2.1, $\bar{R}$ is again a CI SAGA presented by quadrics and $[\bar{z}_2], \ldots, [\bar{z}_k] \in \mathbb{P}(\bar{R}^1)$ are distinct nilpotent elements. Hence, they are in general position so one can iterate the process k times in order to conclude. □

By considering a CI SAGA R presented by quadrics, it seems to be not so easy to control the behaviour of the elements of $\mathcal{L}^c$ when passing to a quotient $\bar{R} = R/(z)$ of R, where $z \in \mathcal{L}^c \setminus \mathcal{N}_2$. Namely, it is not clear if their classes still represent non-Lefschetz elements in such a quotient and whether new non-Lefschetz or nilpotent elements can arise in $\bar{R}$. Answers to this kind of questions could give a refinement of the above corollary and could allow to follow a strategy to study the weak Lefschetz property in degree 2, which a priori does not stop after passing to successive quotients over ideals generated by all the nilpotent elements one has at the beginning of the process. Some problems of this type can be observed in the following example.

EXAMPLE 2.3. Consider the smooth cubic surface $X = V(f)$ where

$$f = (x_0^3 + x_1^3 + x_2^3 + 3x_0x_1x_2) + x_3^3.$$

The Jacobian ring R of X has non-Lefschetz locus $\mathcal{L}^c$ with two components: an irreducible plane cubic C in the plane $\mathbb{P}(\langle x_0, x_1, x_2 \rangle)$ and the point $\{[x_3]\}$. Moreover one has $\mathcal{N}_2 = \{[x_3]\}$. If one sets

$$z = x_0 - x_1, \quad w = x_0 + x_1 - x_2, \quad t = x_0 + 3x_2 \quad \text{and} \quad a = -4x_0 + 2x_2$$

one can see that the following relations hold in R:

$$z, w, t, a \neq 0 \qquad zw = 0, \qquad at \neq 0 \qquad \text{and} \qquad at \in (z).$$

Hence $[z], [w] \in \mathcal{L}^c$ while one can show that $[a]$ and $[t]$ are Lefschetz elements. On the other hand $[\bar{a}], [\bar{t}]$ are non-Lefschetz elements in $\bar{R} = R/(z)$. Clearly, $\bar{x}_3$ is a non-trivial nilpotent element in $\bar{R}$ but two "new" nilpotents arise. Indeed, one can see that $\bar{R} = R/(z)$ is the Jacobian ring of a Fermat curve so there are exactly 3 independent elements in $\bar{R}$ whose square is 0.

Our aim for this section is to give a generalization of claim (a) in Theorem 2.1 for a CI SAGA $R = S/I$ presented by quadrics. First of all, let us observe that for the above result, we considered non-Lefschetz elements (in degree 1), i.e. elements of $R^1 \setminus \{0\}$ whose associated multiplication map $R^1 \to R^2$ is not injective. By Proposition 1.5, if $z \in R^d$ is not 0, we have $\dim(K_z^1) \leq d$. In particular, $z \in R^1$ is a non-Lefschetz element exactly when K_z^1 has maximal dimension. We will focus our attention on the next case where this property is realized, namely we will analyze non-trivial elements in degree 2 whose associated kernels have dimension 2 (which is the maximum possible by Proposition 1.5).

To do this, let us firstly present an interesting correspondence between such elements in R^2 and pencils of quadrics in I whose base locus contains an $(n-2)$-plane.

More precisely, given a 2-dimensional subspace W of I_2 with base locus containing a $(n-2)$-plane Π, we would like to obtain an element in R^2, whose construction is realized as follows.
Let $\{L_0, L_1\} \subseteq S^1$ such that $V(L_0, L_1) = \Pi$. Then, we can chose a basis $\{q_0, q_1\}$ of W of the form

$$\begin{cases} q_0 = A_{00}L_0 + A_{01}L_1 \\ q_1 = A_{10}L_0 + A_{11}L_1 \end{cases} \tag{2.1}$$

where $A_{ij} \in S^1$. One can then define the matrix $M = [A_{ij}] \in \mathrm{Mat}_2(S^1)$ and the vectors $\underline{L} \in (S^1)^{\oplus 2}, \underline{q} \in (S^2)^{\oplus 2}$ in the obvious way in order to write the above relation in a compact form as $M\underline{L} = \underline{q}$. Although M depends on $\underline{L}$ and on the choice of $\underline{q}$ it is easy to see that $\Delta_{W,\Pi} = \det(M)$ is well defined[1] up to multiplication by elements of $\mathbb{K}^*$ and up to sums by elements in I. We define $\mathcal{P}$ to be the following set:

$$\mathcal{P} = \left\{ (W, \Pi) \;\middle|\; \begin{array}{c} W \subseteq I_2, \dim(W) = 2, \Pi \in \mathrm{Gr}(n-2, \mathbb{P}^n) \\ \text{with } \Delta_{W,\Pi} \notin I \text{ and } \Pi \subseteq \mathrm{BL}(|W|) \end{array} \right\}$$

By the above discussion, for any pair $(W, \Pi) \in \mathcal{P}$ we have a well defined element $[\Delta_{W,\Pi}] \in \mathbb{P}(R^2)$.

THEOREM 2.4. *Set* $\mathcal{X} = \{[\eta] \in \mathbb{P}(R^2) \mid \dim(K_\eta^1) = 2\}$. *For any* $[\eta] \in \mathcal{X}$ *denote by* $(K_\eta^1)_S$ *the ideal of* S *spanned by* $K_\eta^1 \subseteq R^1 = S^1$. *Then*

(a) *For all* $[\eta] \in \mathcal{X}$, $(K_\eta^1)_S \cap I_2$ *has dimension* 2*;*
(b) *The map* $\Delta : \mathcal{P} \to \mathcal{X}$ *such that* $\Delta : (W, \Pi) \mapsto [\Delta_{W,\Pi}]$ *is a bijection;*

In particular, the inverse φ *of* Δ *is such that* $\varphi([\eta]) = ((K_\eta^1)_S \cap I_2, V(K_\eta^1))$.

PROOF. It is clear that $\dim((K_\eta^1)_S \cap I_2) \geq 2$ since $[\eta] \in \mathcal{X}$. Claim (a) follows easily since I_2 is generated, as $\mathbb{K}$-vector space, by a regular sequence. The last claim follows easily from (a) and (b) so we can focus on (b). We denote by $\{g_0, \dots, g_n\}$ a regular sequence of elements of S^2 which generates I.

[1]The description $M\underline{L} = \underline{q}$ is not unique, even if we fix $\underline{L}$ and $\underline{q}$. Indeed, different descriptions of q_0 and q_1 as S-combinations of L_0 and L_1 are possible. More precisely, one can show that all the possible ways to present q_i are of the form $q_i = (A_{i0} + \lambda L_1)L_0 + (A_{i1} - \lambda L_0)L_1$ with $\lambda \in \mathbb{K}$.

Claim B_1: Δ is well defined. Let $(W, \Pi) \in \mathcal{P}$ and consider $M \in \mathrm{Mat}_2(S^1)$ and the vectors $\underline{L} \in (S^1)^{\oplus 2}, \underline{q} \in (S^2)^{\oplus 2}$ such that $M\underline{L} = \underline{q}$, as when we have defined $\Delta_{W,\Pi}$. By assumption $\Delta_{W,\Pi} \notin I$ so $\eta := (\Delta_{W,\Pi} + I) \in R^2$ is well defined and determined, up to scalars, by (W, Π). We want to show that $\eta L_i = 0$ in R. This can be done by multiplying both sides of the equation $M\underline{L} = \underline{q}$ by the adjoint M^* of M. Indeed, one obtains

$$\Delta_{W,\Pi}\underline{L} = \det(M)\underline{L} = M^*\underline{q}$$

so $\Delta_{W,\Pi}L_i$ is a S-linear combination of q_0 and q_1 and then it is in I. Hence, in R, we have $\eta L_i = 0$ so $\langle L_0, L_1\rangle \subseteq K^1_\eta$. On the other hand, since $\eta \neq 0$ by assumption, K^1_η has dimension at most 2 by Proposition 1.5, so $K^1_\eta = \langle L_0, L_1\rangle$ is 2-dimensional as claimed.

Claim B_2: There exists a section $\varphi : \mathcal{X} \to \mathcal{P}$ for Δ. Let $[\eta] \in \mathbb{P}(R^2)$ and assume that $K^1_\eta = \langle L_0, L_1\rangle$ is of dimension 2. We have an exact sequence

$$0 \to \bigwedge^2 I_2 \xrightarrow{\alpha} S^2 \otimes_{\mathbb{K}} I_2 \xrightarrow{\beta} I_4 \to 0 \tag{2.2}$$

which can be obtained from the Koszul sequence associated to the locally free sheaf $E = I_2 \otimes \mathcal{O}_{\mathbb{P}^n}$ and its section $(g_i)_{i=0}^n \in H^0(E) = \bigoplus_{i=0}^n I_2 \otimes H^0(\mathcal{O}_{\mathbb{P}^n})$. In the above sequence, β is simply the evaluation map of the tensor product and α is such that $\alpha(f_1 \wedge f_2) = f_1 \otimes f_2 - f_2 \otimes f_1$.

By assumption we have

$$\eta L_i = \sum_{j=0}^n a_{ij} g_j \tag{2.3}$$

for suitable $a_{ij} \in S^1$. Since $0 = (\eta L_0)L_1 = (\eta L_1)L_0$, we have $\sum_{j=0}^n (L_1 a_{0j} - L_0 a_{1j})g_j = 0$ and so

$$\psi = \sum_{j=0}^n (L_1 a_{0j} - L_0 a_{1j}) \otimes g_j$$

belongs to $\ker(\beta) = \mathrm{Im}(\alpha)$. Let W be the vector space generated by the quadrics $q_j = (L_1 a_{0j} - L_0 a_{1j})$ with $j = 0, \dots, n$. Using the exact sequence (2.2) it is easy to see that W is a subspace of I_2 and all its elements vanish on the $(n-2)$-plane $\Pi = V(L_0, L_1) = V(K^1_\eta)$.

We claim now that $\dim(W) = 2$. Indeed, it has dimension at most 2 since $W \subseteq I_2$ and I is generated by a regular sequence of quadrics. Moreover, if one assumes that $\dim(W) = 0$ we have $q_j = 0$ so $L_1 a_{0j} = L_0 a_{1j}$ for all j. On the other hand, since $a_{0j}, a_{1j}, L_0, L_1 \in S^1$ and L_0, L_1 are linearly independent we have that $a_{0j} = \sigma_j L_0$ and $a_{1j} = \sigma_j L_1$ with a suitable $\sigma_j \in \mathbb{K}$. Then, from Equation (2.3) we have

$$\eta L_0 = \sum_{j=0}^n a_{0j} g_j = \sum_{j=0}^n \sigma_j g_j L_0$$

so η is a $\mathbb{K}$-linear combination of the g_i's. This is impossible since, in this case, we would have $\eta = 0$ in R. Finally, assume that $\dim(W) = 1$. Hence, there exists $q \in I^2$ and $\delta_j \in \mathbb{K}$ such that $\delta_j q = q_j$ for all j. Then we can write ψ as $q \otimes g$ where $g = \sum_{j=0}^n \delta_j g_j$. Since $q \cdot g = \beta(\psi) = 0$ by construction we have either $q = 0$ or $g = 0$. In both cases, one gets that the dimension of W is 0 which is impossible.

It is easy to see that W does not depend on the choice of the expression (2.3): different choices of the linear forms a_i yield to the same vector space W. We set $\varphi([\eta]) = (W, V(K_\eta^1))$. In order to show that the map φ is well defined, we need only to prove that $\Delta_{W,\Pi} \notin I$

Up to a change of indices, we can assume that $W = \langle q_0, q_1\rangle$. The matrix M defined starting from $\underline{L} = [L_0, L_1]^T$ and $\underline{q} = [q_0, q_1]^T$ has determinant $\det(M) = a_{00}a_{11} - a_{01}a_{10}$. Hence we need to prove that $\det(M) \notin I$.

It is easy to see that up to changing the representative of $[\eta] \in R$ and up to a change of the regular sequence defining I, we can assume that

$$\eta L_0 = a_{00}g_0 + a_{01}g_1 \qquad \eta L_1 = a_{10}g_0 + a_{11}g_1, \tag{2.4}$$

Then, in a compact way, we can write $\eta\underline{L} = N\underline{g}$ where $N = [a_{ij}]_{i,j=0,1}$. In particular, notice that $\det(N) = \det(M)$.

Assume now that $\det(M) = \det(N) \in I$. If N^* is the adjoint of N we obtain $\eta N^*\underline{L} = N^*N\underline{g} = \det(M)\underline{g}$ which gives

$$\eta\gamma_0 = \det(M)g_0 \qquad \text{and} \qquad \eta\gamma_1 = \det(M)g_1 \tag{2.5}$$

with $\gamma_i \in S^2$. Since $g_0, g_1, \det(M) \in I$, we have that η cannot divide one of them. Hence we have that η is not irreducible: $\eta = xy$ for suitable $x, y \in S^1$. By the first of the above relations, we can assume $x|\det(M)$ and $y|g_0$. Since $\eta \not| \det(M)$ from the second relation, we get $y|g_1$. Hence g_0 and g_1 have a common linear factor, which is impossible since they are part of a regular sequence. This shows that φ is well-defined. It is easy to see, by Proposition 1.5, that $\Delta \circ \varphi = \mathrm{Id}_{\mathcal{X}}$.

Claim B_3: φ is an inverse for Δ. We only need to prove that $\varphi\circ\Delta = \mathrm{Id}_{\mathcal{P}}$. If $(W, \Pi) \in \mathcal{P}$ one can consider the matrix M and the vectors $\underline{L}, \underline{q}$ such that $M\underline{L} = \underline{q}$. We have $[\eta] = [\Delta(W, \Pi)] = [\det(M)]$. Since q_0, q_1 are independent and in I_2, we can complete them to a regular sequence $\{q_0, q_1, g_2, \ldots, g_n\}$ spanning I. Then the relations (2.3) used in order to define $\varphi([\eta]) = (W', \Pi')$ can be obtained by $\eta\underline{L} = M^*\underline{q}$. From this description one easily obtains that $(W', \Pi') = (W, \Pi)$. □

Finally let us now prove the natural generalization of Theorem 2.1(a).

Corollary 2.5. *Assume that $R = S/I$ is a CI SAGA presented by quadrics of codimension $n+1$. Assume that there exists an element $[\eta] \in \mathbb{P}(R^2)$ such that $\dim(K_\eta^1) = 2$. Let (K_η^1) be the ideal of R generated by K_η^1. Then $\bar{R} = R/(K_\eta^1)$ is a CI SAGA of codimension $n-1$ presented by quadrics and $(0 : \eta) = (K_\eta^1)$.*

Proof. Let us firstly denote by L_0, L_1 the two generators of K_η^1. By construction we have $(L_0, L_1) \subseteq (0 : \eta)$, so we can define an epimorphism of graded $\mathbb{K}$-algebras

$$\varphi : \bar{R} := R/(L_0, L_1) \to \tilde{R} := R/(0 : \eta).$$

By [**FP22**, Lemma 2.3], we get that $\tilde{R}$ is a SAGA of codimension $n-1$ and with socle in degree $\tilde{N} = n-1$. On the other hand, $\bar{R}$ is clearly an Artinian standard algebra of codimension $n-1$: to conclude the proof we need to show that it is also a CI SAGA presented by quadrics.

By Theorem 2.4 we know that we can associate to $[\eta]$ a 2-dimensional subspace $W \subseteq I_2 \cap (L_0, L_1)$. Denote by q_{n-1} and q_n two generators for W and complete $\{q_{n-1}, q_n\}$ to a basis of I_2 by adding $q_0, \ldots, q_{n-2}$. By construction, $q_i \notin (L_0, L_1)$ since $\{q_0, \ldots, q_n\}$ is a regular sequence and $q_{n-1}, q_n \in (L_0, L_1)$. Moreover it is easy to see that the reductions $\{\bar{q}_i\}_{i=0,\ldots,n-2}$ of the q_i's modulo (L_0, L_1) form a regular sequence in $\bar{S} = S/(L_0, L_1)$. Hence

$$\bar{R} = R/(L_0, L_1) = \frac{S/I}{(L_0, L_1)} \simeq \frac{\bar{S}}{(\bar{q}_0, \cdots, \bar{q}_{n-2})}$$

so $\bar{R}$ is a CI SAGA presented by quadrics. It has codimension $n-1$ and thus socle in degree $\bar{N} = n-1 = \tilde{N}$.

Assume, by contradiction, that φ is not injective: there is $x \in \bar{R}^k$ with $2 \leq k \leq n-1$ such that $\varphi(x) = 0$. Since $\bar{R}$ is a SAGA, there exists $y \in \bar{R}^{n-1-k}$ such that xy is a non-zero multiple of the generator $\bar{\sigma}$ of the socle $\bar{R}^{n-1}$. Since φ preserves the degrees and it is an epimorphism, the image of $\bar{\sigma}$ is a non-zero multiple of the generator $\tilde{\sigma}$ of $\tilde{R}^{n-1}$, i.e. the socle of $\tilde{R}$. Then $0 = \varphi(xy) = \varphi(\bar{\sigma}) = \lambda\tilde{\sigma}$ which is impossible. In particular, φ is an isomorphism and $(K^1_\eta) = (0 : \eta)$. □

Corollary 2.5 is a generalization of Theorem 2.1(a). We conjecture that also Theorem 2.1(b) can be generalized as follows:

CONJECTURE 2.6. *Let R and η be as in Corollary 2.5. If $\bar{R}$ satisfies WLP_2 the same holds for R.*

Actually, we expect that liftings for some Lefschetz properties in higher degree can be realized by exploiting the existence of an element η with K^1_η of maximal dimension and by studying the corresponding quotient.

3. Cubics and hessians via SAGAs

We now deal with the relation between linear factors of the hessian of a smooth cubic hypersurface and nilpotent elements of its Jacobian ring.

PROPOSITION 3.1. *Let $X = V(f)$ be a smooth cubic hypersurface in $\mathbb{P}^n$ and denote with $R = S/J_f$ its Jacobian ring. Then $[L] \in \mathcal{N}_2 \iff L|\operatorname{hess}(f)$.*

PROOF. $\Longrightarrow$) This implication is an easy consequence of the following more general fact. If $f \in S^d$ and $p \in S$ is such that $p^k \in J^{d-1}$ for a suitable $k \geq 2$, then, up to a change of coordinates we have $\partial_{x_0}(f) = p^k$. Hence $\partial_{x_i}\partial_{x_0}(f) = \partial_{x_i}(p^k) = kp^{k-1}\partial_{x_i}(p)$ so the first row of the Hessian matrix is divisible by p^{k-1} and the same holds for the hessian of f.

$\Longleftarrow$) Assume that $L \in S^1$ is such that $L|\operatorname{hess}(f)$. Consider the linear subsystem $|W|$ of $|J^2| \subset |\mathcal{O}_{\mathbb{P}^n}(2)|$ given by

$$W = \langle \partial_v(f) \,|\, L(v) = 0 \rangle.$$

Since $X = V(f)$ is smooth - in particular, it is not a cone - we have that the map $v \mapsto \partial_v(f)$ is injective so $\dim(W) = n$. Being $|W|$ spanned by n quadrics which can be completed to a regular sequence in J^2, we have that the base locus of $|W|$ has dimension 0. Notice that, by construction, all the elements of $|W|$ are singular. On the other hand, by Bertini, the general element is smooth outside $\mathrm{BL}(|W|)$, thus all the elements of $|W|$ are singular in at least a point of $\mathrm{BL}(|W|)$. Since the dimension of $\mathrm{BL}(|W|)$ is 0 there exists a point $[w] \in \mathbb{P}^n$ which is singular for all the elements of $|W|$. Equivalently,

$$\partial_w\partial_v(f) = 0 \qquad \text{for all } [v] \in V(L).$$

This implies that $\partial_v(\partial_w(f)) = 0$ for all $[v] \in V(L)$: $V(\partial_w(f))$ is a quadric whose singular locus contains the hyperplane $V(L)$ so $\partial_w(f)$ and L^2 are proportional. □

We stress some consequences of Proposition 1.6. Since nilpotent elements are in general position, (different) linear factors dividing hess(f) are linearly independent. In particular, we have that $V(f) \subseteq \mathbb{P}^n$ is a Fermat hypersurface if and only if hess(f) is the product of $n+1$ linearly independent factors. One might wonder whether powers of linear factors can divide the hessian of a cubic form and what happens when $\#\mathcal{N}_2 \leq n$. Let us now present some results in this direction.

PROPOSITION 3.2. *Assume that $X = V(f)$ is a smooth cubic hypersurface in $\mathbb{P}^n$ and let $R = S/J_f$ be its Jacobian ring. Then:*

(a) *if $|\mathcal{N}_2| \geq k \geq 1$, then, up to a change of coordinates we can write*
$$f = x_0^3 + \cdots + x_{k-1}^3 + p_3(x_k, \ldots, x_n)$$
where p_3 is a degree 3 polynomial defining a smooth hypersurface in $\mathbb{P}^{n-k}$;

(b) *if $[z] \in \mathcal{N}_2$, then $\bar{R} = R/(z)$ is the Jacobian ring of a smooth cubic $(n-2)$-fold.*

PROOF. Assume that $[L] \in \mathcal{N}_2$. Then, up to a change of coordinates we have $\partial_{x_0}(f) = L^2$. Notice that, in this case, $a = \partial_{x_0}(L) \neq 0$ as, otherwise, we would have ${\partial_{x_0}}^2(f) \equiv 0$ and this is impossible by Lemma 1.7. Then
$$f = \int L^2 dx_0 = \frac{1}{3a}L^3 + p_3(x_1, \ldots, x_n)$$
with p_3 of degree 3. Since the coefficient of x_0 in L is $a \neq 0$ the transformation $y_0 = L, y_i = x_i$ is a linear change of coordinates. It is immediate to show that if $[v_1 : \cdots : v_n] \in \mathbb{P}^{n-1}$ is a singular point for $V(p_3) \subseteq \mathbb{P}^{n-1}$, then $[0 : v_1 : \cdots : v_n]$ is singular for X. This proves claim (a) for $k = 1$.

Now (b) follows easily from the above argument: indeed, if $[z] \in \mathcal{N}_2$, then we can set linear coordinates $x_0 = z, x_1, \ldots, x_n$ such that $f = x_0^3 + p_3(x_1, \ldots, x_n)$ as above. Hence, a regular sequence defining J_f is $\{x_0^2, \partial_{x_1}(p_3), \ldots, \partial_{x_n}(p_3)\}$ so $\bar{R} = R/(z) \simeq \mathbb{K}[x_1, \ldots, x_n]/J_{p_3}$ as claimed.

One can now prove (a) for $k \geq 2$ by an induction process. Indeed, let us recall that by Proposition 1.6 the nilpotent elements $[z_0], \cdots, [z_k] \in \mathbb{P}(R^1)$ are in general position. Then we can write $f = x_0^3 + p_3(x_1, \ldots, x_n)$ with respect to some coordinates $x_0 = z_0, \ldots, x_k = z_k, x_{k+1}, \ldots, x_n$. If we set $\bar{R} = R/(x_0)$ we have that $\bar{R} = \mathbb{K}[x_1, \ldots, x_n]/J_{p_3}$ and the classes of $[z_1], \cdots, [z_k] \in \mathbb{P}(\bar{R}^1)$ are again nilpotent elements (and distinct) and thus in general position. □

COROLLARY 3.3. *Assume that $X = V(f)$ is a smooth cubic hypersurface in $\mathbb{P}^n$ and assume that $|\mathcal{N}_2| \geq n-1$. Then $|\mathcal{N}_2| = n+1$ and f is a Fermat cubic.*

PROOF. Assume that $|\mathcal{N}_2| \geq n-1$. By Proposition 3.2, we have
$$f = x_0^3 + \cdots + x_{n-2}^3 + p_3(x_{n-1}, x_n)$$
where p_3 is a degree 3 polynomial whose zero locus is given by 3 distinct points in $\mathbb{P}^1$ (since $V(p_3)$ is smooth). It is well known that such a binary form of degree 3 has Waring rank 2, i.e. it can be written as $L_0^3 + L_1^3$ with L_0 and L_1 independent linear forms. One can prove this directly as follows. Up to a change of coordinates, one can assume $p_3 = y_0 y_1(y_0 + y_1)$. By an easy and direct computation, one can

check that $\text{hess}(p_3) = -4(y_0^2+y_1^2+y_0y_1)$ which splits as product of non-proportional linear factors. Thus, $\text{hess}(f) = \Pi_{i=0}^{n-2} x_i \cdot \text{hess}(p_3)$ splits as a product of $n+1$ linearly independent factors. Then one concludes using Proposition 3.1 □

One has that the above result is sharp: indeed, Example 2.3 shows that a Jacobian ring can have $n-2$ nilpotent elements without being the Jacobian ring of a cubic Fermat hypersurface.

Finally, one can use similar arguments in order to construct examples of CI SAGAs presented by quadrics which are not Jacobian rings.

EXAMPLE 3.4. In $S = \mathbb{K}[x_0, \dots, x_n]$ consider the ideal I spanned by the regular sequence $\{x_0^2, \dots, x_{n-1}^2, x_n^2 + 2x_0x_1\}$. One can easily see that in $R = S/I$ there are exactly n distinct elements in $\mathcal{N}_2$: R can't be the Jacobian ring of a smooth cubic hypersurface by Corollary 3.3.

References

[AAI+23] N. Abdallah, N. Altafi, A. Iarrobino, A. Seceleanu, and J. Yaméogo, *Lefschetz properties of some codimension three Artinian Gorenstein algebras*, J. Algebra **625** (2023), 28–45, DOI 10.1016/j.jalgebra.2023.03.005. MR4567483 ↑126

[AR19] A. Alzati and R. Re, *Complete intersections of quadrics and the weak Lefschetz property*, Collect. Math. **70** (2019), no. 2, 283–294, DOI 10.1007/s13348-018-0230-1. MR3941957 ↑126, 128, 129, 131

[BF22] D. Bricalli and F. Favale, *Lefschetz properties for jacobian rings of cubic fourfolds and other Artinian algebras*, Collectanea Mathematica, posted on 2022, DOI 10.1007/s13348-022-00382-5. ↑126, 128, 129, 130, 131

[BFP23a] D. Bricalli, F. F. Favale, and G. P. Pirola, *A theorem of Gordan and Noether via Gorenstein rings*, Selecta Math. (N.S.) **29** (2023), no. 5, Paper No. 74, 25, DOI 10.1007/s00029-023-00882-7. MR4651997 ↑126, 128, 129, 131

[BFP23b] D. Bricalli, F. Favale, and G. P. Pirola, *On the Hessian of cubic hypersurfaces*, arXiv:2305.10871 (2023). ↑130

[BMMRN18] M. Boij, J. Migliore, R. M. Miró-Roig, and U. Nagel, *The non-Lefschetz locus*, J. Algebra **505** (2018), 288–320, DOI 10.1016/j.jalgebra.2018.03.006. MR3789915 ↑126

[CG80] J. A. Carlson and P. A. Griffiths, *Infinitesimal variations of Hodge structure and the global Torelli problem*, Journées de Géometrie Algébrique d'Angers, Juillet 1979/Algebraic Geometry, Angers, 1979, Sijthoff & Noordhoff, Alphen aan den Rijn—Germantown, Md., 1980, pp. 51–76. MR605336 ↑126

[CGGH83] J. Carlson, M. Green, P. Griffiths, and J. Harris, *Infinitesimal variations of Hodge structure. I*, Compositio Math. **50** (1983), no. 2-3, 109–205. MR720288 ↑126

[DGI20] A. Dimca, R. Gondim, and G. Ilardi, *Higher order Jacobians, Hessians and Milnor algebras*, Collect. Math. **71** (2020), no. 3, 407–425, DOI 10.1007/s13348-019-00266-1. MR4129535 ↑126

[FP22] F. F. Favale and G. P. Pirola, *Infinitesimal variation functions for families of smooth varieties*, Milan J. Math. **90** (2022), no. 1, 209–228, DOI 10.1007/s00032-022-00353-2. MR4443117 ↑134

[GN76] P. Gordan and M. Nöther, *Ueber die algebraischen Formen, deren Hesse'sche Determinante identisch verschwindet* (German), Math. Ann. **10** (1876), no. 4, 547–568, DOI 10.1007/BF01442264. MR1509898 ↑131

[Gri68] P. A. Griffiths, *Periods of integrals on algebraic manifolds. I. Construction and properties of the modular varieties*, Amer. J. Math. **90** (1968), 568–626, DOI 10.2307/2373545. MR229641 ↑126

[GZ18] R. Gondim and G. Zappalà, *Lefschetz properties for Artinian Gorenstein algebras presented by quadrics*, Proc. Amer. Math. Soc. **146** (2018), no. 3, 993–1003, DOI 10.1090/proc/13822. MR3750213 ↑126

[HMM+13] T. Harima, T. Maeno, H. Morita, Y. Numata, A. Wachi, and J. Watanabe, *The Lefschetz properties*, Lecture Notes in Mathematics, vol. 2080, Springer, Heidelberg, 2013, DOI 10.1007/978-3-642-38206-2. MR3112920 ↑126, 127, 128

[HMNW03] T. Harima, J. C. Migliore, U. Nagel, and J. Watanabe, *The weak and strong Lefschetz properties for Artinian K-algebras*, J. Algebra **262** (2003), no. 1, 99–126, DOI 10.1016/S0021-8693(03)00038-3. MR1970804 ↑126

[Mac94] F. S. Macaulay, *The algebraic theory of modular systems*, Cambridge Mathematical Library, Cambridge University Press, Cambridge, 1994. Revised reprint of the 1916 original; With an introduction by Paul Roberts. MR1281612 ↑127

[MW09] T. Maeno and J. Watanabe, *Lefschetz elements of Artinian Gorenstein algebras and Hessians of homogeneous polynomials*, Illinois J. Math. **53** (2009), no. 2, 591–603. MR2594646 ↑127, 130, 131

[MMRO13] E. Mezzetti, R. M. Miró-Roig, and G. Ottaviani, *Laplace equations and the weak Lefschetz property*, Canad. J. Math. **65** (2013), no. 3, 634–654, DOI 10.4153/CJM-2012-033-x. MR3043045 ↑126

[MMRN11] J. C. Migliore, R. M. Miró-Roig, and U. Nagel, *Monomial ideals, almost complete intersections and the weak Lefschetz property*, Trans. Amer. Math. Soc. **363** (2011), no. 1, 229–257, DOI 10.1090/S0002-9947-2010-05127-X. MR2719680 ↑126

[MN13] J. Migliore and U. Nagel, *Gorenstein algebras presented by quadrics*, Collect. Math. **64** (2013), no. 2, 211–233, DOI 10.1007/s13348-012-0076-x. MR3041764 ↑126

Dipartimento di Matematica, Università degli Studi di Pavia, Via Ferrata, 5 I-27100 Pavia, Italy

Email address: davide.bricalli@unipv.it

Dipartimento di Matematica, Università degli Studi di Pavia, Via Ferrata, 5 I-27100 Pavia, Italy

Email address: filippo.favale@unipv.it

Contemporary Mathematics
Volume **805**, 2024
https://doi.org/10.1090/conm/805/16131

Decompositions and Terracini loci of cubic forms of low rank

Luca Chiantini and Fulvio Gesmundo

Abstract. We study Waring rank decompositions for cubic forms of rank $n+2$ in $n+1$ variables. In this setting, we prove that if a concise form has more than one non-redundant decomposition of length $n+2$, then all such decompositions share at least $n-3$ elements, and the remaining elements lie in a special configuration. Following this result, we give a detailed description of the $(n+2)$-th Terracini locus of the third Veronese embedding of n-dimensional projective space.

1. Introduction

For an integer n, let V be a complex vector space of dimension $n+1$ and let $\mathbb{P}^n = \mathbb{P}V$ be the projective space of lines in V. Identify V with the space of linear forms on V^* and let S^dV be the space of homogeneous polynomials of degree d on V^*. The Waring rank of a form $F \in S^dV$ is

$$\mathrm{R}(F) = \min\{r : F = L_1^d + \cdots + L_r^d \text{ for some } L_1, \ldots, L_r \in V\}.$$

The d-th Veronese embedding of $\mathbb{P}V$ is the map $v_d : \mathbb{P}V \to \mathbb{P}S^dV$ defined by $v_d([L]) = [L^d]$. If A is a set of r points in $\mathbb{P}V$, we say that A has length r, and we write $\ell(A) = r$. Geometrically, the Waring rank of F can be equivalently defined as

$$\mathrm{R}(F) = \min\{r : [F] \in \langle v_d(A)\rangle \text{ for a set of points } A \subseteq \mathbb{P}V \text{ with } \ell(A) = r\}.$$

Here $\langle - \rangle$ denotes the (projective) linear span. A finite set A with $\ell(A) = r$ such that $[F] \in \langle \nu_d(A)\rangle$ is a *decomposition* of F of length r. If $A = \{[L_1], \ldots, [L_r]\}$, this is equivalent to the existence of an expression

$$F = \alpha_1 L_1^d + \cdots + \alpha_r L_r^d \tag{1}$$

for some $\alpha_1, \ldots, \alpha_r \in \mathbb{C}$. We say that the decomposition is *non-redundant* (or irredundant) if there is no proper subset $A' \subsetneq A$ such that $[F] \in \langle v_d(A')\rangle$; this is equivalent to saying that $L_1^d, \ldots, L_r^d$ are linearly independent and no coefficient α_i in (1) is 0. We say that the decomposition is *minimal* if its length is the minimal possible, namely $r = \mathrm{R}(F)$. If F is a form of rank r which has a unique minimal decomposition of length r, we say that F is identifiable.

2020 *Mathematics Subject Classification.* Primary 14N07, 14N05, 15A69.

Key words and phrases. Waring rank, tensor decomposition, Hilbert function, Terracini locus.

We say that $F \in S^dV$ is concise if there is no proper subspace $V' \subsetneq V$ such that $F \in S^dV'$; equivalently, there is no change of coordinates in V after which F can be written in fewer than $n+1$ variables. It is a classical fact, implicitly appearing in **[Syl52]**, that F is concise if and only if its first order partial derivatives are linearly independent; equivalently F is concise if and only if its partial derivatives of order $d-1$ span V; we refer to **[IK99, Car06]** for a modern discussion on this topic. If F is concise, then every decomposition A of F satisfies $\langle A \rangle = \mathbb{P}V$; in other words, no decomposition of F is contained in a hyperplane. In particular $\mathrm{R}(F) \geq n+1$. A partial converse of this property in low rank is given in Lemma 2.7.

For a subset $A \subseteq \mathbb{P}V$ with $\ell(A) = r$, we say that A is linearly independent if $\dim\langle A \rangle = \ell(A) - 1$, which is the maximum possible value; here dim is the dimension in projective space. The Kruskal rank of A is the largest integer k such that every subset of k elements of A is linearly independent. We say that A is in *linear general position* (LGP) if any subset of (at most) $n+1$ elements of A is linearly independent, namely the Kruskal rank of A is $\min\{\ell(A), n+1\}$, which is the maximum possible value.

In this work, we study non-redundant decompositions of length $n+2$ for concise cubic forms in $n+1$ variables. This is the first case where Kruskal's criterion does not guarantee uniqueness of the decomposition and in fact there are examples of forms having multiple decompositions, see e.g. **[Der13]**. However, our main result shows that if the form is concise, then non-uniqueness of a decomposition can only occur under very special conditions, and in particular very low Kruskal rank.

THEOREM 1.1. *Let V be a vector space of dimension $n+1$, with $n \geq 1$. Let $F \in S^3V$ be concise with* $\mathrm{R}(F) \leq n+2$. *Then one of the following occurs:*

(I) *there is a unique non-redundant decomposition A of length $n+2$ for F; in this case the Kruskal rank of A is at least* 4;
(II) *there are infinitely many non-redundant decompositions of F of length $n+2$; any two decompositions A, B of length $n+2$ intersect in at least $n-3$ points; moreover $(A \setminus B) \cup (B \setminus A)$ lies in the union of two lines; in particular the Kruskal rank of any decomposition is* 2;
(III) *there are infinitely many non-redundant decompositions of F of length $n+2$; any two decompositions A, B of length $n+2$ intersect in at least $n-2$ points; moreover $(A \setminus B) \cup (B \setminus A)$ is contained in the union of two planes; in particular the Kruskal rank of any decomposition is at most* 3.

Theorem 1.1 is built on an induction argument for $n \geq 4$. The base of the induction is given by the cases $n = 1, 2, 3$, where special configurations of points yield some pathological behaviour. These special configurations already appear in the literature **[CS11, BGI11, BB13, Bal17]** and in some cases they already appear implicitly in the classical **[Syl52, Ter11]**. The general cases are obtained as extensions of a special configuration on a subspace; this result is original to the best of our knowledge.

As a consequence of Theorem 1.1, in Section 5, we obtain a detailed description of the $(n+2)$-th Terracini locus of the Veronese variety $v_3(\mathbb{P}^n)$ in the concise setting.

The study of identifiability of homogeneous polynomials, and more generally of tensors, is a classical topic in algebraic geometry and invariant theory. Generic forms in S^dV are identifiable only for few combination of degree d and number of variables $n+1$, namely $(d, n) = (2k-1, 1), (3, 3), (5, 2)$: identifiability in these cases was known classically **[Syl52, Hil]**; the full proof that these are the only cases where generic identifiability holds is recent [**GM19**]. On the other hand, generic forms of subgeneric rank are identifiable with only few exceptions [**BCO14, COV17**]. In general, however, little is known about the explicit genericity condition providing identifiability. The most general tool in this context is Kruskal's criterion [**Kru77**], which guarantees uniqueness of decomposition under the hypothesis of large enough *Kruskal rank*; a variant of Kruskal criterion is given in [**LP23**]. A geometric study of the subvarieties of non-identifiable forms, in special cases, is given in [**ACV18, AC20, CO23**], which are based on properties of the Hilbert function of points [**Chi19**] and on apolarity theory [**IK99**].

Acknowledgments. This work is partially supported by the Thematic Research Programme "Tensors: geometry, complexity and quantum entanglement", University of Warsaw, Excellence Initiative – Research University and the Simons Foundation Award No. 663281 granted to the Institute of Mathematics of the Polish Academy of Sciences for the years 2021-2023. We thank IMPAN and the AGATES organizers for providing an excellent research environment throughout the semester.

2. Preliminaries

Let $X \subseteq \mathbb{P}^N$ be an irreducible projective variety. The r-th secant variety of X is

$$\sigma_r(X) = \overline{\{p \in \mathbb{P}^N : p \in \langle x_1, \dots, x_r \rangle \text{ for some } x_1, \dots, x_r \in X\}};$$

here the closure can be taken equivalently in the Zariski or the Euclidean topology. The r-th symmetric product of X is $X^{(r)} = X^{\times r}/\mathfrak{S}_r$, where the symmetric group $\mathfrak{S}_r$ acts by permuting the factors of $X^{\times r}$. One can verify that $X^{(r)}$ is a projective variety. The r-th abstract secant variety of X is

$$A\sigma_r(X) = \overline{\{((x_1, \dots, x_r), p) : p \in \langle x_1, \dots, x_r \rangle\}} \subseteq X^{(r)} \times \mathbb{P}^N.$$

There are two natural projections $\pi_X : A\sigma_r(X) \to X^{(r)}$, which is surjective, and $\pi_\sigma : A\sigma_r(X) \to \mathbb{P}^N$, which surjects onto the secant variety $\sigma_r(X)$. It is easy to verify that $A\sigma_r(X)$ is irreducible and $\dim A\sigma_r(X) = r \dim X + (r-1)$; in fact, $A\sigma_r(X)$ is birational to a projective bundle over $X^{(r)}$ whose fibers are copies of $\mathbb{P}^{r-1}$. We say that $\sigma_r(X)$ is non-defective, or that X is not r-defective, if $\dim \sigma_r(X) = \min\{N, r \dim X + (r-1)\}$: if $\sigma_r(X) \subsetneq \mathbb{P}^N$, this is equivalent to the condition that π_σ has generically finite fibers.

DEFINITION 2.1. Let $X \subseteq \mathbb{P}^N$ be an irreducible variety and let $r \geq 1$ be an integer such that $N \geq r \dim X + (r-1)$. The r-th Terracini locus of X is

$$\mathbb{T}_r(X) = \overline{\left\{ (x_1, \dots, x_r) : \begin{array}{c} x_i \in X^{\mathrm{sm}} \text{ are linearly independent,} \\ T_{x_1}X, \cdots, T_{x_r}X \text{ are not linearly independent} \end{array} \right\}} \subseteq X^{(r)},$$

where X^{sm} is the open set of the smooth points of X and T_xX denotes the affine tangent space to X at x.

This object was introduced in [**BC21, BBS23**] as a tool to study identifiability of tensors and singularities of secant varieties. Terracini's Lemma [**BCC**$^+$**18**, Lemma 1] guarantees that if $\sigma_r(X)$ is non-defective and it does not fill the space, then $\mathbb{T}_r(X)$ is a proper subvariety of $X^{(r)}$.

The preimage $\pi_X^{-1}(\mathbb{T}_r(X)) \subseteq A\sigma_r(X)$ is contained in the locus where the differential of π_σ drops rank. If a point $q \in \mathbb{P}^N$ is in the image $\pi_\sigma \pi_X^{-1}(\mathbb{T}_r(X)) \subseteq \sigma_r(X)$ then either $\pi_\sigma^{-1}(q)$ is positive dimensional, or q is a cuspidal singularity of $\sigma_r(X)$.

By definition, the Terracini locus is the sets of points $\{x_1, \ldots, x_r\}$ with the property that they impose independent conditions on $\mathcal{O}_X(1)$ but the double points $\{2x_1, \ldots, 2x_r\}$ do not impose independent conditions on $\mathcal{O}_X(1)$. In fact, with this formulation, one can give the definition of Terracini locus for an abstract variety and any (ample) line bundle, see [**BC21**, **BV23**].

An important tool in the study of sets of point in projective space is given by the Hilbert function. Let $\mathbb{C}[V] = \mathrm{Sym}(V^*)$ be the homogeneous coordinate ring of $\mathbb{P}V$, which is a polynomial ring in $n+1$ variables. If $Z \subseteq \mathbb{P}V$ is a subvariety with homogeneous ideal $I(Z)$, the Hilbert function of Z is

$$\begin{aligned} h_Z : \mathbb{N} &\to \mathbb{N} \\ t &\mapsto \dim(\mathbb{C}[V]_t / I(Z)_t). \end{aligned}$$

If Z is a set of points, $Z = \{[v_1], \ldots, [v_r]\}$ for $v_j \in V$, let $\mathrm{ev}_i : \mathbb{C}[V] \to \mathbb{C}$ be the evaluation map at v_i. For every t, the restriction $\mathrm{ev}_i : \mathbb{C}[V]_t \to \mathbb{C}$ is linear, and the Hilbert function of Z is characterized as

$$h_Z(t) = \mathrm{rank}((\mathrm{ev}_1, \ldots, \mathrm{ev}_r) : \mathbb{C}[V] \to \mathbb{C}^r).$$

The *first difference* of the Hilbert function of Z is $Dh_Z(t) = h_Z(t) - h_Z(t-1)$. It is often identified simply with the sequence of its non-zero values, which is called the h-vector of Z. We record three main properties of the Hilbert function and the h-vector, which will be useful in the following. We refer to [**Chi19**] for details and proofs.

PROPOSITION 2.2. *Let $Z' \subseteq Z \subseteq \mathbb{P}V$ be finite sets of points. Then*

(i) *$Dh_{Z'}(t) \leq Dh_Z(t)$ for every t; in particular $h_{Z'}(t) \leq h_Z(t)$ for every t.*
(ii) *There is an integer $\tau \geq 0$ such that $h_Z(t)$ is strictly increasing for $t \leq \tau$ and $h_Z(t) = \ell(Z)$ for $t \geq \tau$. In particular $Dh_Z(t) > 0$ if $t \leq \tau$ and $Dh_Z(t) = 0$ if $t > \tau$.*
(iii) *If $Dh_Z(t) \leq t$, then $Dh_Z(t+1) \leq Dh_Z(t)$.*

Understanding what functions h can occur as Hilbert functions of sets of points is the object of a long line of research. Macaulay characterized the *maximal growth* of h_Z. In [**BGM94**], strong consequences of Macaulay's result are given; we state here a restricted version of [**BGM94**, Theorem 3.6], extending a result from [**Dav85**] in the case of $\mathbb{P}^2$.

THEOREM 2.3. *Let $Z \subseteq \mathbb{P}V$ be a finite set of points. If $s := Dh_Z(t_0) = Dh_Z(t_0+1) \leq t_0$ for some t_0 then there exists a reduced curve $C \subseteq \mathbb{P}V$ of degree s such that, setting $Z' = Z \cap C$, one has*

- *$Dh_{Z'}(t) = Dh_C(t)$ for $t \leq t_0 + 1$;*
- *$Dh_{Z'}(t) = Dh_Z(t)$ for $t \geq t_0$.*

Intuitively, Theorem 2.3 says that if Dh_Z is constant and small on an interval, then a subset of points of Z lies on a low degree curve. This is useful to deduce pathologies of certain sets of points and prove that under suitable genericity assumptions only certain Hilbert functions are possible.

Sets of points defining non-redundant decompositions satisfy particular conditions that are reflected in their Hilbert function. An important one concerns the

Cayley-Bacharach property. A finite set $Z \subseteq \mathbb{P}V$ satisfies the Cayley-Bacharach property in degree t, denoted $CB(t)$, if, for every $p \in Z$, every form of degree d vanishing on $Z \setminus \{p\}$ vanishes at p as well; in other words, Z satisfy $CB(t)$ if for every $p \in Z$, $I(Z \setminus \{p\})_t = I(Z)_t$.

The role of the Cayley-Bacharach property in the study of decompositions of homogeneous polynomials is stated in the following result.

PROPOSITION 2.4 ([**AC20**], Proposition 2.25). *Let $F \in S^dV$ and let $A, B \subseteq \mathbb{P}V$ be non-redundant disjoint decompositions of F. The $A \cup B$ satisfies the Cayley-Bacharach property in degree d.*

In particular, the failure to satisfy the Cayley-Bacharach property allows one to rule out the existence of certain decompositions. This will be achieved using the fact that the Cayley-Bacharach property poses strong conditions on the Hilbert function of a set of points. More precisely, the following holds, see [**ACV18**, Thm. 4.9].

THEOREM 2.5. *Let $Z \subseteq \mathbb{P}V$ be a set of points satisfying $CB(t)$. Then, for every $s \leq t+1$*

$$Dh_Z(0) + \cdots + Dh_Z(s) \leq Dh_Z(t+1) + \cdots + Dh_Z(t+1-s).$$

2.1. First results. In this section we prove some simple technical results toward the proof Theorem 1.1. The first immediate observation is that $n+2$ points in LGP in $\mathbb{P}V$ can be normalized via the action of $\mathrm{GL}(V)$. We record a slightly more general fact in the following result, see e.g. [**Har92**, Ch. 10].

LEMMA 2.6. *Let V be a vector space with $\dim V = n+1$ and let $e_0, \dots, e_n$ be a basis of V. Let*

$$\Omega = \{A \subseteq \mathbb{P}V : \ell(A) = n+2, \langle A \rangle = \mathbb{P}V\} \subseteq (\mathbb{P}V)^{(n+2)}.$$

Then Ω is Zariski open in $(\mathbb{P}V)^{(n+2)}$; the action of $\mathrm{GL}(V)$ on Ω has exactly n orbits $K_2, \dots, K_{n+1}$. The elements of K_r are subsets having Kruskal rank exactly r and they are all equivalent to

$$A_r = \{[e_0], \dots, [e_n], [e_0 + \cdots + e_{r-1}]\}.$$

Moreover, $K_r \subseteq \overline{K_{r+1}}$ and for every r, $\overline{K_r}$ is irreducible of dimension $n(n+1)+r-1$.

PROOF. The first part of the statement is classical. If $A \in (\mathbb{P}V)^{(n+2)}$ is a set of points having Kruskal rank r and such that $\langle A \rangle = \mathbb{P}V$, then we may assume $A = \{[v_0], \dots, [v_n], [v_{n+1}]\}$ with $v_0, \dots, v_n$ linearly independent and $v_{n+1} \in \langle v_0, \dots, v_{r-1} \rangle$. Since $v_0, \dots, v_n$ are linearly independent, there is an element $g \in \mathrm{GL}(V)$ such that $gv_i = e_i$ for $i = 0, \dots, n$; as a result $gv_{n+1} = \lambda_0 e_0 + \cdots + \lambda_{r-1} e_{r-1}$ for some $\lambda_j \in \mathbb{C} \setminus \{0\}$. Define an element $h \in \mathrm{GL}(V)$ by $he_i = \lambda_i^{-1} e_i$ for $i \leq r-1$ and $he_i = e_i$ for $i \geq r$. We conclude that $hgA = A_r$.

The condition that $K_r \subseteq \overline{K_{r+1}}$ is clear; moreover $\overline{K_r}$ is irreducible, because it is the closure of an orbit under the action of $\mathrm{GL}(V)$. Finally, to compute the dimension of $\overline{K_r}$, consider the diagram

$$\begin{array}{ccc} & (\mathbb{P}V)^{n+2} & \\ {\scriptstyle \pi_{\mathfrak{S}}} \swarrow & & \searrow {\scriptstyle \pi_{n+2}} \\ (\mathbb{P}V)^{(n+2)} & & (\mathbb{P}V)^{n+1} \end{array}$$

where $\pi_{\mathfrak{S}}$ is the projection onto the quotient $(\mathbb{P}V)^{(n+2)} = (\mathbb{P}V)^{n+2}/\mathfrak{S}_{n+2}$ and π_{n+2} is the projection onto the first $n+1$ factors. The preimage $\pi_{\mathfrak{S}}(K_r)$ is the union of several irreducible components, all isomorphic and all surjecting onto K_r; one of these components is

$$J_r = \left\{ ([v_0], \ldots, [v_{n+1}]) \in (\mathbb{P}V)^{n+2} : \begin{array}{c} \mathbb{P}V = \langle [v_0], \ldots, [v_n] \rangle \\ v_{n+1} \in \langle v_0, \ldots, v_{r-1} \rangle \end{array} \right\}$$

Since $\pi_{\mathfrak{S}}$ is generically finite, $\dim K_r = \dim J_r$. The restriction $\pi_{n+2} : J_r \to (\mathbb{P}V)^{n+1}$ is dominant; the fiber over a generic element $([v_0], \ldots, [v_n])$ is the linear span $\mathbb{P}\langle v_0, \ldots, v_{r-1} \rangle$ which has dimension $r-1$. We conclude

$$\dim K_r = \dim J_r = \dim((\mathbb{P}V)^{n+1}) + \dim \mathbb{P}^{r-1}(r-1) = n(n+1) + r - 1.$$

This concludes the proof. □

We observe that forms admitting non-redundant decompositions of length $n+2$ are necessarily concise:

LEMMA 2.7. *Let $F \in S^d V$ be a homogeneous form with* $\dim V = n+1$. *Let $A \subseteq \mathbb{P}V$ be a set of $n+2$ points with $\langle A \rangle = \mathbb{P}V$. If A is a non-redundant decomposition of F then F is concise.*

PROOF. By Lemma 2.6, we may assume $A \in K_r$ for some r and we can easily reduce to the case $r = n$. Write $L = x_0 + \cdots + x_n$ and assume

$$F = \alpha_0 x_0^d + \cdots + \alpha_n x_n^d + \alpha_{n+1} L^d$$

where $\alpha_j \neq 0$ for every j.

Let H be the space of $(d-1)$-th order partial derivatives of F. Up to scaling, we have

$$\frac{\partial^{d-1}}{\partial x_0^{d-2} \partial x_1} F = \alpha_{n+1} L;$$

hence $L \in H$. Moreover, $\frac{\partial^{d-1}}{\partial x_j^d} F = \alpha_j x_j + \alpha_{n+1} L$, showing $x_j \in H$ for every j. This shows $H = V$, therefore F is concise. □

Similarly, forms admitting non-redundant decompositions of length exactly $n+1$ are concise, and their decomposition is unique. The uniqueness of the decomposition can be obtained via an easy calculation with the Hilbert function, but it is also a consequence of the classical Kruskal criterion for tensors, [**Kru77**].

LEMMA 2.8. *Let $F \in S^d V$ be a homogeneous form, with* $\dim V = n+1$ *with $d \geq 3$. Let $A \subseteq \mathbb{P}V$ be a non-redundant decomposition of F with A linearly independent in $\mathbb{P}V$ and $\ell(A) = n+1$. Then F is concise,* $\mathrm{R}(F) = n+1$ *and A is the unique minimal decomposition of F.*

PROOF. Since A is linearly independent with $\ell(A) = n+1$, we may normalize it via the action of $\mathrm{GL}(V)$ and assume $A = \{[x_0], \ldots, [x_n]\}$. Therefore $\langle v_d(A) \rangle = \langle [x_0^d], \ldots, [x_n^d] \rangle$ and $F = x_0^d + \cdots + x_n^d$. This shows that F is concise with $\mathrm{R}(F) = n+1$.

The uniqueness of the decomposition follows by Kruskal's criterion, see [**Kru77**]. The proof follows from the immediate fact that if $A \subseteq \mathbb{P}V$ is a set of linearly elements with $\ell(A) = n+1$, then A is in LGP and in particular it has maximal Kruskal ranks. □

A stronger result holds for forms admitting two non-redundant decompositions of length $n+1$ and $n+2$, respectively.

LEMMA 2.9. *Let $F \in S^dV$ be a concise form with $d \geq 3$. Let $A \subseteq \mathbb{P}V$, $B \subseteq \mathbb{P}V$ be non-redundant decompositions of F of length $n+1$ and $n+2$ respectively. Then $d = 3$, $\ell(A \cap B) \geq n-1$ and $(A \setminus B) \cup (B \setminus A)$ is collinear.*

PROOF. We proceed by induction on n. If $n = 1$ the statement is clearly true. Since 5 points on a rational normal curve of degree $d \geq 4$ are linearly independent, we immediately have $d = 3$. In this case, every concise $F \in S^3V$ has at most one decomposition of length 2 and a two-parameter family of decompositions of length 3. All decompositions are collinear because $\mathbb{P}V$ is a line.

Assume $n \geq 2$. First assume $A \cap B = \emptyset$. Let $Z = A \cup B$. Since F is concise, we have $Dh_A = (1, n)$, which implies $Dh_Z(0) + Dh_Z(1) = n+1$; by the Cayley-Bacharach property of Z, we have $Dh_Z(d+1) + Dh_Z(d) \geq n+1$. This implies $Dh_Z(2) \leq 1$, which arithmetically is a contradiction by Proposition 2.2.

Therefore $A \cap B \neq \emptyset$. Without loss of generality, assume $A = \{[x_0], \ldots, [x_n]\}$ and $F = x_0^d + \cdots + x_n^d$. Since $A \cap B \neq \emptyset$, assume $[x_n] \in B$ and write $B = \{[L_0], \ldots, [L_n], [x_n]\}$, so that $F = L_0^d + \cdots + L_n^d + \beta x_n^d$ for some nonzero $\beta \in \mathbb{C}$. Consider

$$F' = F - \beta x_n^d = x_0^d + \cdots + x_{n-1}^d + (1-\beta)x_n^d = L_0^d + \cdots + L_n^d. \tag{2}$$

Let $B' = \{[L_0], \ldots, [L_n]\}$; clearly $[x_n] \notin B'$. If $\beta \neq 1$, then F' is concise and (2) gives two non-redundant decompositions F' of length $n+1$. By Corollary 2.8, the two decompositions are the same, in contradiction with the fact that $[x_n] \notin B'$. Therefore $\beta = 1$ and F' is non-concise. Moreover, if B' was linearly independent in $\mathbb{P}V$, Corollary 2.8 would imply that F' is concise; hence B' is not linearly independent.

Let $V' = \langle x_0, \ldots, x_{n-1} \rangle$. Notice $V' = \langle B' \rangle$ and F' is concise in S^dV'. Let $A' = \{[x_0], \ldots, [x_{n-1}]\}$. Now F' is concise and it has two decompositions A', B' of length n and $n+1$ respectively. By the inductive hypothesis, we deduce $d = 3$, $\ell(A' \cap B') \geq n-2$ and B' has three collinear points. Since $A = A' \cup \{[x_n]\}$ and similarly for B, we deduce that the statement holds for A and B as well. This concludes the proof. □

In the study of identifiability of forms, it is often useful to consider the case where two potential decompositions of a form are disjoint. We illustrate a general method to reduce the case of non-disjoint decompositions to the one of disjoint decompositions:

REMARK 2.10. Let $F \in S^dV$ and suppose $A, B \subseteq \mathbb{P}V$ give two non-redundant decompositions of F. Write $A = \{[L_1], \ldots, [L_r]\}$ and $B = \{[M_1], \ldots, [M_s]\}$. Then

$$F = \alpha_1 L_1^d + \cdots + \alpha_r L_r^d = \beta_1 M_1^d + \cdots + \beta_s M_s^d.$$

After possibly reordering the elements of A and B, assume $L_i = M_i$ for $i = 1, \ldots, p$, where $p = \ell(A \cap B)$. In this case, define

$$F' = F - (\alpha_1 L_1^d + \cdots + \alpha_p L_p^d) = (\beta_1 - \alpha_1)M_1^d + \cdots + (\beta_p - \alpha_p)M_p^d + \beta_{p+1}M_{p+1} + \cdots + \beta_s M_s^d;$$

note

$$F' = \alpha_{p+1} L_{p+1}^d + \cdots + \alpha_r L_r^d.$$

Let $A' = A \setminus B$; let $B' = \{[M_i] : \alpha_i \neq \beta_i, i = 1, \dots, p\} \cup \{[M_{p+1}], \dots, [M_s]\}$; then A', B' are two disjoint non-redundant decompositions of F'. Moreover, if A is a minimal decomposition for F, then A' is a minimal decomposition for F'.

We conclude this section with a classical result for which we do not know an explicit reference, in the form that we need in the following. In the case $n = 1$, the result dates back to [**Syl52**]; analogous statements, in a more general setting, are discussed in [**BBCG19**, Section 7.1].

PROPOSITION 2.11. *Let F be a concise form in $S^d V$, with $\dim V = n+1$. Let A, B be non-redundant decompositions of F of length r and s respectively. Then $r + s \geq d + 2n$.*

PROOF. Let p be the length of $A \cap B$. If $p = 0$, set $F' = F$. If $p > 0$, consider the form F' with disjoint decompositions A', B' of length r', s' respectively, as constructed in Remark 2.10. Notice that $r - r' = p \geq s - s'$; specifically $s - s'$ is the number of elements of $A \cap B$ appearing with the same coefficient in the two decompositions of F. Observe $\dim\langle B' \rangle \geq n - (s - s')$: indeed F is concise, which implies that B spans $\mathbb{P}V$; since $\dim\langle B \setminus B' \rangle \leq s - s' - 1$, we deduce $\dim\langle B' \rangle \geq n - (s - s')$.

Set $Z = A' \cup B'$ and let $n' = \dim\langle Z' \rangle$. We have $n' \geq \dim\langle B' \rangle \geq n - (s - s')$, so $n - n' \leq s - s' \leq r - r'$. Since Z is union of two disjoint decompositions of a form of degree d, we have $Dh_{Z'}(d+1) \geq 1$, which implies $Dh_{Z'}(t) \geq 1$ for $t \leq d+1$. Moreover, $Dh_{Z'}(1) = \dim\langle Z' \rangle = n'$. Finally, Proposition 2.4 implies that the Cayley-Bacharach property holds for Z' in degree d, and Theorem 2.5 implies $Dh_{Z'}(d+1) + Dh_{Z'}(d) \geq Dh_{Z'}(0) + Dh_{Z'}(1) \geq 1 + n'$.

We deduce $Dh_{Z'} = (1, n', \delta_2, \dots, \delta_{d+1}, \dots)$ with $\delta_t \geq 1$ for $t \leq d+1$ and $\delta_{d+1} + \delta_d \geq 1 + n'$. We conclude $r' + s' = \ell(Z') \geq \sum_{t=0}^{d+1} \delta_t \geq 2 + 2n' + (d-2) = d + 2n'$. Therefore

$$r + s = (r - r') + (s - s') + (r' + s') \geq (n - n') + (n - n') + d + 2n' = d + 2n$$

as desired. □

The bound of Proposition 2.11 is sharp. An example where the bound is attained can be constructed, for every n and d, as follows. When $d = 2$, it is a classical fact [**Ter11**] that general forms of maximal rank have many decompositions of length $n + 1$, which attains the bound.

For higher values of d, write $V = V' \oplus V''$ where with $\dim V' = 2$, $\dim V' = n - 1$. Let let $F' \in S^d V'$ be a binary form having two non-redundant disjoint decompositions A', B' of length r', s' with $r' + s' = d + 2$; it is a classical result that such forms exist [**Syl52, CS11**]. Let $F'' = L_1^d + \cdots + L_{n-1}^d$ be a sum of $n - 1$ d-powers of generic linear forms. Then $F = F' + F''$ is concise and it has two distinct non-redundant decompositions $A = A' \cup \{L_1, \dots, L_{n-1}\}$ and $B = B' \cup \{L_1, \dots, L_{n-1}\}$ with $\ell(A) = r' + (n - 1)$, $\ell(B) = s' + (n - 1)$ and we have $\ell(A) + \ell(B) = r' + s' + 2(n - 1) = d + 2n$.

3. First cases

In this section, we characterize decompositions of length $n+2$ in $\mathbb{P}^n$ for $n = 1, 2$.

Case $n = 1$.

This case is classical, see e.g. [**CS11, BGI11**]. Let V be a vector space with $\dim V = 2$. The generic Waring rank in $\mathbb{P}S^3V$ is 2. In this case Theorem 1.1 is easily verified. Let $F \in S^3V$ be a concise form. If $\mathrm{R}(F) = 2$, then F has a unique decomposition of length 2 by Kruskal criterion, and in particular Corollary 2.8; moreover F has a 2-dimensional family of decompositions of length 3. If $\mathrm{R}(F) = 3$, then $[F]$ is an element of $\tau(v_3(\mathbb{P}^1)) \setminus v_3(\mathbb{P}^1)$, where $\tau(v_3(\mathbb{P}^1)) = \{[L_1^2L_2] \in \mathbb{P}S^3V : [L_1], [L_2] \in \mathbb{P}^1\}$; in this case F has a two dimensional family of decompositions of length 3.

Since $\mathbb{P}V = \mathbb{P}^1$, all decompositions are defined by collinear points. In particular, both case (II) and case (III) of Theorem 1.1 trivially hold.

Case $n = 2$. Let V be a vector space with $\dim V = 3$. The generic Waring rank in $\mathbb{P}S^3V$ is 4. Let $F \in S^3V$ be a concise form. It is known that $3 \le \mathrm{R}(F) \le 5$ [**LT10**] and we are interested in the dense subset of forms satisfying $\mathrm{R}(F) \le 4$. Notice $\dim A\sigma_4(v_3(\mathbb{P}^2)) = 12$. Therefore the projection map $A\sigma_4(v_3(\mathbb{P}^2)) \to \mathbb{P}S^3V$ has generic fibers of dimension 2. This guarantees that if $\mathrm{R}(F) \le 4$, then F has at least a 2-dimensional family of decompositions of length 4.

In particular, case (III) of Theorem 1.1 trivially holds. We provide an explicit geometric characterization of the decompositions, which implicitly appears in the proof of [**BGI11**, Thm. 5].

THEOREM 3.1. *Let $F \in S^3V$ be a concise form with* $\dim V = 3$. *Let A, B be non-redundant decompositions of F, with $\ell(A) = \ell(B) = 4$. Then one of the following holds:*

- *$A \cap B = \emptyset$ and there is a conic passing through $A \cup B$; if A has three collinear points, so does B.*
- *$\ell(A \cap B) \ge 1$ and all non-redundant decompositions of F of length 4 have three collinear points, on a line Λ.*

PROOF. The statement follows from Proposition 3.2 and Proposition 3.3 below. Proposition 3.2 analyzes the case where A and B are disjoint and Proposition 3.3 the one where they are not. □

PROPOSITION 3.2. *Let $F \in S^3V$ be a concise form with* $\dim V = 3$. *Let A, B be non-redundant decompositions of F, with $\ell(A) = \ell(B) = 4$. If A and B are disjoint then there is a unique conic C vanishing on $A \cup B$. Moreover, if Λ is a line with $\ell(A \cap \Lambda) \ge 3$, then $C = \Lambda \cup \Lambda'$ for a line Λ' and $\ell(B \cap \Lambda') = 3$.*

PROOF. Since F is concise, there are no linear forms vanishing on A. Therefore the h-vector of A is $Dh_A = (1, 2, 1)$ and there is a pencil of conics vanishing on A.

Let $Z = A \cup B$; since $A \cap B = \emptyset$, $\ell(Z) = 8$. Since $[F] \in \langle v_3(A) \rangle \cap \langle v_3(B) \rangle$, we have that $v_3(Z)$ is not linearly independent, namely $\dim\langle v_3(Z)\rangle < \ell(Z) - 1 = 7$. In particular, $Dh_Z(4) > 0$. Since $Dh_A = (1, 2, 1)$, we have $Dh_Z(0) + Dh_Z(1) \ge 3$. Since A, B are non-redundant, they satisfy the Cayley-Bacharach property in degree 3 by Proposition 2.4: therefore Theorem 2.5 provides $Dh_Z(3) + Dh_Z(4) \ge 3$, so $Dh_Z(2) \le 2$. If $Dh_Z(2) \le 1$, by Proposition 2.2, we have $Dh_Z(i) \le 1$ for

$i \geq 2$, which provides a contradiction; therefore $Dh_Z(2) = 2$. We obtain $Dh_Z = (1,2,2,2,1)$. In particular, $h_Z(2) = 1+2+2 = 5$, which implies that there is a unique conic vanishing on Z.

Now suppose A contains three collinear points, lying on a line $\Lambda \subseteq \mathbb{P}V$. Since $A \subseteq C$, we deduce that the conic C passing through Z is reducible with $C = \Lambda \cup \Lambda'$ for some other line $\Lambda' \subseteq \mathbb{P}V$. We are going to show that $B \cap \Lambda'$ consists of three points. Suppose by contradiction $B \cap \Lambda'$ contains fewer than three points; since $B \subseteq C$, we deduce that $B \cap \Lambda$ contains at least two points. Notice $h_Z(3) = 1+2+2+2 = 7 = \ell(Z) - 1$: this guarantees $\dim(\langle v_3(A)\rangle \cap \langle v_3(B)\rangle) = 0$ therefore the two linear spans only intersect at $[F]$. However,

$$\ell(\Lambda \cap Z) = \ell(\Lambda \cap A) + \ell(\Lambda \cap B) \geq 3 + 2 = 5;$$

therefore $v_3(\Lambda \cap Z)$ is linearly dependent. This implies $\langle v_3(A\cap\Lambda)\rangle \cap \langle v_3(B\cap\Lambda)\rangle \neq \emptyset$, in contradiction with the fact that the two spans only intersect at $[F]$, which is not an element of $\langle v_3(\Lambda)\rangle$ because F is concise. This shows that $B \cap \Lambda'$ contains three points. □

The case where A and B are not disjoint is characterized in the following result.

PROPOSITION 3.3. *Let $F \in S^3V$ be a concise form with $\dim V = 3$. Let A, B be non-redundant decompositions of F, with $\ell(A) = \ell(B) = 4$. If $\ell(A \cap B) \geq 1$, then there is a line $\Lambda \subseteq \mathbb{P}V$ such that $A \setminus \Lambda = B \setminus \Lambda$ consists of exactly one point. In particular, both A and B have three collinear points on Λ.*

PROOF. Let $p = \ell(A \cap B)$. Let F', A', B' be constructed from F, A, B as in Remark 2.10. Then A' is a non-redundant decomposition of F' of length $4 - p \leq 3$ and B' is a non-redundant decomposition of F' of length at most 4.

Suppose A' is not contained in a line. In particular, $\ell(A') = 3$ and F' is concise by Corollary 2.8. Moreover A' is the unique decomposition of length 3. This contradicts the existence of B'.

Therefore there exists $V' \subseteq V$, with $\dim V' = 2$ and $A' \subseteq \Lambda = \mathbb{P}V'$. In particular $F' \in S^3V'$ is not concise in S^3V. If F' is not concise in S^3V', then $F' = L^3$ for some $L \in V$; therefore $\{[L]\}$ is a decomposition of F' and by Proposition 2.11, any other non-redundant decomposition of F' must have length at least 4. On the other hand $\ell(B') \leq 3$, because if $\ell(B') = 4$, then $B = B'$, and F' would be concise by Lemma 2.7. Therefore, the condition $\ell(B') \leq 3$ provides a contradiction with Proposition 2.11. This guarantees F' is concise in S^3V'.

To conclude, we analyze two possibilities. If $\ell(A') = 2$, then A' is the unique decomposition of F' of length 2 and $\ell(B') = 3$. In this case $\ell(A \cap B) = 2$ and it consists of one point on Λ and one not in Λ, because B' contains three of the four points of B and $B' \subseteq \Lambda$. In particular $A \setminus \Lambda = B \setminus \Lambda = \{[L]\}$ with $F = F' + L^3$. If $\ell(A') = 3 = \ell(B') = 3$, then $A \cap B = A \setminus \Lambda = B \setminus \Lambda = \{[L]\}$ with $F = F' + L^3$. □

4. The general case

In this section, we give the proof of Theorem 1.1. This will follow from an induction argument, where the cases $n = 1, 2, 3$ give the base of the induction. We provide first a result which analyzes the case $n = 3$.

If $\dim V = 4$, the generic Waring rank in $\mathbb{P}S^3V$ is 5. Sylvester's Pentahedral Theorem [**Syl52, Cle61**] guarantees that a generic form in S^3V has a unique decomposition. We first study the situation of two disjoint decompositions, see also [**Bal17**].

PROPOSITION 4.1. *Let* $\dim V = 4$. *Let* $F \in S^3V$ *be a concise form. Let* $A, B \subseteq \mathbb{P}V$ *be two disjoint non-redundant decompositions of* F *with* $\ell(A), \ell(B) = 5$. *Then* $A \cup B$ *is contained in the union of two lines.*

PROOF. Since $A \cap B = \emptyset$, the union $Z = A \cup B$ is a set of length 10. Since F is concise, the decomposition A is not contained in a hyperplane; therefore $Dh_A(0) = 1$, $Dh_A(1) = 3$. Since A, B are non-redundant, by Proposition 2.4 Z satisfies the Cayley-Bacharach property in degree 3. Therefore $Dh_Z(4) + Dh_Z(3) \geq Dh_Z(0) + Dh_Z(1) \geq 4$. We deduce $Dh_Z(2) \leq 2$, which implies $Dh_Z(3), Dh_Z(4) \leq 2$ by Proposition 2.2(iii). This guarantees $Dh_Z = (1, 3, 2, 2, 2)$.

We apply Theorem 2.3: we have $Dh_Z(3) = Dh_Z(4) = 2 \leq 2$. Therefore there is a curve C of degree 2 such that, setting $Z' = Z \cap C$, one has $Dh_{Z'}(t) = Dh_C(t)$ for $t \leq 4$. Since $\deg(C) = 2$, C is either a plane conic or a union of two lines. If C is a plane conic, we have $Dh_{Z'}(t) = (1, 2, 2, 2, 2, \delta_5)$, with $\delta_5 = 0, 1$; in this case $Z' = Z \cap C$ contains at least 9 points, therefore either $A \subseteq C$ or $B \subseteq C$, in contradiction with the conciseness of F, which implies $\langle A \rangle = \langle B \rangle = \mathbb{P}V$. Therefore C is union of two lines, and we have $Dh_{Z'} = (1, 3, 2, 2, 2)$; this guarantees $Z = Z' \subseteq C$, as desired. $\square$

On the other hand, when $n \geq 4$, two distinct decompositions always intersect, as shown in the next result.

PROPOSITION 4.2. *Let* $n \geq 4$ *and let* V *be a vector space with* $\dim V = n + 1$. *Let* $F \in S^3V$ *be a concise form. Let* A *be a non-redundant decomposition of* F *of length* $n + 2$ *and let* B *be a non-redundant decomposition of* F *of length* $s \leq n + 2$. *Then* $A \cap B \neq \emptyset$.

PROOF. Assume $B \cap A = \emptyset$, so that $Z = A \cup B$ is a set of length at most $2n+4$. Since F is concise, then A is not contained in a hyperplane; therefore $Dh_A(0) = 1$, $Dh_A(1) = n$. Since A, B are non-redundant, Z satisfies the Cayley-Bacharach property; therefore $Dh_Z(4) + Dh_Z(3) \geq Dh_Z(0) + Dh_Z(1) \geq n + 1$. We deduce $Dh_Z(2) \leq 2$, which implies $Dh_Z(3), Dh_Z(4) \leq 2$; this provides a contradiction, for $n \geq 4$. $\square$

The proof of Theorem 1.1 will follow because in the setting of Proposition 4.2 one can guarantee that the intersection between two decompositions must be surprisingly large, similarly to Lemma 2.9.

PROPOSITION 4.3. *Let* V *be a vector space with* $\dim V = n + 1$. *Let* $F \in S^3V$ *be a concise form. Let* A *be a non-redundant decomposition of* F *of length* $n + 2$ *and let* B *be a non-redundant decomposition of* F *of length* $s \leq n + 2$. *Then one of the following holds:*

- $\ell(A \cap B) \geq n - 2$, *and the points of* $(A \setminus B) \cup (B \setminus A)$ *are contained in the union of two planes;*
- $\ell(A \cap B) \geq n - 3$ *and the points of* $(A \setminus B) \cup (B \setminus A)$ *are contained in the union of two lines.*

PROOF. If $s = n+1$, then the statement follows immediately from Lemma 2.9. Therefore assume $s = n+2$.

Let $A = \{[L_1], \ldots, [L_{n+2}]\}$, $B = \{[M_1], \ldots, [M_{n+2}]\}$. We proceed by induction on n. The case $n = 1$ and $n = 2$ are clearly verified. So assume $n \geq 3$.

If $n = 3$ and $A \cap B = \emptyset$, then by Proposition 4.1 condition (ii) is verified. If $n > 3$, then $A \cap B \neq \emptyset$ by Proposition 4.2. Therefore assume $A \cap B \neq \emptyset$ with $[L_{n+2}] = [M_{n+2}] \in A \cap B$.

Write, as usual,

$$\begin{aligned} F &= \alpha_1 L_1^3 + \cdots + \alpha_{n+1} L_{n+1}^3 + \alpha_{n+2} L_{n+2}^3 = \\ &= \beta_1 M_1^3 + \cdots + \beta_{n+1} M_{n+1}^3 + \beta_{n+2} L_{n+2}^3. \end{aligned}$$

Let $F' = F - \beta_{n+2} L_{n+2}^3$. Then F' has a decomposition $B' = B \setminus \{[L_{n+2}]\}$ of length $n+1$.

If B' is linearly independent, by Corollary 2.8, we deduce that F' is concise and B' is its unique decomposition of length $n+1$. If $\alpha_{n+2} \neq \beta_{n+2}$, then A defines a non-redundant decomposition of F' of length $n+2$; in this case Lemma 2.9 applies, and we deduce $\ell(A \cap B') \geq n-1$ and $A \setminus B'$ is collinear. Since $A \setminus B' = A \setminus B$, condition (ii) holds. If $\alpha_{n+2} = \beta_{n+2}$, then $A' = A \setminus \{[L_{n+2}]\}$ is a decomposition of F' of length $n+1$; by Kruskal's criterion Corollary 2.8, we conclude $A' = B'$ which implies $A = B$ and the statement follows.

Now assume B' is not linearly independent, which implies that F' is not concise. Let $V' \subseteq V$ be the subspace in which F' is concise. Since F is concise, we have $\dim V' = \dim V - 1 = n$, and $\langle B' \rangle = V'$. Now, if $\alpha_{n+2} \neq \beta_{n+2}$, A defines a non-redundant decomposition of F' of length $n+2$. Moreover, $\langle A \rangle = V$. By Lemma 2.7, this would imply that F' is concise, which is a contradiction. Therefore $\alpha_{n+2} = \beta_{n+2}$.

We reduced to the following setting. The form F' is concise in $S^3 V'$, and A', B' are two non-redundant decompositions of F' of length $n+1$. Therefore the induction hypothesis applies. We deduce that condition (i) or condition (ii) hold for A' and B'. Since $A = A' \cup \{[L_{n+2}]\}$ and $B = B' \cup \{[L_{n+2}]\}$, we deduce that condition (i) or condition (ii) holds for A and B as well. □

The proof of Theorem 1.1 is obtained directly from the results of this section.

PROOF OF THEOREM 1.1. If F has a unique decomposition A of length $n+2$, then the Kruskal rank of A must be at least 4. Indeed, let $A' \subseteq A$ be a minimal linearly dependent set, so that the Kruskal rank of A equals $\ell(A') - 1$. Let $A = A' \cup A''$: A'' is linearly independent and $\langle v_3(A') \rangle \cap \langle v_3(A'') \rangle = \emptyset$. Correspondingly $F = F' + F''$, with A' non-redundant decomposition of F' and A'' non-redundant decomposition of F''. If $\ell(A') \leq 4$, then F' has many decompositions of length $\ell(A')$, by the discussion on the case $n = 2$. This implies that A is not the unique decomposition of F, in contradiction with the assumption.

If F does not have a unique decomposition, then either case (II) or case (III) occur, by Proposition 4.3. □

We record two easy consequence of Theorem 1.1.

COROLLARY 4.4. *Let $F \in S^3V$ be a concise form admitting at least two distinct non-redundant decompositions of length $n+2$. Then the Kruskal rank of any non-redundant decomposition of length $n+2$ is at most 3. Moreover, the family of decompositions of F of length $n+2$ has dimension at least 2.*

PROOF. Since F has at least two non-redundant decompositions of length $n+2$, either case (II) or case (III) of Theorem 1.1 is verified. This guarantees every non-redundant decomposition of F of length $n+2$ has Kruskal rank at most 3.

Write $A = \{[L_0], \ldots, [L_{n+1}]\}$ for a non-redundant decomposition of length $n+2$ and assume $[L_0], \ldots, [L_3]$ are linearly dependent. Write $F = L_0^3 + \cdots + L_{n+1}^3$ and define $F' = L_0^3 + \cdots + L_3^3$. Then $F' \in S^3V'$ with $\dim V' = 3$ and it has a non-redundant decomposition of length 4. From the discussion on the case $n = 2$, we deduce that F' has a 2-dimensional family of decomposition of length 4. For every such decomposition B', define $B = B' \cup \{[L_4], \ldots, [L_{n+1}]\}$. Then B is a non-redundant decomposition of F, showing that the family of decompositions of F has dimension at least 2. □

As a byproduct of these results, we completely characterize the genericity condition of Sylvester's Pentahedral Theorem. Theorem 4.5 can be regarded as an extension of Theorem 3.1 and of some of the results of [**BGI11**].

THEOREM 4.5. *Let $\dim V = 4$ and let $F \in S^3V$ be a concise form. Then the following are equivalent:*

(i) *F has at least two decompositions as sum of five powers of linear forms;*
(ii) *F has a 2-dimensional family of decompositions as sum of five powers of linear forms;*
(iii) *$F = F' + L^3$ for $F' \in S^3V'$ where $V' \subseteq V$ is a subspace of dimension 3 and L is a linear form with $L \notin V'$.*

PROOF. *(i) ⇒ (iii).* Since the decomposition is not unique, either condition (I) or condition (II) of Theorem 1.1 hold. If condition (II) holds, then $F = F' + L^3$ as desired. If condition (I) holds, then $F = F'' + L_1^3 + L_2^3$ where $F'' \in S^3V''$ for a subspace $V'' \subseteq V$ of dimension 2; setting $F' = F'' + L_1^3$ we obtain the desired expression.

(iii) ⇒ (ii) Let $F = F' + L^3$. Every non-redundant decomposition of F' provides a non-redundant decomposition of F. Since $\dim V' = 3$, F' has at least a 2-dimensional family of decompositions.

(ii) ⇒ (i) This is trivially satisfied. □

COROLLARY 4.6. *Let* $\dim V = 4$. *A form $F \in S^3V$ has a unique decomposition as sum of five powers of linear forms if and only if*

$$[F] \notin \mathbf{J}(\mathrm{Sub}_3, v_3(\mathbb{P}V))$$

where Sub_3 *denotes the variety of non-concise forms and* $\mathbf{J}(-,-)$ *denotes the geometric join of two varieties.*

The variety Sub_3 is called *subspace variety* in [**Lan12**, Ch. 3] and it is defined by the vanishing of 4×4 minors of the first partial derivative map, see, e.g., [**Car06**]. We refer to [**Har92**, Ch. 8] for the definition and the basic properties of the join of two varieties.

5. Characterization of the Terracini locus

In this section, we characterize a subvariety of the $(n+2)$-th Terracini locus of the third Veronese embedding of $\mathbb{P}^n$. Define the *r-th concise Terracini locus* to be the subvariety of the Terracini locus arising from points in $\mathbb{P}^n$ whose span is the entire space. More precisely, for a vector space V of dimension $n+1$, define

$$\mathbb{T}_r^{\text{con}}(v_3(\mathbb{P}V)) = \overline{\mathbb{T}_r(v_3(\mathbb{P}V)) \cap \{v_3(A) : A \in \mathbb{P}V^{(r)} \text{ and } \langle A \rangle = \mathbb{P}V\}}.$$

It is clear that if $\mathbb{T}_r^{\text{con}}(v_3(\mathbb{P}V))$ is non-empty, then it is (union of) irreducible components of $\mathbb{T}_r(v_3(\mathbb{P}V))$. Theorem 1.1 allows us to describe $\mathbb{T}_{n+2}^{\text{con}}(v_3(\mathbb{P}V))$ as the closure of the set $v_3(K_3) \subseteq \mathbb{P}V^{(n+2)}$, where K_3 is the orbit, for the action of $\mathrm{GL}(V)$, described in Lemma 2.6. In particular, this guarantees that it is irreducible.

THEOREM 5.1. *The concise Terracini locus $\mathbb{T}_{n+2}^{\text{con}}(v_3(\mathbb{P}V))$ is the closure of*

$$v_3(K_3) = \{v_3(A) \in v_3(\mathbb{P}V)^{(n+2)} : A \text{ concise with four coplanar points}\}. \tag{3}$$

In particular, $\mathbb{T}_{n+2}^{\text{con}}(v_3(\mathbb{P}V))$ is irreducible of dimension $n(n+1)+2$.

The proof of Theorem 5.1 is based on Lemma 2.6 and two technical results presented below, Lemma 5.2 and Lemma 5.3.

LEMMA 5.2. *Let $d \geq 3$. Let $\mathbb{P}V_1, \mathbb{P}V_2 \subseteq \mathbb{P}V$ be disjoint linear spaces. For $i = 1, 2$, let*

$$H_i = \langle T_{[M^d]} v_d(\mathbb{P}V) : M \in V_i \rangle.$$

Then $\mathbb{P}H_1 \cap \mathbb{P}H_2 = \emptyset$.

PROOF. We have $T_{[M^d]} v_d(\mathbb{P}V) = V \cdot M^{d-1}$. Therefore

$$H_i = \langle T_{[M^d]} v_d(\mathbb{P}V) : M \in V_i \} \rangle = V \cdot S^{d-1} V_i;$$

since $V_1 \cap V_2 = 0$ and $d \geq 3$, we deduce $\mathbb{P}H_1 \cap \mathbb{P}H_2 = 0$. □

The next result is essentially contained in [**CC10**, Lemma 5.10].

LEMMA 5.3. *Let $V' \subseteq V$ be a proper subspace of V and let $L_1, \dots L_k \in V'$. If the tangent spaces to $v_d(\mathbb{P}V)$ at $[L_1^d], \dots, [L_k^d]$ are linearly dependent, so are the tangent spaces to $v_d(\mathbb{P}V')$ to $[L_1^d], \dots, [L_k^d]$.*

PROOF. If $[L_1^{d-1}], \dots, [L_k^{d-1}]$ are linearly dependent then the statement is clearly satisfied. Therefore suppose they are independent. If $T_{[L_1^d]} v_d(\mathbb{P}V), \dots, T_{[L_k^d]} v_d(\mathbb{P}V)$ are linearly dependent, then there exist $M_1, \dots, M_k \in V$, not all zero, such that

$$M_1 L_1^{d-1} + \dots + M_k L_k^{d-1} = 0.$$

In fact, we show $M_j \in V'$ for every j. To see this, let V'' be a complement to V', so that $V = V' \oplus V''$; write $M_j = M_j' + M_j''$ for some $M_j' \in V'$, $M_j'' \in V''$. We have

$$\begin{aligned} &M_1 L_1^{d-1} + \dots + M_k L_k^{d-1} = \\ &(M_1' L_1^{d-1} + \dots + M_k' L_k^{d-1}) + (M_1'' L_1^{d-1} + \dots + M_k'' L_k^{d-1}) = 0. \end{aligned}$$

Since $L_j \in V'$, the two summands in the expression above are elements of $S^d V'$ and $V'' \cdot S^{d-1} V'$ respectively; since these two subspaces are linearly independent, each summand must vanish. Since $V' \cap V'' = 0$, the elements of $V'' \cdot S^{d-1} V'$ can be regarded as elements of $V'' \otimes S^{d-1} V' \subseteq S^d V$; in particular, we have $M_1'' \otimes L_1^{d-1} + \dots + M_k'' \otimes L_k^{d-1} = 0$. If some M_j'' is nonzero, we deduce that $L_1^{d-1}, \dots, L_k^{d-1}$ are linearly dependent, which contradicts the initial assumption. Therefore $M_j'' = 0$ for every j and $M_j = M_j'$. □

We can now provide a complete proof of Theorem 5.1.

PROOF OF THEOREM 5.1. Let $\Omega \subseteq \mathbb{P}V^{(n+2)}$ be the subset of sets of points spanning the entire $\mathbb{P}V$. By definition of concise Terracini locus, $v_3(\Omega)\cap\mathbb{T}^{\mathrm{con}}_{n+2}(v_3(\mathbb{P}V))$ is Zariski open in $\mathbb{T}^{\mathrm{con}}_{n+2}(v_3(\mathbb{P}V))$ and non-empty if $\mathbb{T}^{\mathrm{con}}_{n+2}(v_3(\mathbb{P}V))$ is non-empty. We prove that $v_3(\Omega) \cap \mathbb{T}^{\mathrm{con}}_{n+2}(v_3(\mathbb{P}V)) = v_3(K_3)$.

Let $A \in \Omega$ and write $A = \{[L_1], \dots, [L_{n+2}]\}$. Let r be the Kruskal rank of A: by Lemma 2.6, $A \in K_r$ and it can be normalized as $A = \{[x_0], \dots, [x_n], [x_0 + \cdots + x_{r-1}]\}$. Write $T_j = T_{[x_j]}v_3(\mathbb{P}V)$, $T_+ = T_{[x_0+\cdots+x_{r-1}]}v_3(\mathbb{P}V)$; let $V' = \langle x_0, \dots, x_{r-1}\rangle$ and let $T_j' = T_{[x_j]}v_3(\mathbb{P}V')$, $T_+' = T_{[x_0+\cdots+x_{r-1}]}v_3(\mathbb{P}V')$.

If $v_3(A)$ is an element of the Terracini locus, then $T_0, \dots, T_n, T_+$ are linearly dependent. By Lemma 5.2, the same must hold for the tangent spaces $T_0, \dots, T_{r-1}, T_+$. Further, by Lemma 5.3, the same must hold for $T_0', \dots, T_{r-1}', T_+'$. In particular, it is enough to $A' \subseteq A$ defined by $A' = \{[x_0], \dots, [x_{r-1}], [x_0 + \cdots + x_{r-1}]\} \in (\mathbb{P}V')^{(r+2)}$.

We deduce that $v_3(A')$ belongs to the concise Terracini locus $\mathbb{T}^{\mathrm{con}}_{r+2}(v_3(\mathbb{P}V'))$. By the action of $\mathrm{GL}(V')$, the same holds for every set of $r + 2$ points in linear general position in $\mathbb{P}V'$. Passing to the closure, we obtain $v_3(\mathbb{P}V')^{(r+2)}$ is entirely contained in the Terracini locus. By Terracini's Lemma, we deduce that all fibers of $\pi_\sigma : A\sigma_{r+2}(v_3(\mathbb{P}V')) \to \mathbb{P}S^3V'$ have positive dimension. Theorem 1.1 guarantees that this is the case only if $r \leq 3$. This shows that $A \subseteq K_3$, or equivalently that it has four coplanar points. We obtained $v_3(\Omega) \cap \mathbb{T}^{\mathrm{con}}_{n+2}(v_3(\mathbb{P}V)) \subseteq v_3(K_3)$.

The reverse inclusion holds because of Theorem 1.1 and this concludes the proof of the equality in (3).

Finally, by Lemma 2.6, we conclude that $\mathbb{T}^{\mathrm{con}}_{n+2}(v_3(\mathbb{P}^n)) = \overline{v_3(K_3)}$ is irreducible, and of dimension $n(n + 1) + 2$. □

References

[AC20] E. Angelini and L. Chiantini, *On the identifiability of ternary forms*, Linear Algebra Appl. **599** (2020), 36–65, DOI 10.1016/j.laa.2020.03.042. MR4083806

[ACV18] E. Angelini, L. Chiantini, and N. Vannieuwenhoven, *Identifiability beyond Kruskal's bound for symmetric tensors of degree 4*, Atti Accad. Naz. Lincei Rend. Lincei Mat. Appl. **29** (2018), no. 3, 465–485, DOI 10.4171/RLM/817. MR3819100

[Bal17] E. Ballico, *The stratification by rank for homogeneous polynomials with border rank 5 which essentially depend on five variables*, Acta Math. Vietnam. **42** (2017), no. 3, 509–531, DOI 10.1007/s40306-017-0211-6. MR3667462

[BB13] E. Ballico and A. Bernardi, *Stratification of the fourth secant variety of Veronese varieties via the symmetric rank*, Adv. Pure Appl. Math. **4** (2013), no. 2, 215–250, DOI 10.1515/apam-2013-0015. MR3069955

[BBCG19] E. Ballico, A. Bernardi, M. Christandl, and F. Gesmundo, *On the partially symmetric rank of tensor products of W-states and other symmetric tensors*, Atti Accad. Naz. Lincei Rend. Lincei Mat. Appl. **30** (2019), no. 1, 93–124, DOI 10.4171/RLM/837. MR3941503

[BBS23] E. Ballico, A. Bernardi, and P. Santarsiero, *Terracini locus for three points on a Segre variety*, Asian J. of Mathematics, 27(3):375–404, 2023. DOI 10.4310/AJM.2023.v27.n3.a3.

[BC21] E. Ballico and L. Chiantini, *On the Terracini locus of projective varieties*, Milan J. Math. **89** (2021), no. 1, 1–17, DOI 10.1007/s00032-020-00324-5. MR4277359

[BCC+18] A. Bernardi, E. Carlini, M. Catalisano, A. Gimigliano, and A. Oneto. *The Hitchhiker Guide to: Secant Varieties and Tensor Decomposition.* Mathematics, 6(12):314, 2018. DOI 10.3390/math6120314.

[BCO14] C. Bocci, L. Chiantini, and G. Ottaviani, *Refined methods for the identifiability of tensors*, Ann. Mat. Pura Appl. (4) **193** (2014), no. 6, 1691–1702, DOI 10.1007/s10231-013-0352-8. MR3275258

[BGI11] A. Bernardi, A. Gimigliano, and M. Idà, *Computing symmetric rank for symmetric tensors*, J. Symbolic Comput. **46** (2011), no. 1, 34–53, DOI 10.1016/j.jsc.2010.08.001. MR2736357

[BGM94] A. Bigatti, A. V. Geramita, and J. C. Migliore, *Geometric consequences of extremal behavior in a theorem of Macaulay*, Trans. Amer. Math. Soc. **346** (1994), no. 1, 203–235, DOI 10.2307/2154949. MR1272673

[BV23] E. Ballico and E. Ventura, *A note on very ample Terracini loci*, Rend. Circ. Mat. Palermo (2) **72** (2023), no. 8, 4115–4118, DOI 10.1007/s12215-023-00886-4. MR4670931

[Car06] E. Carlini, *Reducing the number of variables of a polynomial*, Algebraic geometry and geometric modeling, Math. Vis., Springer, Berlin, 2006, pp. 237–247, DOI 10.1007/978-3-540-33275-6_15. MR2279854

[CC10] L. Chiantini and C. Ciliberto, *On the dimension of secant varieties*, J. Eur. Math. Soc. (JEMS) **12** (2010), no. 5, 1267–1291, DOI 10.4171/JEMS/229. MR2677616

[Chi19] L. Chiantini, *Hilbert functions and tensor analysis*, Quantum physics and geometry, Lect. Notes Unione Mat. Ital., vol. 25, Springer, Cham, 2019, pp. 125–151. MR3890657

[Cle61] A. Clebsch, *Zur Theorie der algebraischen Flächen*, J. Reine Angew. Math., 58:93–108, 1861. DOI 10.1515/9783112368480-003.

[CO23] L. Chiantini and G. Ottaviani, *A footnote to a footnote to a paper of B. Segre*, The art of doing algebraic geometry, Trends Math., Birkhäuser/Springer, Cham, [2023] ©2023, pp. 93–117, DOI 10.1007/978-3-031-11938-5_4. MR4592538

[COV17] L. Chiantini, G. Ottaviani, and N. Vannieuwenhoven, *On generic identifiability of symmetric tensors of subgeneric rank*, Trans. Amer. Math. Soc. **369** (2017), no. 6, 4021–4042, DOI 10.1090/tran/6762. MR3624400
label=CS11,

[CS11] G. Comas and M. Seiguer, *On the rank of a binary form*, Found. Comput. Math. **11** (2011), no. 1, 65–78, DOI 10.1007/s10208-010-9077-x. MR2754189

[Dav85] E. D. Davis, *Complete intersections of codimension* 2 *in* $\mathbf{P}^r$*: the Bezout-Jacobi-Segre theorem revisited*, Rend. Sem. Mat. Univ. Politec. Torino **43** (1985), no. 2, 333–353 (1986). MR859862

[Der13] H. Derksen, *Kruskal's uniqueness inequality is sharp*, Linear Algebra Appl. **438** (2013), no. 2, 708–712, DOI 10.1016/j.laa.2011.05.041. MR2996363

[GM19] F. Galuppi and M. Mella, *Identifiability of homogeneous polynomials and Cremona transformations*, J. Reine Angew. Math. **757** (2019), 279–308, DOI 10.1515/crelle-2017-0043. MR4036576

[Har92] J. Harris, *Algebraic geometry*, Graduate Texts in Mathematics, vol. 133, Springer-Verlag, New York, 1992. A first course, DOI 10.1007/978-1-4757-2189-8. MR1182558

[Hil] D. Hilbert. *Lettre adresseé à M. Hermite*. Gesam. Abh., II:148–153.

[IK99] A. Iarrobino and V. Kanev, *Power sums, Gorenstein algebras, and determinantal loci*, Lecture Notes in Mathematics, vol. 1721, Springer-Verlag, Berlin, 1999. Appendix C by Iarrobino and Steven L. Kleiman, DOI 10.1007/BFb0093426. MR1735271

[Kru77] J. B. Kruskal, *Three-way arrays: rank and uniqueness of trilinear decompositions, with application to arithmetic complexity and statistics*, Linear Algebra Appl. **18** (1977), no. 2, 95–138, DOI 10.1016/0024-3795(77)90069-6. MR444690

[Lan12] J. M. Landsberg, *Tensors: geometry and applications*, Graduate Studies in Mathematics, vol. 128, American Mathematical Society, Providence, RI, 2012, DOI 10.1090/gsm/128. MR2865915

[LP23] B. Lovitz and F. Petrov, *A generalization of Kruskal's theorem on tensor decomposition*, Forum Math. Sigma **11** (2023), Paper No. e27, 40, DOI 10.1017/fms.2023.20. MR4572077

[LT10] J. M. Landsberg and Z. Teitler, *On the ranks and border ranks of symmetric tensors*, Found. Comput. Math. **10** (2010), no. 3, 339–366, DOI 10.1007/s10208-009-9055-3. MR2628829

[Syl52] J. J. Sylvester. *On the principles of the calculus of forms.* Cambridge and Dublin Math. J., 7:52–97, 1852.

[Ter11] A. Terracini, *Sulle v_k per cui la varietà degli $s_h(h+1)$-seganti ha dimensione minore dell'ordinario.* Rend. Circ. Mat., 31:392–396, 1911. DOI 10.1007/BF03018812.

(L. Chiantini) Dipartimento di Ingegneria dell'Informazione e Scienze Matematiche, Università di Siena, Italy
Email address: luca.chiantini@unisi.it

(F. Gesmundo) Saarland University, Saarbrücken 66123, Germany

(F. Gesmundo, current address) Institut de Mathématiques de Toulouse; UMR5219 – Université de Toulouse; CNRS – UPS, F-31062 Toulouse Cedex 9, France
Email address: fgesmund@math.univ-toulouse.fr

Contemporary Mathematics
Volume **805**, 2024
https://doi.org/10.1090/conm/805/16132

Almost finitely generated inverse systems and reduced k-algebras

Joan Elias and Maria Evelina Rossi

Abstract. The purpose of this paper is to characterize one-dimensional local domains, or more in general reduced, in terms of its Macaulay's inverse system. This leads to study almost finitely generated modules in the divided power ring. We specialize the results to a numerical semigroup ring by computing explicitly its inverse system. In the graded case we characterize reduced arithmetically Gorenstein 0-dimensional schemes. Several examples are given.

1. Introduction

Let $(R, \mathfrak{m}, \mathbf{k})$ be a complete Noetherian local ring where $\mathbf{k} = R/\mathfrak{m}$ denotes the residue field and let $E_R(\mathbf{k})$ be the injective envelope. Given an R-module M, the Matlis dual $M^\vee = \operatorname{Hom}_R(M, E_R(\mathbf{k}))$ defines a controvariant, additive and exact functor from the category of the R-modules into itself. In particular, by Matlis duality, the functor $(-)^\vee$ is an anti-equivalence between the category of the finitely generated R-modules and the category of the Artinian R-modules, [**2**, Theorem 3.2.13].

Let R be the power series ring $\mathbf{k}[\![x_1, \dots, x_n]\!]$ or the polynomial ring $\mathbf{k}[x_1, \dots, x_n]$ over a field $\mathbf{k}$. We denote by $\mathfrak{m}$ the maximal ideal of R generated by $x_1, \dots, x_n$. It is known that the injective envelope $E_R(\mathbf{k})$ is isomorphic as R-module to the divided power ring $\Gamma = \mathbf{k}_{DP}[y_1, \dots, y_n]$ (see [**13**], [**21**]) also denoted in the literature by $\mathbf{k}[x_1^{-1}, \dots, x_n^{-1}]$.

Macaulay in [**19**, IV] proved a particular case of Matlis duality, called Macaulay's correspondence, between the ideals $I \subseteq R$ such that R/I is an Artinian local ring and R-submodules $I^\perp$ of Γ of R which are finitely generated. For more detailed information concerning Macaulay's Inverse System see also [**5**], [**10**], [**11**], [**17**, Appendix A]. This theory was recently extended by Kleiman and Kleppe in [**18**] to the more general situation of A-algebras, where A is any Noetherian ring.

Macaulay's correspondence is an effective method for computing Gorenstein Artinian rings, see [**3**], Section 1, [**16**], [**14**] and [**17**]. An Artinian Gorenstein $\mathbf{k}$-algebra $A = R/I$ of socle degree s corresponds to a cyclic R-submodule of Γ generated by a polynomial $F \neq 0$ of degree s.

2020 *Mathematics Subject Classification.* Primary 13H10; Secondary 13H15, 14C05.

The first author was partially supported by PID2022-137283NB-C22.
The second author was partially supported by PRIN-MIUR 2020355B8Y..

The authors extended Macaulay's correspondence to d-dimensional local Gorenstein $\mathbf{k}$-algebras proving that they are in bijective correspondence with suitable submodules of Γ, called G-admissible, see [**8**]. This result was extended by S. Masuti, P. Schulze and L.Tozzo to any Cohen-Macaulay $\mathbf{k}$-algebra (local or graded), given the dimension of the socle of a suitable Artinian reduction, see [**20**], [**22**].

Despite several advances on the topic, many basic problems remain open. For instance, in the Artinian case, it is not known which polynomials F in Γ correspond to the ideals of R which are a complete intersection, see [**17**, pag. 261]. In positive dimension, it would be interesting to describe Macaulay's inverse system of domains and, more in general, of reduced $\mathbf{k}$-algebras. The last question is the purpose of this paper.

We briefly describe the organization of the paper. In Section 2 we present the main tools concerning Macaulay's Inverse System that will be used in the paper. In Section 3, inspired by results of W.D. Weakley [**24**], we prove that the inverse system of a 1-dimensional local domain is an almost finitely generated R-submodule of Γ and conversely, see Theorem 3.3. As a consequence, taking advantage of the properties of the G-admissible R-submodules of Γ, see Definition 2.1, we characterize the inverse system of one dimensional local Gorenstein domains, see Proposition 3.6. In Section 4, Theorem 4.1, we explicitly describe the generators of the almost finitely generated dual module of a numerical semigroup ring, see also [**12**, Theorem 2.1] for analogous results recently obtained by K. Eto and K. Watanabe. The defining ideal of a numerical semigroup ring is an homogeneous ideal in a weighted polynomial ring, hence in the process we need to study the inverse system in the non-standard graded case. In Section 5 we study the inverse system of one dimensional reduced $\mathbf{k}$-algebras, see Proposition 5.1. In the graded setting, this is the case of the coordinate ring of a set X of distinct points in the projective space. In Theorem 5.4 we present conditions on the inverse system of any Gorenstein zero-dimensional scheme X for being a reduced scheme. This results completes Theorem 3.14 in [**7**]. In particular we translate the problem in terms of the identifiability (in the sense of [**1**]) of a specific polynomial in the inverse system of X.

Throughout the paper several examples are given. The computations are performed by using the computer algebra system Singular [**4**] and in particular the Singular library INVERSE-SYST.lib, [**6**].

2. Preliminaries

We recall that $\Gamma = \mathbf{k}_{DP}[y_1, \dots, y_n]$ is an R-module acting R by *contraction* denoted by $\circ$. Given $\alpha, \beta \in \mathbb{N}^n$, we denote by $x^\alpha = x_1^{\alpha_1} \cdots x_n^{\alpha_n} \in R$ and $y^\beta = y_1^{\beta_1} \cdots y_n^{\beta_n} \in \Gamma$, then

$$x^\alpha \circ y^\beta = \begin{cases} y^{\beta-\alpha} & \text{if } \beta_i \geq \alpha_i \text{ for all } i = 1, \dots, n \\ 0 & \text{otherwise} \end{cases}$$

If the characteristic of field $\mathbf{k} = R/\mathfrak{m}$ is zero then there is a natural isomorphism of R-algebras between $(\Gamma, \circ)$ equipped with an internal product and the polynomial ring replacing the contraction with the partial derivatives. This action is sometimes called the "apolarity" action of R on Γ defined by

$$\begin{array}{rccc} \circ : & R \times \Gamma & \longrightarrow & \Gamma \\ & (f, g) & \to & f \circ g = f(\partial_{y_1}, \dots, \partial_{y_n})(g) \end{array}$$

where ∂_{y_i} denotes the partial derivative with respect to y_i.

In this paper we assume $\mathbf{k}$ of any characteristic, unless otherwise specified.

In both cases (contraction or derivation) the action of R on Γ lowers degree. Thus, Γ is not a finitely generated R-module. Let R_i (resp. Γ_i) be the sub-$\mathbf{k}$-vector space of R (resp. of Γ) generated by the standard monomials of degree $i \geq 0$. Notice that the apolarity action induces a non-singular $\mathbf{k}$-bilinear pairing:

$$\circ : \quad R_j \times \Gamma_j \quad \longrightarrow \quad \mathbf{k}$$

for every $j \geq 0$. The standard grading will be extended to a weighted grading in Section 3. However, unless explicitly stated otherwise, a polynomial ring has the standard grading in which all the variables have weight 1.

If $I \subset R$ is an ideal of R, then $(R/I)^\vee = \operatorname{Hom}_R(R/I, E_R(\mathbf{k}))$ is the R-submodule of Γ

$$I^\perp = \{F \in \Gamma \mid I \circ F = 0\ \}.$$

This submodule of Γ is called the *Macaulay's inverse system of* I. Remark that $I^\perp$ is also an R/I-module since $I \circ I^\perp = 0$.
Conversely, given an R-submodule W of Γ, the dual $W^\vee = \operatorname{Hom}_R(W, E_R(\mathbf{k}))$ is the ring $R/\operatorname{Ann}_R(W)$ where

$$\operatorname{Ann}_R(W) = \{g \in R \ \mid \ g \circ F = 0 \text{ for all } g \in W\}$$

is an ideal of R. Macaulay's correspondence in [**19**, IV] gives a correspondence between the ideals $I \subseteq R$ such that R/I is an Artinian local ring and R-submodules of Γ of R which are finitely generated. In particular Macaulay proved that Artinian Gorenstein $\mathbf{k}$-algebras $A = R/I$ of socle degree s correspond to cyclic R-submodules of Γ generated by a polynomial $F \neq 0$ of degree s.

In [**8**] the authors extended Macaulay's correspondence to the d-dimensional Gorenstein $\mathbf{k}$-algebras R/I characterizing the generators of the R-submodules of Γ (not finitely generated), called G-admissible, in correspondence with R/I. We present here the definition of G-admissible for $d = 1$.

DEFINITION 2.1. An R-submodule M of Γ is called **G-admissible** if it admits a countable system of generators $\{H_l\}_{l \in \mathbb{N}_+}$ satisfying the following conditions

(1) There exists a linear form $z \in R$ such that for all $l \in \mathbb{N}_+$

$$z \circ H_l = \begin{cases} H_{l-1} & \text{if } l > 1 \\ 0 & \text{otherwise.} \end{cases}$$

(2) $\operatorname{Ann}_R(\langle H_l \rangle) \circ H_{l+1} = \langle H_1 \rangle$ for all $l \in \mathbb{N}_+$.

If this is the case, we say that $M = \langle H_l, l \in \mathbb{N}_+ \rangle$ is a G-admissible R-submodule of Γ with respect to the linear form $z \in R$.

We present the main result of [**8**] in the one-dimensional case.

THEOREM 2.2 ([**8**], Theorem 3.8). *There is a one-to-one correspondence $\mathcal{C}$ between the following sets:*

(i) *one-dimensional Gorenstein* $\mathbf{k}$*-algebras* $A = R/I$,
(ii) *non-zero G-admissible R-submodules* $M = \langle H_l, l \in \mathbb{N}_+ \rangle$ *of* Γ.

In particular, given an ideal $I \subset R$ with $A = R/I$ satisfying (i) and z a linear regular element modulo I, then

$$\mathcal{C}(A) = I^{\perp} = \langle H_l, l \in \mathbb{N}_+ \rangle \subset S \quad \text{with} \quad \langle H_l \rangle = (I + (z^l))^{\perp}$$

is G-admissible. Conversely, given an R-submodule M of Γ satisfying (ii), then

$$\mathcal{C}^{-1}(M) = R/I \quad \text{with} \quad I = \operatorname{Ann}_R(M) = \bigcap_{l \in \mathbb{N}_+} \operatorname{Ann}_R(\langle H_l \rangle).$$

3. Inverse systems of local domains

We recall the definition of almost finitely generated A-module, where A is a Noetherian local ring, see [**24**], [**15**].

DEFINITION 3.1. An A-module M is **almost finitely generated**, a.f.g. for short, if M is not finitely generated, but any proper sub-A-module N of M is finitely generated. Moreover, M is a **divisible** A-module if for all $a \in R \setminus \{0\}$ it holds $M = aM$.

As before, let R be the power series ring $R = \mathbf{k}[\![x_1, \dots, x_n]\!]$. The following remark is an straightforward consequence of the definition of a.f.g. R-module that will be used systematically along this section.

LEMMA 3.2. *Let $I \subsetneq J$ be ideals of R such that* $\dim R/J > 0$. *Then $I^{\perp}$ is not an a.f.g. R-module.*

PROOF. Notice that $J^{\perp}$ is not finitely generated because R/J is not Artinian. Since $J^{\perp} \subsetneq I^{\perp}$, then $I^{\perp}$ is not an a.f.g. R-module. □

In the next result we characterize prime ideals in terms of their inverse systems. We remark that $I^{\perp}$ is an R-module, but also an R/I-module since $I \subseteq Ann_R(I^{\perp})$. The following result can eventually be deduced from [**24**, Proposition 2.6], here we give a short and straight proof.

THEOREM 3.3. *Let I be an ideal of R such that $A = R/I$ is one-dimensional. The following conditions are equivalent:*

(i) *I is a prime ideal,*
(ii) *$I^{\perp}$ is an a.f.g. A-module,*
(iii) *$I^{\perp}$ is a divisible A-module.*

PROOF. Assume that I is a prime ideal. Since R/I is not an Artinian ring, $I^{\perp}$ is not a finitely generated R-module. Let $N \subsetneq I^{\perp}$ be an R-submodule of $I^{\perp}$. Since I is a prime ideal and $I \subsetneq J := \operatorname{Ann}_R(N)$, we have that R/J is Artinian and hence $J^{\perp} = N$ is finitely generated. Hence $I^{\perp}$ is an a.f.g. R-module and also a.f.g. A-module.

Assume that $I^{\perp}$ is an a.f.g. A-module. From [**24**], Proposition 1.1 (3), we get that $I^{\perp}$ is divisible.

If $I^{\perp}$ is a divisible A-module then for all $a \in A \setminus \{0\}$ it holds $I^{\perp} = a \circ I^{\perp}$, i.e. the morphism of R-modules

$$I^{\perp} \xrightarrow{a\circ} I^{\perp}$$

is an epimorphism. By Matlis duality we deduce that the product by $A \xrightarrow{a\cdot} A$ is a monomorphism, i.e. a is a non-zero divisor. □

REMARK 3.4. Notice that if $I^\perp$ is a divisible A-module, then $I^\perp = \mathfrak{m} \circ I^\perp$. This condition does not imply $I^\perp = 0$ since $I^\perp$ is not, in general, a finitely generated A-module, hence Nakayama Lemma does not hold in this situation.

In the next result we characterize the G-admissible sets which are divisible.

LEMMA 3.5. *Let $A = R/I$ be a one-dimensional Gorenstein ring. Let $\{H_t; t \geq 1\}$ be a G-admissible system of generators of $I^\perp$. Then for all $H \in I^\perp$ there exists $r \in \mathbb{N}$ and $F \in R$ such that $H = F \circ H_r$.*

PROOF. It follows easily from the condition (1) of G-admissibility, Definition 2.1, because if $t > r$, then $H_t = z^{t-r} \circ H_r$. □

PROPOSITION 3.6. *Let I be an ideal of R such that $A = R/I$ is one-dimensional and Gorenstein. Let $I^\perp = \langle H_t; t \geq 1\rangle$, where $\{H_t; t \geq 1\}$ is a G-admissible system of generators with respect to $z \in R$. Then I is prime if and only if for all $a \in A \setminus \{0\}$ and for all $t \geq 1$ there exist $F \in R$ and $r \geq t$ such that*

$$H_t = a \circ (F \circ H_r).$$

PROOF. Assume that I is a prime ideal. From Theorem 3.3, $I^\perp$ is a divisible A-module. Then for all $a \in A \setminus \{0\}$ and for all $t \geq 1$ there exist $G \in I^\perp$ such that $H_t = a \circ G$. From Lemma 3.5 there exists $F \in R$ such that $G = F \circ H_r$ for some $r \in \mathbb{N}$. Hence

$$H_t = a \circ G = a \circ (F \circ H_r).$$

It is easy to prove that $r \geq t$. Let $\{H_t; t \geq 1\}$ a G-admissible system of generators with respect to z. If $r < t$ then

$$H_1 = z^{t-1} \circ H_t = (Fa) \circ (z^{t-1} \circ H_r) = (Fa) \circ 0 = 0$$

which is not possible.

Assume now that for all $a \in A \setminus \{0\}$ and for all $t \geq 1$ there exists $F \in R$ and $r \geq t$ such that

$$H_t = a \circ (F \circ H_r).$$

From Theorem 3.3 we only have to prove that $I^\perp$ is divisible, i.e. for all $a \in A \setminus \{0\}$ and $H \in I^\perp$ there exist $L \in I^\perp$ such that $H = a \circ L$.

Let $0 \neq a \in A$ be an element of A and $H \in I^\perp$. Then there exist $q \in R$ and $w \in \mathbb{N}$ such that

$$H = q \circ H_w.$$

From the hypothesis there exist $F \in R$ and $r \geq w$ such that $H_w = a \circ (F \circ H_r)$, so

$$H = q \circ H_w = a \circ ((qF) \circ H_r) = a \circ L.$$

□

REMARK 3.7. Let $A = R/I$ be a local ring with maximal ideal $\mathfrak{m}$. We recall that the valuation with respect to $\mathfrak{m}$ is the function

$$\begin{array}{rccl} v_{\mathfrak{m}}: & A \setminus \{0\} & \longrightarrow & \mathbb{N} \\ & a & \mapsto & v_{\mathfrak{m}}(a) = \operatorname{Max}\{u \mid a \in \mathfrak{m}^u\} \end{array}$$

and $v_{\mathfrak{m}}(0) = +\infty$. It is easy to see that in Proposition 3.6, instead of all the elements $a \in A$, we may restrict the control to the elements $a \in A$ such that $v_{\mathfrak{m}}(a) \leq e - 1$

where e is the multiplicity of $A = R/I$. In fact, since A is a one-dimensional Cohen-Macaulay local ring we have

$$\mathfrak{m}^{e-1+u} = z^u \mathfrak{m}^{e-1}.$$

for every $u \geq 1$. Hence if $v_{\mathfrak{m}}(a) = e-1+u > e-1$, then $a = z^u c$ with $c \in \mathfrak{m}^{e-1} \setminus \mathfrak{m}^e$, in particular $v_{\mathfrak{m}}(c) \leq e-1$. In the proof it is enough to take $L = F \circ H_{r+u}$ then we have

$$a \circ L = a \circ (F \circ H_{r+u}) = (z^u c) \circ (F \circ H_{r+u}) = c \circ (F \circ H_r) = H.$$

4. Inverse system of numerical semigroup rings

Let $1 \leq a_1 \leq \cdots \leq a_n$ be an n-ple of positive integers and let $\omega = (a_1, \dots, a_n)$. Consider the corresponding ring $R = \mathbf{k}[\![x_1, \dots, x_n]\!]$ where $\deg_\omega(x_i) = a_i$, $i = 1, \dots, n$. Denote $\omega(b_1, \dots, b_n) := \sum_{i=1}^n b_i a_i$. If $x^K = x_1^{k_1} \dots x_n^{k_n}$, then $\deg_\omega(x^K) = \omega(K)$ for $K = (k_1, \dots, k_n) \in \mathbb{N}^n$. Denote by $R_{\omega,j}$ (resp. $\Gamma_{\omega,j}$) the sub-$\mathbf{k}$-vector space of R (resp. Γ) generated by the monomials of degree j with respect to ω.

Then the apolarity action induces a non-singular $\mathbf{k}$-bilinear pairing:

$$\circ: \quad R_{\omega,j} \times \Gamma_{\omega,j} \longrightarrow \mathbf{k}$$

for every $j \geq 0$, where $\circ$ is the contraction. As for the standard graded case we have that if I is an homogeneous ideal in the ω-weighted ring R, then $I^\perp$ is homogeneous in the ω-weighted divided power ring Γ and $I^\perp = \oplus_j I^\perp_{\omega,j}$ where $I^\perp_{\omega,j} = \{F \in \Gamma_{\omega,j} \mid g \circ F = 0 \text{ for all } g \in I\}$, i.e. $(I^\perp)_{\omega,j} = I^\perp_{\omega,j}$. In fact we can repeat the same proof as in the standard case, see [**14**], Proposition 2.5.

We consider now local rings defined by numerical semigroup rings and we compute their inverse system. Notice a description of the inverse system of numerical semigroup rings was also the main task of [**12**, Theorem 2.1]. Here we prove the result in an easier way and we observe, following the definition, that it is an almost finitely generated R-module.

Given the integers $1 \leq a_1 < \cdots < a_n$ with $\gcd(a_1, \dots, a_n) = 1$, we denote by $A(a_1, \dots, a_n) := R/I(a_1, \dots, a_n)$ the ring associated to the monomial curve with parameterization

$$\begin{array}{cccc} \phi: & R & \longrightarrow & \mathbf{k}[\![t]\!] \\ & x_i & \to & t^{a_i} \end{array}$$

i.e. $I(a_1, \dots, a_n) = \ker(\phi)$. If we denote $\omega = (a_1, \dots, a_n)$ as before, then the ideal $I(a_1, \dots, a_n)$ is generated by the binomials $x^K - x^L$ with $K, L \in \mathbb{N}^n$ such that $\deg_\omega(K) = \deg_\omega(L)$, see for instance [**23**, Lemma 4.1].

We denote by $\mathcal{J}$ the set of $j \geq 0$ such that there exists $K \in \mathbb{N}^n$ with $\deg_\omega(K) = j$. For all $j \in \mathcal{J}$ we define the following homogeneous form with respect to ω as follows:

$$L_{\omega,j} = \sum_{\deg_\omega(K)=j} y^K.$$

Notice that $x_i \circ L_{\omega,j} = L_{\omega,j-a_i}$ for all $j \in \mathcal{J}$ and $i = 1, \dots, n$.

THEOREM 4.1. *Given the integers $1 \leq a_1 < \cdots < a_n$ with $\gcd(a_1, \dots, a_n) = 1$, then*

$$I(a_1, \dots, a_n)^\perp = \bigcup_{j \in \mathcal{J}} \langle L_{\omega,j} \rangle$$

and the R-module $I(a_1, \dots, a_n)^\perp$ is an a.f.g. R-module.

PROOF. For all $j \in \mathcal{J}$, by the non-singular $\mathbf{k}$-bilinear pairing induced by $\circ$ we get

$$\dim_{\mathbf{k}} I^{\perp}_{\omega,j} = \dim_{\mathbf{k}} A(a_1,\dots,a_n)_{\omega,j} = \dim_{\mathbf{k}}(\mathbf{k}[\![t]\!]_j) = 1,$$

so we only have to prove that $L_{\omega,j} \in I^{\perp}_{\omega,j}$. This is equivalent to show that for every $j \in \mathcal{J}$,

$$(x^K - x^L) \circ L_{\omega,j} = 0$$

for all $K, L \in \mathbb{N}^n$ such that $\deg(K) = \deg(L)$. Furthermore, this is equivalent to prove that for all $y^\alpha \in \Gamma_{\omega,j}$ such that $x^K \circ y^\alpha \neq 0$ there exists a unique $y^\beta \in \Gamma_{\omega,j}$ with $x^K \circ y^\alpha = x^L \circ y^\beta$. Since $x^K \circ y^\alpha \neq 0$ we have $\alpha - K \in \mathbb{N}^n$, so $\beta = L + (\alpha - K) \in \mathbb{N}^n$.

Since $I(a_1,\dots,a_n)$ is a prime ideal of R, by Theorem 3.3 the R-module $I(a_1,\dots,a_n)^{\perp}$ is an a.f.g. R-module. □

COROLLARY 4.2. *Given the integers* $1 \le a_1 < \dots < a_n$ *with* $\gcd(a_1,\dots,a_n) = 1$, *it holds*

$$(I(a_1,\dots,a_n) + (x_1^t))^{\perp} = \langle L_{\omega,j} \mid \deg_{y_1}(L_{\omega,j}) \le t-1 \rangle$$

for all $t \ge 1$.

PROOF. By Matlis' duality we know that

$$(I(a_1,\dots,a_n) + (x_1^t))^{\perp} = I(a_1,\dots,a_n)^{\perp} \cap (x_1^t)^{\perp}.$$

Being (x_1^t) monomial, $(x_1^t)^{\perp}$ is homogeneous with respect the grading defined by ω. Hence

$$(I(a_1,\dots,a_n) + (x_1^t))^{\perp}_{\omega,j} = I(a_1,\dots,a_n)^{\perp}_{\omega,j} \cap (x_1^t)^{\perp}.$$

From this identity and the previous result we get the claim. □

EXAMPLE 4.3. Let us consider the integers $a_1 = 5, a_2 = 6, a_3 = 9$. The corresponding monomial curve C is a complete intersection, so C is Gorenstein. Moreover, the ideal defining C is minimally generated by two elements: $I(5,6,9) = (x_1^3 - x_2x_3, x_2^3 - x_3^2)$. Then by using Corollary 4.2 we get:

$I = I(5,6,9)$	generator $L_{\omega,j}$	j
$(I + (x_1))^{\perp}$	$y_3^2 + y_2^3$	$j = 18$
$(I + (x_1^2))^{\perp}$	$y_1y_3^2 + y_1y_2^3$	$j = 23$
$(I + (x_1^3))^{\perp}$	$y_1^2y_3^2 + y_1^2y_2^3$	$j = 28$
$(I + (x_1^4))^{\perp}$	$y_2y_3^3 + y_2^4y_3 + y_1^3y_3^2 + y_1^3y_2^3$	$j = 33$

EXAMPLE 4.4. Let us consider the integers $a_1 = 5, a_2 = 6, a_3 = 7$. The corresponding monomial curve C is not a complete intersection and the Cohen-Macaulay type of C is two. The ideal defining C is minimally generated by three elements: $I(5,6,7) = (x_1^4 - x_2x_3^2, x_2^2 - x_1x_3, x_1^3x_2 - x_3^3)$. Then by using Corollary 4.2 we get:

$I = I(5,6,7)$	generators $L_{\omega,j}$	j
$(I + (x_1))^{\perp}$	y_2y_3, y_3^2	$j = 13, 14$
$(I + (x_1^2))^{\perp}$	$y_2^3 + y_1y_2y_3, y_2^2y_3 + y_1y_3^2$	$j = 18, 19$
$(I + (x_1^3))^{\perp}$	$y_1y_2^3 + y_1^2y_2y_3, y_2^4 + y_1^2y_2y_3 + y_1^2y_3^2$	$j = 23, 24$
$(I + (x_1^4))^{\perp}$	$y_3^4 + y_1^2y_2^3 + y_1^3y_2y_3, y_1y_2^4 + y_1^2y_2^2y_3 + y_1^3y_3^2$	$j = 28, 29$

EXAMPLE 4.5. Let us consider the integers $a_1 = 6, a_2 = 7, a_3 = 11, a_4 = 15$. The corresponding monomial curve C is not a complete intersection: the ideal defining C is minimally generated by five elements: $I(6,7,11,15) = (x_4^2 - x_1^2x_2x_3, x_3x_4 - x_1^2x_2^2, x_1x_4 - x_2^3, x_3^2 - x_2x_4, x_2x_3 - x_1^3)$. Then:

$I = I(6,7,11,15)$	generators $L_{\omega,j}$	j
$(I+(x_1))^\perp$	$y_2^2, y_3^2 + y_2y_4$	$j = 14, 22$
$(I+(x_1^2))^\perp$	$y_1y_2^2, y_1y_3^2 + y_1y_2y_4 + y_2^4$	$j = 20, 28$
$(I+(x_1^3))^\perp$	$y_3y_4 + y_1^2y_2^2, y_1^2y_3^2 + y_1^2y_2y_4 + y_1y_2^4$	$j = 26, 34$
$(I+(x_1^4))^\perp$	$y_1y_3y_4 + y_2^3y_3 + y_1^3y_2^2, y_2y_3^3 + y_2^2y_3y_4 + y_1^3y_3^2 + y_1^3y_2y_4 + y_1^2y_2^4$	$j = 32, 40$

Hence $A(6,7,11,15)/(x_1)$ is level of Cohen-Macaulay type two. In [**20**], Example 1, it is shown that $A(6,7,11,15)/(x_1)$ is level and $A(6,7,11,15)/(x_1+x_2)$ does not. Notice that x_1 is homogeneous with respect to ω but $x_1 + x_2$ is not homogeneous with respect to ω.

5. Inverse systems of reduced rings

As a consequence of Theorem 3.3, we present a characterization of the radical ideals $I \subseteq R = \mathbf{k}[\![x_1, \ldots, x_n]\!]$ in terms of a.f.g. modules in the case $\dim R/I = 1$.

PROPOSITION 5.1. *Let I be an ideal of R such that* $\dim R/I = 1$. *The following conditions are equivalent:*

(i) *I is a radical ideal,*
(ii) *there exist $M_1, \ldots, M_r$ a.f.g. sub-R-modules of $I^\perp$ such that $I^\perp = M_1 + \cdots + M_r$.*

PROOF. Assume that I is a radical ideal. Then $I = p_1 \cap \cdots \cap p_r$ where $p_1, \ldots, p_r$ are prime ideals of R such that $\dim R/p_i = 1$ for all i. Then the modules $M_i = p_i^\perp$ are a.f.g by Theorem 3.3. and $I^\perp = M_1 + \cdots + M_r$. Hence (ii) is proved.

Assume now (ii). We define $J_i = \operatorname{Ann}(M_i)$; by Theorem 3.3 J_i is a prime ideal. Since $I \subset J_i$ the ideal J_i has height at least $n-1$. On the other hand, M_i is not f.g. so J_i is a height $n-1$ prime ideal of R. By Matlis duality we get $I = J_1 \cap \cdots \cap J_r$, so I is radical. $\square$

In the following we give an example of a reduced 1-dimensional local ring obtained as a monomial curve linked to a straight line.

EXAMPLE 5.2. Let us consider the non-Gorenstein monomial curve C of Example 4.4; its defining ideal is $I_1 = (x_1^4 - x_2x_3^2, x_2^2 - x_1x_3, x_1^3x_2 - x_3^3) \subset R = \mathbf{k}[\![x_1, x_2, x_3]\!]$. Let L be the straight line defined by the ideal $I_2 = (x_2, x_3) \subset R$.

The union $D = C \cup L$ is a complete intersection defined by the radical ideal $J = I_1 \cap I_2 = (x_3^3 - x_1^3x_2, x_2^2 - x_1x_3)$; in other words: C is linked to L by the complete intersection D.

We know that $J^\perp = I_1^\perp + I_2^\perp$. Since I_1, I_2 are prime ideals, from Theorem 3.3 the R-modules $I_1^\perp$ and $I_2^\perp$ are a.f.g., see Proposition 5.1.

Notice that $I_1^\perp$ is generated by $L_{\omega,j}$, $j \geq 0$, with $\omega = (5,6,7)$, see Theorem 4.1 and Example 4.4. By a straightforward computation we get that $I_2^\perp$ is generated by y_1^n, $n \geq 0$. The ring R/J is Gorenstein, so by using Singular library [**6**], $J^\perp$ can be generated by the following G-admissible system of generators with respect to x_1: $H_1 = y_2y_3^2, H_2 = y_1H_1 + y_2^3y_3, H_3 = y_1H_2 + y_2^5, H_4 = y_1H_3 + y_3^5, \ldots$.

Next, it could be interesting to see explicitly how some generators of $J^\perp$ belong to $I_1^\perp + I_2^\perp$ and vice-versa. For instance, let us consider the generator H_2 of $J^\perp$:
$H_2 = x_3^2 \circ (y_2^3 x_3^3 + x_1 x_2 x_3^4 + y_1^3 y_2^4 + y_1^4 y_2^2 y_3 + y_1^5 y_3^2) - y_1^5 = x_3^2 \circ L_{\omega,39} - y_1^5 \in I_1^\perp + I_2^\perp$.
On the other hand let us consider the generator $L_{\omega,23}$ of $I_1^\perp$:

$$L_{\omega,23} = y_1 y_2^3 + y_1^2 y_2 y_3 = x_3 \circ H_3 \in J^\perp,$$

and the generator y_1^3 of $I_2^\perp$:

$$y_1^3 = x_2 x_3^2 \circ H_4 \in J^\perp.$$

From now on, R will be the polynomial ring $\mathbf{k}[x_1, \ldots, x_n]$ with the standard grading ($\deg x_i = 1$). In this part we use results in [**14**] and [**7**], hence we from now on we assume $\mathbf{k}$ is an algebraically closed field of characteristic zero and $\circ$ **is the derivation**.

Let X be a zero-dimensional scheme, the first purpose is to understand when X is reduced, that is X consists of a set of distinct points. Recall that $I(X)^\perp$ is not a finitely generated graded R-module and $(I(X)^\perp)_j = (I(X)_j)^\perp$ for every $j \geq 0$. Moreover the Hilbert function of $R/I(X)$ is

$$\mathrm{HF}_{R/I(X)}(j) = \dim_{\mathbf{k}}(R_j/I(X)_j) = \dim_{\mathbf{k}}(I(X)^\perp)_j,$$

$j \geq 0$, see [**14**, Proposition 2.5].

Let z be a linear form in R such that $z(P_i) \neq 0$ for every $i = 1, \ldots, r$, that is z is a not zero divisor in $R/I(X)$. Then we will say that $R/I(X) + (z)$ is an Artinian reduction of X. Notice that $h_t = \Delta\mathrm{HF}_X(t) = \mathrm{HF}_X(t) - \mathrm{HF}_X(t-1)$ is the Hilbert function in degree t of any Artinian reduction of X and $h_t = 0$ for every $t > s$ where s is called the **socle degree** of $R/I(X)$. Since $R/I(X)$ is Cohen-Macaulay, we recall that s coincides with the regularity of $R/I(X)$. The vector $(h_0, \ldots, h_s)$ is the h-vector of $R/I(X)$ or, for short, the h-vector of X.

Given a point $P = (a_0, \ldots, a_n) \in \mathbb{P}^n_{\mathbf{k}}$ we define the dual linear form $L := a_0 y_0 + \cdots + a_n y_n$ in Γ and we say that L is the linear form associated to P, actually it generates the dual of $I(P)$.

We present a well known result concerning the inverse system of the ideal of a set of points.

PROPOSITION 5.3 ([**17**], Theorem 1.15(2)). *Let $X = \{P_1, \ldots, P_r\}$ be a set of distinct points of $\mathbb{P}^n_{\mathbf{k}}$ and let $L_1, \ldots, L_r$ be the associated linear forms. Then for all $j \geq 0$*

$$(I(X)^\perp)_j = \langle L_1^j, \ldots, L_r^j \rangle_{\mathbf{k}}.$$

Comparing Proposition 5.3 and Proposition 5.1, it is clear that in the case of zero dimensional schemes $X = \{P_1, \ldots, P_r\}$, the a.f.g. R-submodules of $I^\perp$ are the inverse systems of $I(P_i)$, that is $M_i = \langle L_i^j : j \geq 0 \rangle$.

By using Theorem 3.14 and Remark 3.15 in [**7**], we characterize zero-dimensional Gorenstein schemes which are reduced. The new part in the following result is (3) implies (1).

THEOREM 5.4. *Let $X \subset \mathbb{P}^n_{\mathbf{k}}$ be zero-dimensional Gorenstein scheme of degree r. We denote by s the socle degree of $R/I(X)$. Let $I(X)^\perp = \langle H_t; t \geq 1 \rangle$ be a G-admissible system of generators with respect to a linear form $z \in R$. The* *following conditions are equivalent:*

(1) *X is reduced,*

(2) *there exist (unique up scalars) linear forms $L_1, \dots, L_r \in \Gamma$ pairwise linear independent and unique elements $\alpha_1, \dots, \alpha_r \in \mathbf{k}$ such that for all $t \geq r+1$*

$$H_t = \frac{1}{(t+s-1)!} \sum_{i=1}^{r} \frac{\alpha_i}{z(P_i)} L_i^{t+s-1},$$

(3) *there exist (unique up to scalars) linear forms $L_1, \dots, L_r \in \Gamma$ pairwise linear independent and unique elements $\alpha_1, \dots, \alpha_r \in \mathbf{k}$ such that*

$$H_{r+2} = \frac{1}{(r+s+1)!} \sum_{i=1}^{r} \frac{\alpha_i}{z(P_i)} L_i^{r+s+1}.$$

Under these conditions, $X = \{P_1, \dots, P_r\}$ where P_i is the point defined by the linear form L_i, $i = 1, \dots, r$, and $\alpha_1, \dots, \alpha_r \in \mathbf{k}^*$.

PROOF. Recall that $s = \deg(H_1)$, $\deg(H_t) = s + t - 1$ and $r = \dim_{\mathbf{k}}\langle H_1 \rangle$.

Now (1) implies (2) (and hence trivially (3)) follows by Theorem 3.13 and Remark 3.14 in [**7**]. Assume now (3) and we prove that it implies (1). Let P_i be the point of $\mathbb{P}^n_{\mathbf{k}}$ defined by the linear form L_i, $i = 1, \dots, r$. Since the linear forms are pairwise linear independent we have that $P_i \neq P_j$ if $i \neq j$. We consider the set of points $Y = \{P_1, \dots, P_r\}$, we will prove that $X = Y$.

From (3) and Proposition 5.3 we get

$$\langle H_{r+2} \rangle \subset \langle L_1^{r+s-1}, \dots L_r^{r+s-1} \rangle \subset I(Y)^{\perp},$$

so $I(Y) \subset \operatorname{Ann}(H_{r+2})$. Furthermore, by [**9**, Proposition 2.11] we deduce

$$I(Y)_{\leq r+1} R \subset \operatorname{Ann}(H_{r+2})_{\leq r+1} R = I(X).$$

On the other hand, since the Castelnuovo-Mumford regularity of $I(Y)$ is at most r we get that $I(Y)_{\leq r+1} R = I(Y)$. Hence we have that

$$I(Y) \subset I(X).$$

Since the ground field $\mathbf{k}$ is infinite there is a linear regular element $x \in R_1$ of $R/I(X)$ and $R/I(Y)$. We have $I(Y) + (x) \subset I(X) + (x)$. Recall that both $R/I(X)$ and $R/I(Y)$ are Cohen-Macaulay rings, so

$$\operatorname{Length}(R/I(X) + (x)) = \operatorname{Length}(R/I(Y) + (x)) = r$$

and then $I(Y) + (x) = I(X) + (x)$. In particular the $h-$vector of $R/I(X)$ and $R/I(Y)$ coincides. Being both rings graded and Cohen-Macaulay we get that

$$\operatorname{HF}_{R/I(Y)} = \operatorname{HF}_{R/I(X)}.$$

Since $I(Y) \subset I(X)$ we deduce that $Y = X$.

Finally, if one the equivalent conditions holds, then $\alpha_1, \dots, \alpha_r \in \mathbf{k}^*$ by Theorem 3.13 in [**7**]. □

The condition (3) in Theorem 5.4 translates the problem to be reduced for X in terms of the identifiability of a specific form, see [**1**]. This means that we have to prove that H_{r+2} can be uniquely written as sum of powers of r linear forms $L_1, \dots, L_r$. If we know $\deg X = r$ and the socle degree s of $R/I(X)$, then to verify if X is reduced it consists in an effective computation on $\langle H_{r+2} \rangle = (I(X) + (z^{r+2}))^{\perp}$, where z is a linear regular element of $R/I(X)$.

In the following example we consider a reduced zero-dimensional Gorenstein scheme $X \subset \mathbb{P}^2_{\mathbf{k}}$ of degree r, hence the corresponding H_{r+2} is given by Theorem 5.4(3).

EXAMPLE 5.5. Let us consider the ideal I of $R = \mathbf{k}[x_1, x_2, x_3]$ generated by the forms $x_1^2 - x_1x_3, x_2^2 - x_2x_3$. The projective variety $X = V(I)$ is a complete intersection of $\mathbb{P}^2_{\mathbf{k}}$ defining four points: $(1,0,1)$, $(0,1,1)$, $(0,0,1)$ and $(1,1,1)$. Hence X is a reduced Gorenstein (in fact a complete intersection) set of $r = 4$ points. We can deduce this fact from the last result starting from the defining ideal I. It is easy to prove that x_3 is a non-zero divisor of $A = R/I$, the socle degree $s = 2$, $r + 2 = 6$ and that H_6, i.e. a generator of $(I + (x_3^6))^\perp$, can be computed by [**6**] :

$$\begin{aligned} H_6 = y_1^6y_2+3y_1^5y_2^2+5y_1^4y_2^3+5y_1^3y_2^4+3y_1^2y_2^5+y_1y_2^6+6y_1^5y_2y_3+15y_1^4y_2^2y_3+20y_1^3y_2^3y_3+ \\ 15y_1^2y_2^4y_3 + 6y_1y_2^5y_3 + 15y_1^4y_2y_3^2 + 30y_1^3y_2^2y_3^2 + 30y_1^2y_2^3y_3^2 + 15y_1y_2^4y_3^2 + 20y_1^3y_2y_3^3+ \\ 30y_1^2y_2^2y_3^3 + 20y_1y_2^3y_3^3 + 15y_1^2y_2y_3^4 + 15y_1y_2^2y_3^4 + 6y_1y_2y_3^5 \end{aligned}$$

If $L_1 = y_1 + y_3$, $L_2 = y_2 + y_3$, $L_3 = y_3$ and $L_4 = y_1 + y_2 + y_3$ then

$$H_6 = \frac{1}{7}(-L_1^7 - L_2^7 + L_3^7 + L_4^7)$$

so condition (3) of Theorem 5.4 holds true. Indeed X is a reduced Gorenstein zero-dimensional scheme.

In the following example we consider a non-reduced zero-dimensional Gorenstein scheme $X \subset \mathbb{P}^2_{\mathbf{k}}$ of degree r, hence the corresponding H_{r+2} cannot verify condition (3) of Theorem 5.4.

EXAMPLE 5.6. Let us consider the ideal I of $R = \mathbf{k}[x_1, x_2, x_3]$ generated by the forms $x_1^2 + x_2^2 - x_3^2, x_1^2 - x_2x_3 - x_3^2$. The projective variety $X = V(I)$ is a complete intersection of $\mathbb{P}^2_{\mathbf{k}}$ defining four points: a double point $(0, -1, 1)$ and two simple points: $(1, 0, 1)$ and $(-1, 0, 1)$. Hence X is a non-reduced Gorenstein set of points. In this case $r = 4$ and $s = 2$. We have that x_3 is a non-zero divisor of $A = R/I$ and that H_6, i.e. a generator of $(I + (x_3^6))^\perp$ can be computed by [**6**]:

$$\begin{aligned} H_6 = y_1^7 - 7y_1y_2^6 + 42y_1y_2^5y_3 + 21y_1^5y_3^2 - 105y_1y_2^4y_3^2 \\ + 140y_1y_2^3y_3^3 + 35y_1^3y_3^4 - 105y_1y_2^2y_3^4 + 42y_1y_2x_3^5 \end{aligned}$$

A linear algebra computation by [**4**] shows that there are not $\alpha_1, \dots, \alpha_4 \in \mathbf{k}$ and linear forms $L_1, \dots, L_4$ such that

$$H_6 = \frac{1}{7!}\sum_{i=1}^{4} \alpha_i L_i^7.$$

Hence we recover that I is not radical from the condition (3) of Theorem 5.4.

References

[1] E. Angelini and L. Chiantini, *On the identifiability of ternary forms*, Linear Algebra Appl. **599** (2020), 36–65, DOI 10.1016/j.laa.2020.03.042. MR4083806

[2] W. Bruns and J. Herzog, *Cohen-Macaulay rings*, Cambridge Studies in Advanced Mathematics, vol. 39, Cambridge University Press, Cambridge, 1993. MR1251956

[3] Y. H. Cho and A. Iarrobino, *Inverse systems of zero-dimensional schemes in* $\mathbb{P}^n$, J. Algebra **366** (2012), 42–77, DOI 10.1016/j.jalgebra.2012.04.032. MR2942643

[4] W. Decker, G.-M. Greuel, G. Pfister, and H. Schönemann, SINGULAR *4-3-0–A computer algebra system for polynomial computations*, `http://www.singular.uni-kl.de`, 2022.

[5] D. Eisenbud, *Commutative algebra*, Graduate Texts in Mathematics, vol. 150, Springer-Verlag, New York, 1995. With a view toward algebraic geometry, DOI 10.1007/978-1-4612-5350-1. MR1322960

[6] J. Elias, INVERSE-SYST–*Singular library for computing Macaulay's inverse systems*, `arXiv:1501.01786`, `http://www.ub.edu/C3A/elias/inverse-syst-v.5.2.lib`, 2015.

[7] J. Elias and M. E. Rossi, *Inverse system of Gorenstein points in* $\mathbb{P}^n_k$, to appear in Israel Journal of Math, `arXiv:2301.07056v1`.

[8] J. Elias and M. E. Rossi, *The structure of the inverse system of Gorenstein k-algebras*, Adv. Math. **314** (2017), 306–327, DOI 10.1016/j.aim.2017.04.025. MR3658719

[9] J. Elias and M. E. Rossi, *A constructive approach to one-dimensional Gorenstein* **k**-*algebras*, Trans. Amer. Math. Soc. **374** (2021), no. 7, 4953–4971, DOI 10.1090/tran/8376. MR4273181

[10] J. Emsalem, *Géométrie des points épais* (French, with English summary), Bull. Soc. Math. France **106** (1978), no. 4, 399–416. MR518046

[11] A. Iarrobino and J. Emsalem, *Some zero-dimensional generic singularities; finite algebras having small tangent space*, Compositio Math. **36** (1978), no. 2, 145–188. MR515043

[12] K. Eto and K. Watanabe, *Inverse polynomials of numerical semigroup rings*, 2021, `arXiv:2108.04513`.

[13] P. Gabriel, *Objects injectifs dans les catégories abéliennes*, Séminaire Pierre Dubriel 1958/59, Numdam, 1959, pp. 17–01, 32.

[14] A. V. Geramita, *Inverse systems of fat points: Waring's problem, secant varieties of Veronese varieties and parameter spaces for Gorenstein ideals*, The Curves Seminar at Queen's, Vol. X (Kingston, ON, 1995), Queen's Papers in Pure and Appl. Math., vol. 102, Queen's Univ., Kingston, ON, 1996, pp. 2–114. MR1381732

[15] W. Heinzer and D. Lantz, *Artinian modules and modules of which all proper submodules are finitely generated*, J. Algebra **95** (1985), no. 1, 201–216, DOI 10.1016/0021-8693(85)90101-2. MR797663

[16] A. A. Iarrobino, *Associated graded algebra of a Gorenstein Artin algebra*, Mem. Amer. Math. Soc. **107** (1994), no. 514, viii+115, DOI 10.1090/memo/0514. MR1184062

[17] A. Iarrobino and V. Kanev, *Power sums, Gorenstein algebras, and determinantal loci*, Lecture Notes in Mathematics, vol. 1721, Springer-Verlag, Berlin, 1999. Appendix C by Iarrobino and Steven L. Kleiman, DOI 10.1007/BFb0093426. MR1735271

[18] S. L. Kleiman and J. O. Kleppe, *Macaulay duality and its geometry*, 2022, `arXiv:2210.10934`.

[19] F. S. Macaulay, *The algebraic theory of modular systems*, Cambridge Mathematical Library, Cambridge University Press, Cambridge, 1994. Revised reprint of the 1916 original; With an introduction by Paul Roberts. MR1281612

[20] S. K. Masuti and L. Tozzo, *The structure of the inverse system of level K-algebras*, Collect. Math. **69** (2018), no. 3, 451–477, DOI 10.1007/s13348-018-0212-3. MR3842217

[21] D. G. Northcott, *Injective envelopes and inverse polynomials*, J. London Math. Soc. (2) **8** (1974), 290–296, DOI 10.1112/jlms/s2-8.2.290. MR360555

[22] M. Schulze and L. Tozzo, *Inverse limits of Macaulay's inverse systems*, J. Algebra **525** (2019), 341–358, DOI 10.1016/j.jalgebra.2019.01.024. MR3911647

[23] B. Sturmfels, *Gröbner bases and convex polytopes*, University Lecture Series, vol. 8, American Mathematical Society, Providence, RI, 1996, DOI 10.1090/ulect/008. MR1363949

[24] W. D. Weakley, *Modules whose proper submodules are finitely generated*, J. Algebra **84** (1983), no. 1, 189–219, DOI 10.1016/0021-8693(83)90075-3. MR716778

DEPARTAMENT DE MATEMÀTIQUES I INFORMÀTICA, UNIVERSITAT DE BARCELONA (UB), GRAN VIA 585, 08007 BARCELONA, SPAIN

Email address: `elias@ub.edu`

DIPARTIMENTO DI MATEMATICA, UNIVERSITÀ DI GENOVA, VIA DODECANESO 35, 16146 GENOVA, ITALY

Email address: `rossim@dima.unige.it`

Contemporary Mathematics
Volume **805**, 2024
https://doi.org/10.1090/conm/805/16133

Deformations of local Artin rings via Hilbert-Burch matrices

Roser Homs and Anna-Lena Winz

ABSTRACT. In the local setting, Gröbner cells are affine spaces that parametrize ideals in $\mathbf{k}[\![x,y]\!]$ that share the same leading term ideal with respect to a local term ordering. In particular, all ideals in a cell have the same Hilbert function, so they provide a cellular decomposition of the punctual Hilbert scheme compatible with its Hilbert function stratification. We exploit the parametrization of Gröbner cells via Hilbert-Burch matrices to compute the Betti strata, with hands-on examples of deformations that preserve the Hilbert function, and revisit some classical results along the way. Moreover, we move towards an explicit parametrization of all local Gröbner cells.

1. Introduction

The punctual Hilbert scheme $\mathrm{Hilb}^n(\mathbf{k}[\![x,y]\!])$ is a fundamental object in algebraic geometry that parametrizes ideals of colength n in $\mathbf{k}[\![x,y]\!]$. During the decades of the 70s and 80s of the past century several authors [**4, 5, 8, 10, 12, 16, 22**] studied its geometrical properties. In particular, Briançon [**4**] and Iarrobino [**16**] proved that $\mathrm{Hilb}^n(\mathbf{k}[\![x,y]\!])$ is irreducible of dimension $n-1$ for $\mathbf{k}$ algebraically closed.

A natural approach to understand these schemes better is decomposing them into smaller spaces by considering sets of flat deformations of a given colength n monomial ideal. On one hand, Briançon and Galligo [**5**] introduced the technique of vertical standard basis stratification of $\mathrm{Hilb}^n(\mathbf{k}[\![x,y]\!])$ consisting on the study of the generators of such deformations. Yaméogo [**22**] showed that each vertical stratum of $\mathrm{Hilb}^n(\mathbf{k}[\![x,y]\!])$ is isomorphic to an affine space of dimension given by a very simple formula.

On the other hand, Conca and Valla [**6**] focused on studying the syzygies of the deformations of monomial ideals. For a monomial ideal E, they used Hilbert-Burch matrices to parametrize the so-called Gröbner cell $V_{\mathrm{lex}}(E)$ of all ideals J with leading term ideal $\mathrm{Lt}_{\mathrm{lex}}(J)=E$ with respect to the lexicographical term ordering lex. When the Hilbert scheme is smooth these cells are affine spaces by [**3**]. A term ordering τ determines a $\mathbf{k}^*$-action on the punctual Hilbert scheme with isolated fixed points — the monomial ideals — and the $V_\tau(E)$ are all those points in the Hilbert scheme specialising to the point E. When one has a smooth scheme with a $\mathbf{k}^*$-action with isolated fixed points, one obtains a cellular decomposition of that

2020 *Mathematics Subject Classification.* Primary 13D02, 14C05.

space. This approach has been used to study certain invariants — such as Betti numbers — of Hilbert schemes of points of $\mathbb{A}^2$, $\mathbb{P}^2$ and $(\mathbb{A}^2, 0)$, see [**8**, **11**].

The cellular decomposition of $\mathrm{Hilb}^n(\mathbf{k}[\![x,y]\!])$ resulting from vertical stratification or equivalently Gröbner cells is not compatible with the Hilbert function stratification of $\mathrm{Hilb}^n(\mathbf{k}[\![x,y]\!])$. In other words, colength n ideals with different Hilbert functions may belong to the same Gröbner cell. Building on the works of Rossi and Sharifan [**21**] and Bertella [**1**] on producing zero-dimensional ideals in $\mathbf{k}[\![x,y]\!]$ with a given Hilbert function, the authors of the present paper consider in [**15**] a local term ordering $\overline{\tau}$ to define Gröbner cells that overcome this problem.The parametrization of $V_{\overline{\tau}}(E)$ in terms of Hilbert-Burch matrices has been completely described for a special class of monomial ideals, that includes the well-studied lex-segment ideals, see Theorem 2.8 or [**15**, Theorem 5.7].

The goal of the present contribution is twofold: (1) to exploit the parametrization of Gröbner cells for lex-segment ideals to study the Betti strata and homogeneous subcells of the subscheme $\mathrm{Hilb}^h(\mathbf{k}[\![x,y]\!])$ of $\mathrm{Hilb}^n(\mathbf{k}[\![x,y]\!])$ that parametrizes ideals with a given Hilbert function h; and (2) to move towards a complete description of a cellular decomposition of $\mathrm{Hilb}^n(\mathbf{k}[\![x,y]\!])$ that is compatible with the Hilbert function stratification.

The paper is structured as follows. Section 2 introduces the machinery needed for studying Gröbner cells in the local setting and recalls the fundamental result of [**15**], namely the parametrization of Gröbner cells of lex-segment ideals in terms of canonical Hilbert-Burch matrices. In Section 3, we focus on the use — in both practise and theory — of such matrices to obtain deformations of the lex-segment ideal with desired properties, such as a given minimal number of generators or homogeneity. The main results of this section are Theorem 3.10 and Proposition 3.11, where we parameterize the Betti strata and the homogeneous sub-cell, respectively, of the Gröbner cell for lex-segment ideals. Our journey towards these statements includes revisiting classical results from the perspective of canonical Hilbert-Burch matrices. Finally, Section 4 is devoted to cellular decompositions of the punctual Hilbert scheme $\mathrm{Hilb}^n(\mathbf{k}[\![x,y]\!])$ that are compatible with its Hilbert function stratification. We explicitly compute such a decomposition for $n = 6$ and briefly describe the Julia programming language module we created to compute Gröbner cells of any monomial ideal according to the parametrization in [**15**, Conjecture 5.14]. We provide computational evidence for the correctness of the conjecture.

2. Gröbner cells in the local setting

In this section we will introduce the notation and recall some basics on local term orderings. We will finish with our main result from [**15**], which will be applied in the following sections to prove and reprove some properties of the punctual Hilbert scheme and its points, codimension two ideals in the power series ring.

DEFINITION 2.1. A term ordering τ in the polynomial ring $P = \mathbf{k}[x_1, \ldots, x_n]$ induces a reverse-degree ordering $\overline{\tau}$ in $R = \mathbf{k}[\![x_1, \ldots, x_n]\!]$ such that for any monomials m, m' in R, $m >_{\overline{\tau}} m'$ if and only if

$$\deg(m) < \deg(m') \text{ or } \deg(m) = \deg(m') \text{ and } m >_\tau m'.$$

We call $\overline{\tau}$ the **local term ordering** induced by the global ordering τ.

DEFINITION 2.2. Given an ideal J of R, we define the **leading term ideal** of J as the monomial ideal in P generated by the leading terms with respect to the local degree ordering $\overline{\tau}$, i.e. $\mathrm{Lt}_{\overline{\tau}}(J) = (\mathrm{Lt}_{\overline{\tau}}(f) : f \in J) \subset \mathbf{k}[x_1, \dots, x_n]$.

We call a subset $\{f_1, \dots, f_m\}$ of J a **$\overline{\tau}$-enhanced standard basis of J** if $\mathrm{Lt}_{\overline{\tau}}(J) = (\mathrm{Lt}_{\overline{\tau}}(f_1), \dots, \mathrm{Lt}_{\overline{\tau}}(f_m))$.

From the computational point of view, the tangent cone algorithm given by Mora in [**19**] — a variant of Buchberger's algorithm to compute Gröbner bases of ideals in the polynomial ring — allows us to explicitly obtain $\overline{\tau}$-enhanced standard basis of ideals in $\mathbf{k}[\![x_1, \dots, x_n]\!]$ that are generated by polynomials. See [**13**, Sections 1.6,1.7,6.4] for a detailed treatment and implementation.

From now on we will restrict to the codimension two case. Let $P = \mathbf{k}[x, y]$, $R = \mathbf{k}[\![x, y]\!]$ and fix $x >_\tau y$.

DEFINITION 2.3. Given a zero-dimensional monomial ideal E in R, we denote by $V(E) := \{J \subset R : \mathrm{Lt}_{\overline{\tau}}(J) = E\}$ the **Gröbner cell** of E with respect to the local term ordering $\overline{\tau}$.

As the name suggests these are affine spaces by results about group actions of $\mathbf{k}^*$ on smooth varieties with isolated fixed points of [**3**, Theorem 4.4]. They have been explicitly described — at least for a special class of monomial ideals — for the lexicographical term ordering [**6**], the degree lexicographical term ordering [**7**] and the local term ordering induced by both of them [**15**].

REMARK 2.4. It is important to note that, as opposed to what happens for global term orderings, ideals in a Gröbner cell with respect to a local term ordering preserve the Hilbert function of the monomial ideal, see [**15**, Corollary 5.10]. Indeed, by definition of local term ordering $\mathrm{Lt}_{\overline{\tau}}(f) = \mathrm{Lt}_\tau(f^*)$, where f^* is the initial form of f. In particular, any $\overline{\tau}$-enhanced standard basis $\{f_1, \dots, f_m\}$ of J is also a standard basis of J, namely the initial ideal J^* is generated by the initial forms $f_1^*, \dots, f_m^*$. Therefore, $\mathrm{Lt}_\tau(J^*) = \mathrm{Lt}_{\overline{\tau}}(J)$ and hence $\mathrm{HF}_{R/J} = \mathrm{HF}_{P/J^*} = \mathrm{HF}_{P/\mathrm{Lt}_{\overline{\tau}}(J)}$. So these Gröbner cells are compatible with the Hilbert function stratification of $\mathrm{Hilb}^n(\mathbf{k}[\![x, y]\!])$ and induce a cellular decomposition of the strata.

The strategy used in the description of the Gröbner cells of the three previously mentioned term orderings consists in focusing directly on the syzygies instead of the generators themselves and to that aim it is convenient to define the notion of canonical Hilbert-Burch matrix.

Note that any zero-dimensional monomial ideal in $\mathbf{k}[\![x, y]\!]$ can be written as $E = (x^t, \dots, x^{t-i}y^{m_i}, \dots, y^{m_t})$ with $0 = m_0 < m_1 \leq \cdots \leq m_t$.

DEFINITION 2.5. The **canonical Hilbert-Burch matrix** of the monomial ideal $E = (x^t, \dots, x^{t-i}y^{m_i}, \dots, y^{m_t})$ is the Hilbert-Burch matrix of E of the form

$$H = \begin{pmatrix} y^{d_1} & 0 & \cdots & 0 \\ -x & y^{d_2} & \cdots & 0 \\ 0 & -x & \cdots & 0 \\ \vdots & \vdots & & \vdots \\ 0 & 0 & \cdots & y^{d_t} \\ 0 & 0 & \cdots & -x \end{pmatrix},$$

where $d_i = m_i - m_{i-1}$ for any $1 \le i \le t$. The **degree matrix** U of E is the $(t+1) \times t$ matrix with integer entries $u_{i,j} = m_j - m_{i-1} + i - j$, for $1 \le i \le t+1$ and $1 \le j \le t$.

It follows from the definition that $u_{i,i} = d_i$ and $u_{i+1,i} = 1$, for $1 \le i \le t$.

DEFINITION 2.6. We denote by $\mathcal{M}(E)$ the set of matrices $N = (n_{i,j})$ of size $(t+1) \times t$ with entries in $\mathbf{k}[y]$ such that

- $n_{i,j} = 0$ for any $i \le j$,
- and for $i > j$ the non-zero entries satisfy $u_{i,j} \le \operatorname{ord}(n_{i,j}) \le \deg(n_{i,j}) < d_j$.

For the time being we will focus on a special class of monomial ideals called lex-segment ideals.

DEFINITION 2.7. A monomial ideal $L = (x^t, \dots, x^{t-i}y^{m_i}, \dots, y^{m_t})$ is a **lex-segment ideal** if and only if $0 < m_1 < \cdots < m_t$.

By Macaulay's theorem [**18**], any Hilbert function h has a unique lex-segment ideal $\operatorname{Lex}(h)$, see [**14**, Theorem 6.3.1] for a modern treatment. Consider an ideal J with Hilbert function h. In characteristic 0, after possibly a generic change of coordinates on J, $\operatorname{Lt}_{\overline{\tau}}(J)$ coincides with $\operatorname{Lex}(h)$. In other words, the lex-segment ideal is the generic initial ideal with respect to $\overline{\tau}$. For the reader familiar with generic initial ideals in the global setting, $\operatorname{Gin}_{\overline{\tau}}(J) = \operatorname{Gin}_{\tau}(J^*)$. See [**1**, Section 1.4] for an exposition about the notion of generic initial ideal in the local setting.

THEOREM 2.8. [**15**, Theorem 5.7] *Given a zero-dimensional lex-segment monomial ideal $L \subset R$ with canonical Hilbert-Burch matrix H, the map*

$$\begin{array}{rccc} \Phi_L : & \mathcal{M}(L) & \longrightarrow & V(L) \\ & N & \longmapsto & I_t(H+N) \end{array}$$

is a bijection, where $I_t(M)$ denotes the ideal of t-minors of the matrix M.

This theorem allows us to define the canonical Hilbert-Burch matrix of any zero-dimensional ideal $J \subset R$ that can be obtained as a deformation of a lex-segment ideal L as $H + \Phi_L^{-1}(J)$, where H is the canonical Hilbert-Burch matrix of the monomial ideal $\operatorname{Lt}_{\overline{\tau}}(J)$. In particular, the maximal minors of $H + \Phi_L^{-1}(J)$ form a $\overline{\tau}$-enhanced standard basis of J.

In characteristic 0, the affine space $V(L)$ encodes all ideals with a given Hilbert function up to generic change of coordinates. In other words, the map Φ_L provides a parametrization via canonical Hilbert-Burch matrices of all ideals with Hilbert function $h = \operatorname{HF}_{P/L}$ up to generic change of coordinates. For a formal version of this statement, see [**15**, Corollary 5.10]. See Section 4 for a discussion about the description of Gröbner cells and the notion of canonical Hilbert-Burch matrix beyond the lex-segment case.

3. Parametrizing ideals with a given Hilbert function

In this section we will showcase how to exploit canonical Hilbert-Burch matrices to obtain all ideals with a given Hilbert function, and special additional properties, up to generic change of coordinates. In the first part of the section, we will exhibit hands-on examples on how to refine the parametrization from Theorem 2.8

by adding constraints on the ideals such as the minimal number of generators or homogeneity. The second part is devoted to formalizing these procedures. We will revisit some classical results on admissible numbers of generators and on certain subschemes of the punctual Hilbert scheme $\mathrm{Hilb}^n(\mathbf{k}[\![x,y]\!])$, and give alternative proofs based on our canonical Hilbert-Burch matrices. This will assist us in reaching Theorem 3.10 and Proposition 3.11, where we provide explicit descriptions of the affine and quasi-affine varieties forming the Betti strata and homogeneous cells, respectively.

In what follows h is an admissible Hilbert function for some zero-dimensional ideal in $R = \mathbf{k}[\![x,y]\!]$, namely $h = (1, 2, \ldots, t, h_t, \ldots, h_s, 0)$, with $t \geq h_t \geq \cdots \geq h_s$, see [**18**]. Moreover, we will denote by $\Delta(h)$ the maximal jump in the Hilbert function h, that is $\Delta(h) := \max\{|h_i - h_{i-1}| : 1 \leq i \leq s\}$. For an ideal J we denote by $\mu(J)$ the minimal number of generators of J. We will say that a minimal number of generators d is admissible for a Hilbert function h if there exists an ideal J with Hilbert function h and $\mu(J) = d$.

Let $L = (x^t, x^{t-1}y^{m_1}, \ldots, y^{m_t})$ be the associated lex-segment ideal to the Hilbert function h and let us focus on special subsets of $V(L)$:

DEFINITION 3.1. For any admissible minimal number of generators d of the Hilbert function h, we call $V_d(L) = \{J \in V(L) : \mu(J) = d\}$ the d-**Betti stratum** of $V(L)$. The set $V_2(L)$ is the subset of $V(L)$ consisting of complete intersection ideals. Therefore, we will also refer to it as $V_{\mathrm{CI}}(L)$.

We denote the homogeneous sub-cell of $V(L)$ by $V_{\mathrm{hom}}(L) = \{J \in V(L) : J = J^*\}$.

In [**21**, Remark 4.7], Rossi and Sharifan give a procedure to obtain special instances of ideals in the d-Betti stratum $V_d(L)$. The authors considered very specific deformations of a Hilbert-Burch matrix H of the lex-segment ideal L: whenever $u_{i+2,i} \leq 0$, the entry $n_{i+2,i}$ is set to 1 and all other entries are set to 0. Note that a matrix N obtained this way belongs to $\mathcal{M}(L)$, hence the ideal $I_t(H+N)$ is in $V(L)$. This approach is a particular case of the fact that the minimal number of generators of $J \in V(L)$ decreases as the rank of the matrix $\bar{N}$ formed by the constant terms of $N \in \mathcal{M}(L)$ increases. We reproduce a tailored version of a result of Bertella [**1**, Lemma 2.1] and its proof here for the sake of completeness.

LEMMA 3.2. *Let L be a lex-segment ideal with canonical Hilbert-Burch matrix H and consider $J \in V(L)$. Then the minimal number of generators $\mu(J) = t + 1 - \mathrm{rk}(\bar{N})$, where N is the unique matrix in $\mathcal{M}(L)$ such that $J = I_t(H + N)$.*

PROOF. Since the columns of the matrix $H+N$ encode a system of generators of the syzygies of its maximal minors (that is, of the generators of the ideal J), the sequence $R^t \xrightarrow{H+N} R^{t+1} \longrightarrow J \longrightarrow 0$ is exact. Tensoring with $R/\mathfrak{m}$ over R yields

$$(R/\mathfrak{m})^t \xrightarrow{\bar{N}} (R/\mathfrak{m})^{t+1} \longrightarrow J/\mathfrak{m}J \longrightarrow 0,$$

hence the minimal number of generators of J is $\dim_{R/\mathfrak{m}} J/\mathfrak{m}J = t + 1 - \mathrm{rk}(\bar{N})$. □

In the following example, we will show how the parametrization of the Gröbner cell $V(L)$ via canonical Hilbert-Burch matrices generalizes the constructions of Rossi and Sharifan in [**21**].

EXAMPLE 3.3. Consider the lex-segment ideal $L = (x^4, x^3y, x^2y^5, xy^8, y^{10})$ with Hilbert function $h = (1, 2, 3, 4, 3, 3, 3, 2, 2, 1)$, canonical Hilbert-Burch matrix H and degree matrix U:

$$H = \begin{pmatrix} y & 0 & 0 & 0 \\ -x & y^4 & 0 & 0 \\ 0 & -x & y^3 & 0 \\ 0 & 0 & -x & y^2 \\ 0 & 0 & 0 & -x \end{pmatrix} \qquad U = \begin{pmatrix} 1 & 4 & 6 & 7 \\ 1 & 4 & 6 & 7 \\ -2 & 1 & 3 & 4 \\ -4 & -1 & 1 & 2 \\ -5 & -2 & 0 & 1 \end{pmatrix}.$$

By Theorem 2.8, any $J \in V(L)$ is the ideal of maximal minors of the matrix

$$H + N = \begin{pmatrix} y & 0 & 0 & 0 \\ -x & y^4 & 0 & 0 \\ c_1 & -x + c_2y + c_3y^2 + c_4y^3 & y^3 & 0 \\ c_5 & c_6 + c_7y + c_8y^2 + c_9y^3 & -x + c_{10}y + c_{11}y^2 & y^2 \\ c_{12} & c_{13} + c_{14}y + c_{15}y^2 + c_{16}y^3 & c_{17} + c_{18}y + c_{19}y^2 & -x + c_{20}y \end{pmatrix},$$

for some coefficients $c_1, \dots, c_{20} \in \mathbf{k}$. In other words, $V(L)$ is a 20-dimensional affine space with origin the lex-segment ideal L. Any ideal of the form $J = I_4(H + N)$ can therefore be identified with the point with coordinates $(c_1, \dots, c_{20})$ in $\mathbb{A}^{20}$.

Note that the coordinates $c_2, c_{10}, c_{17}, c_{20}$ play a special role: J is homogeneous if and only if those are the only non-zero coordinates. Indeed, they are the coefficients of degree matching the corresponding entry of the degree matrix. Thus V_{hom} as defined in Definition 3.1 is the subset of $V(L)$ where all other coordinates vanish. Let $\mathcal{I} \subset \mathbf{k}[c_1, \dots, c_{20}]$ be an ideal, then $\mathcal{V}(\mathcal{I}) \subset \mathbb{A}^{20}$ will denote the vanishing set of $\mathcal{I}$. And we can formalize the previous statement to:

$$V_{\text{hom}}(L) = \mathcal{V}(c_1, c_3, c_4, c_5, c_6, c_7, c_8, c_9, c_{11}, c_{12}, c_{13}, c_{14}, c_{15}, c_{16}, c_{18}, c_{19}) \simeq \mathbb{A}^4.$$

Since a $\overline{\tau}$-enhanced standard basis is in particular a standard basis, it offers a very simple procedure for computing the initial ideal J^* for a non-homogeneous J: set all $c_i = 0$ except for $i = 2, 10, 17, 20$, namely project onto $\mathbb{A}^4$.

Let us revisit the specific deformation of L given in [**21**, Example 4.8]. The authors constructed the complete intersection ideal $J = (x^4 - x^2y^2 - x^2y^3 - x^2y^4 + y^6, x^3y - xy^3 - xy^4) \in V(L)$ and calculated J^*. Note that $J = I_4(H + N)$ with $N \in \mathcal{M}(L)$ given by $c_1 = c_6 = c_{17} = 1$ and 0's otherwise. Indeed, J^* is obtained by keeping $c_{17} = 1$ and setting 0's otherwise.

Note that the third column of the canonical Hilbert-Burch matrix of J allows us to express f_4 as a combination of f_2, f_3, the second column yields f_3 as a combination of f_1 and f_2 and the first one gives f_2 as a combination of f_0 and f_1. In fact, any ideal obtained from a matrix satisfying $c_1c_6c_{17} \neq 0$ is a complete intersection generated by f_0 and f_1.

By Lemma 3.2, this is actually the only way to obtain complete intersections in $V(L)$. $V_{\text{CI}}(L)$ (as in Definition 3.1) is a full-dimensional quasi-affine variety obtained by removing the union of the three hyperplanes $\{c_1 = 0\}$, $\{c_6 = 0\}$ and $\{c_{17} = 0\}$ from $\mathbb{A}^{20}$.

Similarly, the sets of ideals in $V(L)$ with exactly 5, 4 and 3 generators can be described as follows:

$$V_5(L) = \mathcal{V}(c_1, c_5, c_6, c_{12}, c_{13}, c_{17}) \simeq \mathbb{A}^{14},$$
$$V_4(L) = \mathcal{V}(c_1c_6, c_1c_{13}, c_1c_{17}, c_5c_{17}, c_6c_{17}, c_5c_{13} - c_6c_{12})\backslash V_5(L),$$
$$V_3(L) = \mathcal{V}(c_1c_6c_{17})\backslash \mathcal{V}(c_1c_6, c_1c_{13}, c_1c_{17}, c_5c_{17}, c_6c_{17}, c_5c_{13} - c_6c_{12}).$$

In particular, homogeneous ideals with Hilbert function h will always have either 4 or 5 generators, since the only non-zero constant term of $H+N$ allowed in $V_{\mathrm{hom}}(L)$ is c_{17}.

We will now consider the closures of the d-Betti strata. The stratum of the maximal admissible minimal number of generators $V_5(L)$ is closed, of codimension 6 inside the 20-dimensional $V(L)$ and irreducible. The closure of the 4-Betti stratum has 3 irreducible components, all of them are reduced of codimension 3. The closure $\overline{V_3(L)}$ has 3 irreducible components, each of codimension 1. The stratum of complete intersection ideals is dense in $V(L)$, thus $\overline{V_{\mathrm{CI}}(L)} = V(L)$.

Next we will display an example where $\Delta(h) > 1$:

EXAMPLE 3.4. Consider the lex-segment ideal $L = (x^4, x^3y^2, x^2y^3, xy^5, y^7)$ with Hilbert function $h = (1, 2, 3, 4, 4, 2, 1)$ and degree matrix U:

$$U = \begin{pmatrix} 2 & 2 & 3 & 4 \\ 1 & 1 & 2 & 3 \\ 1 & 1 & 2 & 3 \\ 0 & 0 & 1 & 2 \\ -1 & -1 & 0 & 1 \end{pmatrix}$$

By Theorem 2.8, $V(L) \simeq \mathbb{A}^{12}$ and any ideal in $V(L)$ is of the form

$$J = I_4 \begin{pmatrix} y^2 & 0 & 0 & 0 \\ -x + c_1y & y & 0 & 0 \\ c_2y & -x & y^2 & 0 \\ c_3 + c_4y & c_5 & -x + c_6y & y^2 \\ c_7 + c_8y & c_9 & c_{10} + c_{11}y & -x + c_{12}y \end{pmatrix},$$

for some coefficients $c_1, \ldots, c_{12} \in \mathbf{k}$.

Note that J is homogeneous if and only if, in each entry (i, j) of the matrix above, coefficients of terms with degree other than $u_{i,j}$ vanish, namely $V_{\mathrm{hom}}(L) = \mathcal{V}(c_4, c_7, c_8, c_9, c_{11}) \simeq \mathbb{A}^7$.

We study the matrix of constant terms

$$\bar{N} = \begin{pmatrix} 0 & 0 & 0 & 0 \\ 0 & 0 & 0 & 0 \\ 0 & 0 & 0 & 0 \\ c_3 & c_5 & 0 & 0 \\ c_7 & c_9 & c_{10} & 0 \end{pmatrix} \tag{3.1}$$

to determine the sets of ideals with a given admissible minimal number of generators:

$$V_5(L) = \mathcal{V}(c_3, c_5, c_7, c_9, c_{10}) \simeq \mathbb{A}^7,$$
$$V_4(L) = \mathcal{V}(c_3c_{10}, c_5c_{10}, c_3c_9 - c_5c_7)\backslash V_5(L),$$
$$V_3(L) = \mathbb{A}^{12}\backslash \mathcal{V}(c_3c_{10}, c_5c_{10}, c_3c_9 - c_5c_7).$$

The stratum $V_5(L)$ with the maximal admissible minimal number of generators is again closed, as in Example 3.3. It is of codimension 5 and irreducible. The closure of $V_4(L)$ has two irreducible components, both of codimension 2. The stratum $V_3(L)$ with the minimal admissible minimal number of generators is dense in $V(L)$.

In this example, we can obtain homogeneous ideals in each of the admissible minimal numbers of generators since $\bar{N}$ is projected into $V_{\text{hom}}(L)$ just by setting $c_7 = c_9 = 0$, hence its rank can still take any value in $\{0, 1, 2\}$. For example, take any $J \in V_3(L)$, then its initial ideal

$$J^* = I_4 \begin{pmatrix} y^2 & 0 & 0 & 0 \\ -x + c_1 y & y & 0 & 0 \\ c_2 y & -x & y^2 & 0 \\ c_3 & c_5 & -x + c_6 y & y^2 \\ 0 & 0 & c_{10} & -x + c_{12} y \end{pmatrix}$$

is minimally generated by a quartic and two quintics and will have the graded resolution

$$0 \longrightarrow P(-6) \oplus P(-8) \longrightarrow P(-4) \oplus P^2(-5) \longrightarrow P \longrightarrow P/J^* \longrightarrow 0,$$

which can be obtained from the resolution of P/L by the only two admissible zero cancellations $(P(-6), P(-6))$ and $(P(-7), P(-7))$, in the notation of [**21**].

As an intermediate step towards proving the main result on Betti strata and homogeneous cells, we will reprove some classical results using the parametrization of Theorem 2.8. We will start with the well-known numerical condition on Hilbert functions that admit complete intersections, that is, there exists a complete intersection ideal with that Hilbert function. This is a result from Macaulay in [**18**], which has been reproved several times with different techniques in [**4, 16, 21**].

From now on we will assume that $\mathbf{k}$ is a field of characteristic 0. This assumption is necessary whenever we want to translate results on $V(\text{Lex}(h))$ to results on $\text{Hilb}^h(\mathbf{k}[\![x, y]\!])$, or from $V_{\text{hom}}(\text{Lex}(h))$ to results on $\text{Hilb}^h_{\text{hom}}(\mathbf{k}[\![x, y]\!])$. The computation of dimensions and strata in the Gröbner cells themselves hold for any arbitrary field $\mathbf{k}$.

PROPOSITION 3.5. *[Hilbert functions that admit a complete intersection] A Hilbert function h admits a complete intersection if and only if $\Delta(h) = 1$.*

PROOF. Combining Lemma 3.2 and the fact that $\text{Lex}(h)$ is the generic initial ideal, h admits a complete intersection if and only if $\text{rk}\,\bar{N}$ can reach $t-1$ for some $J = I_t(H+N)$ in $V(\text{Lex}(h))$, where $N \in \mathcal{M}(\text{Lex}(h))$. This can only occur if $u_{i,i-2} \leq 0$ for all $i = 3, \dots, t+1$. Since $0 < m_1 < \cdots < m_t$ in the lex-segment case, $u_{i,j} \leq 1$ for any $j \leq i-1$ and it is decreasing in the sense that $u_{i,j} \leq u_{i-1,j}, u_{i-1,j+1}, u_{i,j+1}$. More precisely, $u_{i,j} = 1$ iff $u_{i-1,j} = u_{i-1,j+1} = u_{i,j+1} = 1$ for $j \leq i-2$. From the fact that $d_k = 1$ for $k \geq 2$ if and only if there is a jump in h of height at least 2, we see that $u_{i,i-2} \leq 0$ for all $i = 3, \dots, t$ if and only if $\Delta(h) = 1$. □

The main point in the proof of Proposition 3.5 is to show that for a Hilbert function h with maximal jump $\Delta(h) = 1$ there exists $N \in \mathcal{M}(\text{Lex}(h))$ with $\text{rk}\,\bar{N} = t-1$. Similarly, for Hilbert functions h with $\Delta(h) > 1$ we can investigate what is the maximal rank of $\bar{N}$ that can be achieved.

EXAMPLE 3.6. Let us consider a Hilbert function $h = (1, 2, 3, 4, 5, 6, 7, 8, \dots)$ which starts decreasing at $t = 8$ and has maximal jump $\Delta(h) = 3$. We want to know which are the possible ranks of a matrix $\bar{N}$ formed by constant terms of $N \in \mathcal{M}(\mathrm{Lex}(h))$. We claim that this only depends on the size of the maximal jump and, more precisely, $0 \leq \mathrm{rk}\,\bar{N} \leq 5 = 8 - 3$.

To illustrate this, we will consider the following extra assumptions: h has a single jump by 3 (reflected by the fact that there are only two consecutive 1's in the main diagonal of the degree matrix U, namely $d_5 = d_6 = 1$) and a single jump by 2 (namely $d_3 = 1$).

$$U = \begin{pmatrix} d_1 & & & & & & & \\ 1 & d_2 & & & & & & \\ * & 1 & 1 & & & & & \\ * & 1 & 1 & d_4 & & & & \\ * & * & * & \mathbf{1} & \mathbf{1} & \mathbf{1} & & \\ * & * & * & \mathbf{1} & \mathbf{1} & \mathbf{1} & & \\ * & * & * & \mathbf{1} & \mathbf{1} & \mathbf{1} & d_7 & \\ * & * & * & * & * & * & 1 & d_8 \\ * & * & * & * & * & * & * & 1 \end{pmatrix} \qquad \bar{N} = \begin{pmatrix} 0 & 0 & 0 & 0 & 0 & 0 & 0 & 0 \\ 0 & 0 & 0 & 0 & 0 & 0 & 0 & 0 \\ * & 0 & 0 & 0 & 0 & 0 & 0 & 0 \\ * & 0 & 0 & 0 & 0 & 0 & 0 & 0 \\ \boxed{*} & * & * & 0 & 0 & 0 & 0 & 0 \\ * & \boxed{*} & * & 0 & 0 & 0 & 0 & 0 \\ * & * & \boxed{*} & \mathbf{0} & 0 & 0 & 0 & 0 \\ * & * & * & \boxed{*} & * & * & 0 & 0 \\ * & * & * & * & \boxed{*} & * & * & 0 \end{pmatrix}$$

FIGURE 1. Degree matrix U and matrix of constant terms $\bar{N}$.

The asterisks below the diagonal in U represent entries such that $u_{i,j} \leq 0$, hence the corresponding entries in $\bar{N}$ are allowed to be non-zero. Note that only the maximal size of a square of 1's (here 3) determines which is the longest main diagonal of non-zeroes (boxed asterisks) and hence $\mathrm{rk}(\bar{N}) \leq 5$. By considering a matrix N with (possibly zero) constants in the boxed entries and zeroes elsewhere, we can obtain ideals $J = I_8(H + N)$ with Hilbert function h and $4 \leq \mu(J) \leq 9$.

This argument is used in the following proposition on the minimal number of generators of an ideal with a given Hilbert function, reproving the results in [**1**, Theorem 2.3, Theorem 2.4]:

PROPOSITION 3.7 (Admissible number of generators of an ideal with a given Hilbert function). *Given h, any ideal J such that $\mathrm{HF}_{R/J} = h$ satisfies $\Delta(h) < \mu(J) \leq t + 1$, and we can find ideals with any admissible number of generators.*

PROOF. Combining Theorem 2.8 and Lemma 3.2, finding ideals J with $\Delta(h) < \mu(J) \leq t + 1$ is equivalent to proving that for $N \in \mathcal{M}(L)$, with $L = \mathrm{Lex}(h)$, the rank of $\bar{N}$ is in the range $0 \leq \mathrm{rk}\,\bar{N} \leq t - \Delta(h)$ and that for all the values l in this range there exists an $N \in \mathcal{M}(L)$ such that $\mathrm{rk}\,\bar{N} = l$.

In the proof of Proposition 3.5 we saw that $u_{i,j} = 1$ iff $u_{i-1,j} = u_{i-1,j+1} = u_{i,j+1} = 1$ for $j \leq i - 2$ and $d_k = 1$ for $k \geq 2$ iff $\Delta(h) > 1$. Therefore, $\Delta(h)$ is the size of the maximal square of 1's that appears in the degree matrix U, see Figure 1. It can be checked that it is precisely this size what gives an upper bound the maximal rank of $\bar{N}$: the maximal diagonal that admits non-zero entries has exactly $t - \Delta(h)$ entries (see boxed entries in Figure 1). By setting the entries in this diagonal to zero or non-zero values, we obtain matrices with the desired ranks. $\square$

We can also use the parametrization of Gröbner cells of lex-segment ideals to recover classical results on the dimension of special subschemes of the punctual Hilbert scheme $\mathrm{Hilb}^n(\mathbf{k}[\![x,y]\!])$. The dimension of the subscheme $\mathrm{Hilb}^h(\mathbf{k}[\![x,y]\!])$ of $\mathrm{Hilb}^n(\mathbf{k}[\![x,y]\!])$ parametrizing ideals with a given Hilbert function was already given in [**4**, Theorem III.3.1] and [**16**, Theorem 2.12], where $\mathrm{Hilb}^h(\mathbf{k}[\![x,y]\!])$ was denoted by $S(H^\alpha)$ and Z_T, respectively.

PROPOSITION 3.8 (Dimension of the subscheme of the punctual Hilbert scheme with a given Hilbert function). *The dimension of the subscheme of* $\mathrm{Hilb}^n(\mathbf{k}[\![x,y]\!])$ *corresponding to a certain Hilbert function h is*

$$\dim(\mathrm{Hilb}^h(\mathbf{k}[\![x,y]\!])) = n - t - \sum_{l\geq 2} n_l \binom{l}{2}, \tag{3.2}$$

where n_l denotes the number of jumps of height $l \geq 2$ in h.

PROOF. Consider the lex-segment ideal L associated to the Hilbert function h. It is immediate from the parametrization of $V(L)$ in Theorem 2.8 that $\dim V(L) = \sum_{2\leq j+1\leq i\leq t+1}(d_j - \max(u_{i,j},0))$. This expression can be rewritten as $n - t - \#\{u_{i,j} : u_{i,j} = 1 \text{ for } j \leq i-2\}$. Using a similar argument as in the proof of Proposition 3.5, 1's below the second main diagonal arise only when there is a jump of height $l \geq 2$ in the Hilbert function and they occur in U in disjoint lower triangular blocks of size $l-1$. Therefore, we derive that the cardinality of the set $\{u_{i,j} : u_{i,j} = 1 \text{ for } j \leq i-2\}$ is exactly the sum of the number of jumps of height $l \geq 2$ times $l(l-1)/2$. □

We reprove the result on the dimension of the subscheme $\mathrm{Hilb}^h_{\mathrm{hom}}(\mathbf{k}[\![x,y]\!])$ that parametrizes homogeneous ideals with a given Hilbert function presented in [**16**, Theorem 2.12] and [**6**, Corollory 3.1], where $\mathrm{Hilb}^h_{\mathrm{hom}}(\mathbf{k}[\![x,y]\!])$ was denoted by G_T and $\mathbb{G}(h)$ respectively:

PROPOSITION 3.9 (Dimension of the subscheme of homogeneous ideals with a given Hilbert function). *The dimension of the subscheme of* $\mathrm{Hilb}^h(\mathbf{k}[\![x,y]\!])$ *parametrizing homogeneous ideals is*

$$\dim(\mathrm{Hilb}^h_{\mathrm{hom}}(\mathbf{k}[\![x,y]\!])) = t + \sum_{i=t-1}^{s}(h_{i-1}-h_i)(h_i - h_{i+1}). \tag{3.3}$$

PROOF. By Theorem 2.8, the dimension of the affine space $V_{\mathrm{hom}}(L)$ is the sum of the number of 0's in the degree matrix U and the number of 1's in the strictly lower triangle of U for which their corresponding diagonal entry is strictly higher than 1. More precisely, $\dim V_{\mathrm{hom}}(L) = \#\{u_{i,j} : u_{i,j} = 0 \text{ for } j \leq i-2\} + \#\{u_{i,j} : u_{i,j} = 1 \text{ for } j \leq i-1 \text{ such that } d_j \neq 1\}$.

We start by counting 0's in the degree matrix. For any $3 \leq i \leq t+1$, $u_{i,i-2} = 0$ if and only if $d_{i-1} = 2$. Note that $d_k = 2$, for $k \geq 2$, is the indicator that there are two consecutive jumps of height $l_1 \geq 1$ and $l_2 \geq 1$, respectively, in the Hilbert function (after position t). This yields a rectangle of 0's of size $l_1 \times l_2$ and hence the number of 0's is the sum of the area of all such rectangles. Therefore,

$$\#\{u_{i,j} : u_{i,j} = 0 \text{ for } j \leq i-2\} = \sum_{i=t}^{s}(h_{i-1}-h_i)(h_i - h_{i+1}).$$

Let us now count the number of admissible 1's. If $d_1 \neq 1$, there are t such 1's. Indeed, in the case $\Delta(h) = 1$, the only 1's are in the third main diagonal and all of them are admissible. Alternatively, if $\Delta(h) > 1$, then we will have squares of 1's in the degree matrix whose edge lengths are the size of the jumps, but the admissible ones are only those in the first column of each square, so we still obtain t 1's.

However, if $d_1 = 1$, we have to remove the 1's that might be in the first column of U. These 1's occur if and only if $h_t < t$ and there will be $t - h_t$ many such 1's. Since $h_{t-2} - h_{t-1} = (t-1) - t = -1$, the number of admissible ones in U is exactly $t + (h_{t-2} - h_{t-1})(h_{t-1} - h_t) = t - (t - h_t)$ for any value of d_1. Adding the number of 0's to this quantity completes the proof. □

We are now in the position of stating the parametrization of the Betti strata of $\mathrm{Hilb}^h(\mathbf{k}[\![x,y]\!])$ up to generic change of coordinates. Recall that, for the lex-segment ideal L associated to h, the Gröbner cell $V(L)$ is an affine space of dimension $D := n - t - \sum_{l\geq 2} n_l \binom{l}{2}$. For the sake of clarity in the notation of the following theorem, $I_d(\bar{N})$ is an ideal in $\mathbf{k}[c_1, .., c_D]$, where $\bar{N}$ is the matrix obtained by taking only the constant terms of a matrix $N \in \mathcal{M}(L)$ in general form. See (3.1) for an example of $\bar{N}$.

THEOREM 3.10 (Betti strata). *Consider a Hilbert function h and its associated lex-segment ideal L. For any $\Delta(h) + 1 \leq d \leq t + 1$, the d-Betti stratum of the D-dimensional affine space $V(L)$, is the quasi-affine variety*

$$V_d(L) = \mathcal{V}(I_{t+2-d}(\bar{N}))\backslash\mathcal{V}(I_{t+1-d}(\bar{N})) \subset V(L) \simeq \mathbb{A}^D.$$

In particular, $V_{t+1}(L)$ is an affine space and

$$V_{\Delta(h)+1}(L) = \mathbb{A}^D\backslash\mathcal{V}(I_{t-\Delta(h)}(\bar{N}))$$

is a D-dimensional quasi-affine variety.

Note that a consequence of this result is that ideals that have the minimum minimal number of generators admissible for a given h, namely those in $V_{\Delta(h)+1}(L)$, are dense in $\mathrm{Hilb}^h(\mathbf{k}[\![x,y]\!])$, as proved in [**4**, Proposition III.2.1]. In particular, if $\Delta(h) = 1$, then $V_{\mathrm{CI}}(L)$ is the quasi-affine variety obtained by removing $t-1$ hyperplanes from $\mathbb{A}^D$.

PROPOSITION 3.11 (Homogeneous sub-cell). *$V_{\mathrm{hom}}(L)$ is a D_h-dimensional affine space, where D_h is the right-hand side of* (3.3). *Moreover, for any $J = I_t(H+N)$ its initial ideal J^* can be computed by projecting N into that affine space.*

PROOF. The maximal minors $f_0, \ldots, f_t$ of $H + N$ are a $\overline{\tau}$-enhanced standard basis of $J = I_t(H+N)$ by construction. By Remark 2.4, they also form a standard basis of J, namely $J^* = (f_0^*, \ldots, f_t^*)$. Let us call N^* the matrix obtained by only keeping the terms of degree $u_{i,j}$ in each entry $N_{i,j}$. Note that $N^* \in \mathcal{M}(L)$ and $I_t(H + N^*) = J^*$.

We claim that an ideal $J = I_t(H + N)$, for some $N \in \mathcal{M}(L)$, is homogeneous if and only the maximal minors $f_0, \ldots, f_t$ of $H + N$ are homogeneous. Indeed, homogeneity of a system of generators yields homogeneity of J. Conversely, if J is homogeneous then $J = I_t(H+N) = I_t(H+N^*)$, thus $N = N^*$ by Theorem 2.8. □

4. A cellular decomposition of the punctual Hilbert scheme

In this section we leave the class of lex-segment ideals and consider general zero-dimensional monomial ideals E in $\mathbf{k}[\![x,y]\!]$. The collection of all Gröbner cells $V(E)$ forms a cellular decomposition of the punctual Hilbert scheme $\mathrm{Hilb}^n(\mathbf{k}[\![x,y]\!])$. A cellular decomposition which additionally respects the Hilbert function stratification induces a cellular decomposition of $\mathrm{Hilb}^h(\mathbf{k}[\![x,y]\!])$, the stratum of $\mathrm{Hilb}^n(\mathbf{k}[\![x,y]\!])$

with a prescribed Hilbert function h. When working over $\mathbb{C}$ a cellular decomposition of a scheme can be used to calculate its Betti numbers. This was done in [**8**] for the Hilbert schemes of points of $\mathbb{P}^2$, $\mathbb{A}^2$ and the punctual Hilbert scheme $\mathrm{Hilb}^n(\mathbf{k}[\![x,y]\!])$. In [**11**] this result is refined for the stratum $\mathrm{Hilb}^h(\mathbf{k}[\![x,y]\!])$. Both papers use representation theoretical methods without giving explicit parametrizations of the cells. We conjecture what should be the parametrization of Gröbner cells for general E in terms of Hilbert-Burch matrices and give some detailed examples and evidence for our conjecture.

In [**15**, Proposition 5.1/Definition 5.2] we showed that the set of matrices $\mathcal{M}(E)$ from Definition 2.6 parametrizes a subset of $V(E)$, namely those ideals J in $V(E)$ that admit a $\overline{\tau}$-enhanced standard basis that also forms a Gröbner basis of the ideal $J \cap \mathbf{k}[x,y]$ with respect to the lexicographic ordering. We proved that for the class of monomial ideals $E = (x^t, x^{t-1}y^{m_1}, \ldots, y^{m_t})$ that satisfy

$$m_j - j - 1 \leq m_i - i \text{ for all } j < i, \tag{4.1}$$

this is already the whole Gröbner cell $V(E)$. The class includes lex-segment ideals, but for example also allows $m_i = m_{i+1}$ for exactly one i. For general E there exist ideals $J \in V(E)$ which do not admit such a $\overline{\tau}$-enhanced standard basis. What occurs in this situation is that $\mathrm{Lt}_{\mathrm{lex}}(J \cap \mathbf{k}[x,y]) \neq E$, for an example check [**15**, Lemma 5.12, Example 5.13]. In these cases, to obtain J as the ideal of maximal minors of $H + N$ it is no longer enough to consider lower triangular matrices N.

Definition 4.1. For a monomial ideal $E \subset \mathbf{k}[\![x,y]\!]$ we denote by $\mathcal{N}_{<\underline{d}}(E)$ the set of matrices $N = (n_{i,j})$ of size $(t+1) \times t$ with entries in $\mathbf{k}[y]$ such that its non-zero entries satisfy

- $u_{i,j} + 1 \leq \mathrm{ord}(n_{i,j}) \leq \deg(n_{i,j}) < d_i$ for any $i \leq j$,
- $u_{i,j} \leq \mathrm{ord}(n_{i,j}) \leq \deg(n_{i,j}) < d_j$ for any $i > j$.

Note that the set $\mathcal{M}(E)$ is a subset of $\mathcal{N}_{<\underline{d}}(E)$, and for ideals satisfying condition (4.1) $\mathcal{M}(E) = \mathcal{N}_{<\underline{d}}(E)$, since the inequality $u_{i,j} + 1 > d_i$ holds for $i \leq j$, and thence yields $n_{i,j} = 0$ for all $i \leq j$.

Conjecture 4.2. [**15**, Conjecture 5.14] Let $E \subset \mathbf{k}[\![x,y]\!]$ be a zero-dimensional monomial ideal. Then

$$\begin{array}{rccc} \Phi_E : & \mathcal{N}_{<\underline{d}}(E) & \longrightarrow & V(E) \\ & N & \longmapsto & I_t(H+N) \end{array}$$

is a bijection.

By [**15**, Lemma 4.4] it is clear that Φ_E is well-defined and that the minors form a $\overline{\tau}$-enhanced standard basis of $I_t(H+N)$. By Theorem 2.8, Conjecture 4.2 holds for lex-segment ideals. If the conjecture holds for all monomial ideals, we obtain a cellular decomposition of the whole punctual Hilbert scheme and can define a canonical Hilbert-Burch matrix for all ideals without considering a change of coordinates. This decomposition induces a cellular decomposition of $\mathrm{Hilb}^h(\mathbf{k}[\![x,y]\!])$. This is not the case for the cellular decomposition given in [**6**] because the Hilbert function of ideals in a cell can vary, see [**15**, Example 5.11].

Note that even without a proof of Conjecture 4.2, we can still prove a version of Proposition 3.11 for homogeneous sub-cells of Gröbner cells of general E. This result agrees with [**6**, Corollary 1] and shows that even though their Gröbner cells differ, the subset of homogeneous ideals agrees. This is no surprise, since for homogeneous f we have that $\mathrm{Lt}_{\overline{\tau}}(f) = \mathrm{Lt}_{\mathrm{lex}}(f)$ by definition.

LEMMA 4.3. *The homogeneous sub-cell $V_{\text{hom}}(E)$ of $V(E)$ is an affine space of dimension $\#\{(i,j) \mid i > j, 0 \leq u_{i,j} < d_j\}$.*

PROOF. Homogeneous ideals J in $V(E)$ are inside the set of ideals in $V(E)$ admitting a canonical Hilbert-Burch matrix. This is clear since for homogeneous f, the leading terms $\mathrm{Lt}_{\overline{\tau}}(f) = \mathrm{Lt}_{\text{lex}}(f)$ coincide, so a homogeneous $\overline{\tau}$-enhanced standard basis of J will also form a lex-Gröbner basis of $J \cap \mathbf{k}[x,y]$. Thence by [**15**, Proposition 5.1] $V_{\text{hom}}(E) \subset \phi_E(\mathcal{M}(E))$, and there exists a unique N such that $J = I_t(H+N)$. Now J is homogeneous if and only if $N \in \mathcal{M}(E)$ has only terms of degree $u_{i,j}$. Thus $V_{\text{hom}}(E)$ is an affine space of dimension $\#\{(i,j) \mid i > j, 0 \leq u_{i,j} < d_j\}$. □

EXAMPLE 4.4 (A cellular decomposition of $\mathrm{Hilb}^6(\mathbf{k}[\![x,y]\!])$). Let us consider $n = 6$. There are eleven partitions of 6, so the punctual Hilbert scheme $\mathrm{Hilb}^6(\mathbf{k}[\![x,y]\!])$ has a cellular decomposition into eleven cells. There are four possible Hilbert functions $[1,2,3]$, $[1,2,2,1]$, $[1,2,1,1,1]$ and $[1,1,1,1,1,1]$.

Note that this is significantly different from Briançon's table in [**4**] because there the author provides a representative of all possible analytic types of ideals in $\mathrm{Hilb}^6(\mathbf{k}[\![x,y]\!])$, whereas our cells contain ideals with a common leading term ideal but in general different analytic types coexist in the same cell (for example, ideals with different minimal number of generators).

The Hilbert function $[1,2,3]$ is only attained by the ideal $(x^3, x^2y, xy^2, y^3) = (x,y)^3$. By Proposition 3.8 the dimension of this stratum is

$$n - t - \sum_{l \geq 2} n_l \binom{l}{2} = 6 - 3 - 1 \cdot \binom{3}{2} = 0,$$

and the Gröbner cell consists only of the single point corresponding to the monomial ideal itself. It is minimally generated by four elements.

There are six cells with Hilbert function $[1,2,2,1]$, see Figure 2. Admissible parameters to obtain a homogeneous ideal, i.e. those of $V_{\text{hom}}(E)$, are indicated in bold. The Hilbert function has only jumps of height 1, so the maximal dimensional cell among them, corresponding to $m = (2,4)$, has dimension $n - t = 6 - 2 = 4$. General ideals in this cell will be minimally generated by two elements. If $c_2 = 0$ the ideal is generated by three elements.

The other five cells with this Hilbert function are not lex-segment. The cells $[2,2,2]$ and $[3,3]$, corresponding to monomial complete intersection ideals (x^3, y^2) and (x^2, y^3), consist only of complete intersection ideals. The two cells $[1,1,2,2]$ and $[1,1,1,3]$ only contain ideals that are minimally generated by 3 elements while the cell $[1,1,4]$ contains ideals I, J with $\mu(I) = 2$ and $\mu(J) = 3$. Notice that $\dim(V(E)) - \dim(V_{\text{hom}}(E)) = 1$ for all E with Hilbert function $[1,2,2,1]$.

There are two cells with Hilbert function $[1,2,1,1,1]$, see Figure 3. Both contain ideals minimally generated by three elements (in case $c_3 = 0$ for $E = (x^5, xy, y^2)$ or $c_1 = 0$ for $E = (x^2, xy, y^5)$) and complete intersection ideals.

The Hilbert function $[1,1,1,1,1,1]$ is attained by the monomial ideals (x^6, y) and the lex-segment monomial ideal (x, y^6). Both cells consist only of complete intersection ideals and have dimensions 4 and 5, see Figure 4. The cell of $m = [6]$ is the dense open subset of the punctual Hilbert scheme $\mathrm{Hilb}^6(\mathbf{k}[\![x,y]\!])$.

m	boxes	H+N	dimension	μ
		[1,2,3]		
$[1,2,3]$	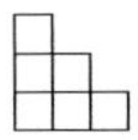	$\begin{pmatrix} y & 0 & 0 \\ -x & y & 0 \\ 0 & -x & y \\ 0 & 0 & -x \end{pmatrix}$	0	4
		[1,2,2,1]		
$[1,1,2,2]$	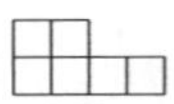	$\begin{pmatrix} y & 0 & 0 & c_1 \\ -x & 1 & 0 & 0 \\ 0 & -x & y & 0 \\ 0 & 0 & -x & 1 \\ 0 & 0 & 0 & -x \end{pmatrix}$	1	3
$[1,1,1,3]$	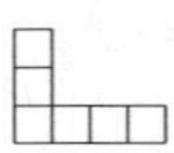	$\begin{pmatrix} y & 0 & c_1 & 0 \\ -x & 1 & 0 & 0 \\ 0 & -x & 1 & 0 \\ 0 & 0 & -x & y^2 \\ 0 & 0 & 0 & -x+\mathbf{c_2}y \end{pmatrix}$	2	3
$[2,2,2]$	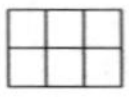	$\begin{pmatrix} y^2 & 0 & c_1 y \\ -x+\mathbf{c_2}y & 1 & 0 \\ 0 & -x & 1 \\ 0 & 0 & -x \end{pmatrix}$	2	2
$[1,1,4]$	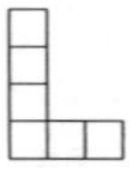	$\begin{pmatrix} y & 0 & 0 \\ -x & 1 & 0 \\ 0 & -x & y^3 \\ \mathbf{c_1} & 0 & -x+\mathbf{c_2}y+c_3y^2 \end{pmatrix}$	3	2,3
$[3,3]$	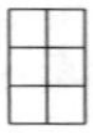	$\begin{pmatrix} y^3 & 0 \\ -x+\mathbf{c_1}y+c_2y^2 & 1 \\ \mathbf{c_3}y^2 & -x \end{pmatrix}$	3	2
$[2,4]$	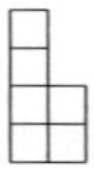	$\begin{pmatrix} y^2 & 0 \\ -x+\mathbf{c_1}y & y^2 \\ \mathbf{c_2}+c_3y & -x+\mathbf{c_4}y \end{pmatrix}$	4	2,3

FIGURE 2. Gröbner cells for Hilbert functions [1,2,3] and [1,2,2,1]

Note that here the difference between the dimension of $V(E)$ and $V_{\text{hom}}(E)$ is again constant for all E with the same Hilbert function: $\dim(V(E))-\dim(V_{\text{hom}}(E))$ is 3 for $[1,2,1,1,1]$ and 4 for $[1,1,1,1,1,1]$.

m	boxes	H+N	dimension	μ
		$[1,2,1,1,1]$		
$[1,1,1,1,2]$	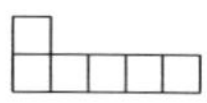	$\begin{pmatrix} y & 0 & c_1 & c_2 & c_3 \\ -x & 1 & 0 & 0 & 0 \\ 0 & -x & 1 & 0 & 0 \\ 0 & 0 & -x & 1 & 0 \\ 0 & 0 & 0 & -x & y \\ 0 & 0 & 0 & 0 & -x \end{pmatrix}$	3	2,3
$[1,5]$		$\begin{pmatrix} y & 0 \\ -x & y^4 \\ c_1 & -x+\mathbf{c_2}y+c_3y^2+c_4y^3 \end{pmatrix}$	4	2,3

FIGURE 3. Gröbner cells with Hilbert function [1,2,1,1,1]

m	boxes	H+N	dimension	μ
		$[1,1,1,1,1,1]$		
$[1]^6$	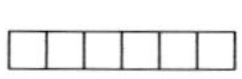	$\begin{pmatrix} y & 0 & c_1 & c_2 & c_3 & c_4 \\ -x & 1 & 0 & 0 & 0 & 0 \\ 0 & -x & 1 & 0 & 0 & 0 \\ 0 & 0 & -x & 1 & 0 & 0 \\ 0 & 0 & 0 & -x & 1 & 0 \\ 0 & 0 & 0 & 0 & -x & 1 \\ 0 & 0 & 0 & 0 & 0 & -x \end{pmatrix}$	4	2
$[6]$	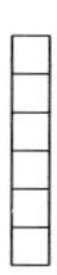	$\begin{pmatrix} y^6 \\ -x+\mathbf{c_1}y+c_2y^2+c_3y^3+c_4y^4+c_5y^5 \end{pmatrix}$	5	2

FIGURE 4. Gröbner cells with Hilbert function [1,1,1,1,1,1]

As stated before, for E not satisfying condition (4.1) the set $\mathcal{N}_{<\underline{d}}(E)$ contains matrices with non-zero entries above the diagonal. This is the case for $m = (1,1,2,2), (1,1,1,3), (2,2,2), (1,1,1,1,2)$ and $(1,1,1,1,1,1)$. For those ideals we checked by comparing to the reduced standard basis that Conjecture 4.2 gives a parametrization of the Gröbner cell.

We can investigate the dimensions of the occurring cells. When we define a_i as the number of cells of dimension i, we obtain

$$a = (a_0, a_1, a_2, a_3, a_4, a_5) = (1, 1, 2, 3, 3, 1).$$

This vector is an invariant of the space.

When a scheme X over $\mathbb{C}$ has a cellular decomposition with dimension vector a as defined in Example 4.4, then all other cellular decompositions of X will have the same dimension vector. It holds that $a_i = b_{2i}(X)$, the $2i$-th Betti number of X, and that the Betti numbers $b_{2i+1}(X) = 0$, see [**3**, Theorem 4.4/4.5] and [**9**, Chapter 19.1].

In [**8**] this method is used to calculate the Betti numbers of the Hilbert scheme of points of the projective plane, the affine plane and, of most interest to us, the punctual Hilbert scheme $\mathrm{Hilb}^n(\mathbf{k}[\![x, y]\!])$. To state their theorem, we need the following definition.

Definition 4.5. Let $l, n \in \mathbb{Z}_{>0}$, then we define $P(n, l)$ as the number of partitions of n bounded by l, i.e. as the number of sequences $0 = m_0 \leq m_1 \leq \cdots \leq m_t \leq l$ such that $\sum_{i=1}^{t} m_i = n$.

Theorem 4.6. [**8**, Theorem 1.1 (iv)] *The non-zero Betti numbers of the punctual Hilbert scheme are*

$$b_{2i}(\mathrm{Hilb}^n(\mathbf{k}[\![x, y]\!])) = P(i, n - i).$$

This result gives us a way of checking Conjecture 4.2 for plausibility by checking if the number of monomial ideals E with $\dim(\mathcal{N}_{<\underline{d}}(E)) = i$ is equal to $P(i, n - i)$.

We created a Julia ([**2**]) module that uses OSCAR.jl ([**20**]) to calculate examples and perform this plausibility check. The module and Jupyter-notebooks with experiments are available at `https://github.com/anelanna/LocalHilbertBurch.jl`.

The module can calculate the (conjectural) cellular decomposition for a given n. For all partitions of n it creates a `Cell` that has the following properties:

- `m` – the partition,
- `E` – its associated monomial ideal,
- `d` – the vector of differences,
- `U` – the degree matrix,
- `hilb` – the Hilbert function of the ideals in the cell,
- `H` – the canonical Hilbert-Burch matrix of the monomial ideal,
- `M` – the general canonical Hilbert-Burch matrix $H + N$,
- `N` – the corresponding element in $\mathcal{N}_{<\underline{d}}(E)$,
- `I` – the maximal minors of $M = H + N$, and
- `dim` – the dimension of the cell.

The vector of dimensions a can be computed with `sorted_celllist(n)`, a function that returns a dictionary mapping an integer i to a vector with the cells of dimension i.

The numbers of bounded partitions can be found in the On-Line Encyclopedia of Integer Sequences, [**17**], as (diagonals in) A008284, or relabelled, and thus serving our purposes a bit better, as (diagonals in) A058398 (see `https://oeis.org/A008284` and `https://oeis.org/A058398`).

For $n \leq 50$ we calculated all dimension vectors of our proposed cellular decomposition and checked that they are the correct ones. A stronger evidence for Conjecture 4.2 would be that $\phi_E : \mathcal{N}_{<\underline{d}}(E) \to V(E)$ is an isomorphism of affine spaces for all $n \leq 40$. We have checked this for $n \leq 7$ where we described $V(E)$ by means of the reduced $\overline{\tau}$-enhanced standard bases.

EXAMPLE 4.7 (Example 4.4 continued). We can use the connection between cellular decompositions and Betti numbers described above to calculate the Betti numbers of $\mathrm{Hilb}^h(\mathbf{k}[\![x,y]\!])$ for $\mathbf{k} = \mathbb{C}$ and h an admissible Hilbert function in the $n = 6$ case. We obtain the following Betti numbers:

$$b_i(\mathrm{Hilb}^{(1,2,3)}(\mathbf{k}[\![x,y]\!])) = \begin{cases} 1, & i = 0; \\ 0, & \text{otherwise.} \end{cases}$$

$$b_i(\mathrm{Hilb}^{(1,2,2,1)}(\mathbf{k}[\![x,y]\!])) = \begin{cases} 1, & i = 2,8; \\ 2, & i = 4,6; \\ 0, & \text{otherwise.} \end{cases}$$

$$b_i(\mathrm{Hilb}^{(1,1,1,2,1)}(\mathbf{k}[\![x,y]\!])) = \begin{cases} 1, & i = 6,8; \\ 0, & \text{otherwise.} \end{cases}$$

$$b_i(\mathrm{Hilb}^{(1,1,1,1,1,1)}(\mathbf{k}[\![x,y]\!])) = \begin{cases} 1, & i = 8,10; \\ 0, & \text{otherwise.} \end{cases}$$

REMARK 4.8. We can describe the subscheme $\mathrm{Hilb}^{(1,1,\ldots,1)}(\mathbf{k}[\![x,y]\!])$ of the punctual Hilbert scheme for any n. There are only two cells with this Hilbert function, namely $L = (x, y^n)$ and $E = (x^n, y)$. The dimension of the cell of the lex-segment ideal L is $n-1$. By considering the reduced $\overline{\tau}$-enhanced standard basis of E we can verify that Conjecture 4.2 holds for E or just directly show that $\dim(V(E)) = n - 2$. So the cellular decomposition has one cell of dimension $n-2$ and one of dimension $n-1$. Hence, the only non-vanishing Betti numbers are $b_{2(n-2)}(\mathrm{Hilb}^{(1,\ldots,1)}(\mathbf{k}[\![x,y]\!])) = b_{2(n-1)}(\mathrm{Hilb}^{(1,\ldots,1)}(\mathbf{k}[\![x,y]\!])) = 1$. This agrees with the observations in [**11**, Remark 2.2a].

Another relevant observation from Example 4.4 is that within each stratum $\mathrm{Hilb}^h(\mathbf{k}[\![x,y]\!])$ of the punctual Hilbert scheme $\mathrm{Hilb}^n(\mathbf{k}[\![x,y]\!])$, the difference of dimensions between any local Gröbner cell and its homogeneous sub-cell is constant. This fact actually follows from a result by Iarrobino:

THEOREM 4.9. [**16**, Theorem 2.11, Theorem 3.14] *The stratrum* $\mathrm{Hilb}^h(\mathbf{k}[\![x,y]\!])$ *is a locally trivial bundle over* $\mathrm{Hilb}^h_{\mathrm{hom}}(\mathbf{k}[\![x,y]\!])$ *having fibre, an affine space, and a global section.*

The cellular decomposition of $\mathrm{Hilb}^h(\mathbf{k}[\![x,y]\!])$ given by all Gröbner cells $V(E)$ where E is a monomial ideal with Hilbert function h, induces a cellular decomposition of $\mathrm{Hilb}^h_{\mathrm{hom}}(\mathbf{k}[\![x,y]\!])$ by taking homogeneous sub-cells $V_{\mathrm{hom}}(E)$. By a combination of Proposition 3.11 and Lemma 4.3 the fibration restricts to $V(E) \to V_{\mathrm{hom}}(E)$. Since we know the dimension of $V_{\mathrm{hom}}(E)$ by Lemma 4.3, we can check (at least in examples) whether the difference $\dim(\mathcal{N}_{<\underline{d}}(E)) - \dim(V_{\mathrm{hom}}(E))$ is constant for all E with the same Hilbert function. If this holds, it implies that our conjectured cells have the right dimension. Using our `Julia` module, we have checked this for all strata $\mathrm{Hilb}^h(\mathbf{k}[\![x,y]\!])$ of $\mathrm{Hilb}^n(\mathbf{k}[\![x,y]\!])$ with $n \leq 50$, thus providing strong evidence for Conjecture 4.2.

Acknowledgments

We want to thank Lars Kastner for helping us to set up the Julia module and helping with technical issues. We thank the two anonymous referees for many useful comments on a preliminary version of this paper. The first author wants to thank Anthony Iarrobino, Pedro Macias Marques, Maria Evelina Rossi, and Jean Vallès, organizers of the AMS-EMS-SMF Special Session on Deformation of Artinian Algebras and Jordan Types, for their invitation to the session.

References

[1] V. Bertella, *Hilbert function of local Artinian level rings in codimension two*, J. Algebra **321** (2009), no. 5, 1429–1442, DOI 10.1016/j.jalgebra.2008.11.023. MR2494398

[2] J. Bezanson, A. Edelman, S. Karpinski, and V. B. Shah, *Julia: a fresh approach to numerical computing*, SIAM Rev. **59** (2017), no. 1, 65–98, DOI 10.1137/141000671. MR3605826

[3] A. Białynicki-Birula, *Some theorems on actions of algebraic groups*, Ann. of Math. (2) **98** (1973), 480–497, DOI 10.2307/1970915. MR366940

[4] J. Briançon, *Description de* $H\mathrm{ilb}^n C\{x,y\}$, Invent. Math. **41** (1977), no. 1, 45–89, DOI 10.1007/BF01390164. MR457432

[5] J. Briançon and A. Galligo, *Déformations distinguées d'un point de* $\mathbf{C}^2$ *ou* $\mathbf{R}^2$ (French), Singularités à Cargèse (Rencontre Singularités Géom. Anal., Inst. Études Sci., Cargèse, 1972), Astérisque, Nos. 7 et 8, Soc. Math. France, Paris, 1973, pp. 129–138. MR361139

[6] A. Conca and G. Valla, *Canonical Hilbert-Burch matrices for ideals of* $k[x,y]$, Michigan Math. J. **57** (2008), 157–172, DOI 10.1307/mmj/1220879402. Special volume in honor of Melvin Hochster. MR2492446

[7] A. Constantinescu, *Parametrizations of ideals in* $K[x,y]$ *and* $K[x,y,z]$, J. Algebra **346** (2011), 1–30, DOI 10.1016/j.jalgebra.2011.07.029. MR2842069

[8] G. Ellingsrud and S. A. Strømme, *On the homology of the Hilbert scheme of points in the plane*, Invent. Math. **87** (1987), no. 2, 343–352, DOI 10.1007/BF01389419. MR870732

[9] W. Fulton, *Intersection theory*, 2nd ed., Ergebnisse der Mathematik und ihrer Grenzgebiete. 3. Folge. A Series of Modern Surveys in Mathematics [Results in Mathematics and Related Areas. 3rd Series. A Series of Modern Surveys in Mathematics], vol. 2, Springer-Verlag, Berlin, 1998, DOI 10.1007/978-1-4612-1700-8. MR1644323

[10] A. Galligo, *À propos du théorème de-préparation de Weierstrass* (French), Fonctions de plusieurs variables complexes (Sém. François Norguet, 1970–1973; à la mémoire d'André Martineau), Lecture Notes in Math., Vol. 409, Springer, Berlin-New York, 1974, pp. 543–579. Thèse de 3ème cycle soutenue le 16 mai 1973 à l'Institut de Mathématique et Sciences Physiques de l'Université de Nice. MR402102

[11] L. Göttsche, *Betti numbers for the Hilbert function strata of the punctual Hilbert scheme in two variables*, Manuscripta Math. **66** (1990), no. 3, 253–259, DOI 10.1007/BF02568495. MR1031195

[12] M. Granger, *Géométrie des schémas de Hilbert ponctuels* (French, with English summary), Mém. Soc. Math. France (N.S.) **8** (1983), 84. MR716783

[13] G.-M. Greuel and G. Pfister, *A Singular introduction to commutative algebra*, Second, extended edition, Springer, Berlin, 2008. With contributions by Olaf Bachmann, Christoph Lossen and Hans Schönemann; With 1 CD-ROM (Windows, Macintosh and UNIX). MR2363237

[14] J. Herzog and T. Hibi, *Monomial ideals*, Graduate Texts in Mathematics, vol. 260, Springer-Verlag London, Ltd., London, 2011, DOI 10.1007/978-0-85729-106-6. MR2724673

[15] R. Homs and A.-L. Winz, *Canonical Hilbert-Burch matrices for power series*, J. Algebra **583** (2021), 1–24, DOI 10.1016/j.jalgebra.2021.04.021. MR4261197

[16] A. A. Iarrobino, *Punctual Hilbert schemes*, Mem. Amer. Math. Soc. **10** (1977), no. 188, viii+112, DOI 10.1090/memo/0188. MR485867

[17] OEIS Foundation Inc., *The on-line encyclopedia of integer sequences*, Published electronically at https://oeis.org., 2023.

[18] F. S. Macaulay, *Some Properties of Enumeration in the Theory of Modular Systems*, Proc. London Math. Soc. (2) **26** (1927), 531–555, DOI 10.1112/plms/s2-26.1.531. MR1576950

[19] T. Mora, *An algorithm to compute the equations of tangent cones*, Proceedings of the European Computer Algebra Conference on Computer Algebra (Berlin, Heidelberg), EUROCAM '82, Springer-Verlag, 1982.

[20] *Oscar – open source computer algebra research system, version 0.11.3-dev*, 2023.

[21] M. E. Rossi and L. Sharifan, *Consecutive cancellations in Betti numbers of local rings*, Proc. Amer. Math. Soc. **138** (2010), no. 1, 61–73, DOI 10.1090/S0002-9939-09-10010-2. MR2550170

[22] J. Yaméogo, *Sur l'alignement dans les schémas de Hilbert ponctuels du plan. La famille des N-uplets de $\mathbf{P}^2$ contenant au moins r points sur une droite* (French), Math. Ann. **285** (1989), no. 3, 511–525, DOI 10.1007/BF01455071. MR1019716

Centre de Recerca Matemàtica, Campus de Bellaterra, Edifici C, 08193 Bellaterra, Barcelona

Email address: `rhoms@crm.cat`

Freie Universität Berlin, Arnimallee 3, 14195 Berlin, Germany

Email address: `winz.math@posteo.de`

Contemporary Mathematics
Volume **805**, 2024
https://doi.org/10.1090/conm/805/16134

The Hilbert function of local Artinian level algebras of codimension 3 and type 2

Shreedevi K. Masuti

Abstract. Classification of Hilbert functions of Artinian algebras with additional properties, like Gorenstein or more generally level algebras, plays an important role in singularity theory and in algebraic geometry. In this paper, we investigate the structure of Hilbert functions of quadratic Artinian level algebras of type 2 and codimension 3. We observe that if $h = (1, 3, 3, h_3, \ldots, 2)$ is the Hilbert function of a level algebra, then $|h_i - h_{i-1}| \leq 4$ for all $i \geq 3$. Moreover, we show that if $\max\{|h_i - h_{i-1}| : i \geq 1\} \leq 2$, then h is a level sequence of type 2. Using these results, we give a complete list of $(1, 3, 3)$ level sequences of type 2 and socle degree at most 6.

1. Introduction

The purpose of this paper is to investigate the Hilbert functions of Artinian level K-algebras. Artinian level algebras have been studied extensively in different contexts, especially in the graded case, see for instance [**Boi94**],[**Boi99**], [**Boi00**], [**BI09**], [**GHMS07**]. They play an important role in the study of the irreducibility and the smoothness of the punctual Hilbert scheme $\mathrm{Hilb}_d(\mathbb{P}^n_K)$ parameterizing zero-dimensional subschemes in $\mathbb{P}^n_K$ of degree d, see among others [**Poo08**], [**EV08**], [**CEVV09**], [**CENR13**],[**CJN15**].

Let $(A, \mathfrak{m})$ be an Artinian local K-algebra where $K \simeq A/\mathfrak{m}$. Let s be the socle degree of A, that is the maximum integer j such that $\mathfrak{m}^j \neq 0$. One of the most important numerical invariant associated to A is its Hilbert function. By definition the Hilbert function $H(A) = (h_i(A))_{i \geq 0}$, where $h_i = h_i(A) := \dim_K \mathfrak{m}^i/\mathfrak{m}^{i+1}$, is the Hilbert function of the associated graded ring $gr_{\mathfrak{m}}(A) := \oplus_{i \geq 0} \mathfrak{m}^i/\mathfrak{m}^{i+1}$. In [**Mac27**] Macaulay characterized the possible Hilbert functions of A. We call a sequence of positive integers $h = (h_0, h_1, \ldots, h_s)$ satisfying Macaulay's criterion, that is

$$h_0 = 1 \text{ and } h_{i+1} \leq h_i^{\langle i \rangle}, \tag{1.1}$$

2020 *Mathematics Subject Classification.* Primary 13D40, 13H10; Secondary 13A30, 13J05, 13F20.

Key words and phrases. Hilbert functions, level rings, Macaulay inverse system.

SKM is supported by INSPIRE faculty award funded by Department of Science and Technology, Govt. of India. She is also supported by CRG grant CRG/2022/007572 and MATRICS grant MTR/2022/000816 funded by SERB, Govt. of India.

for $i = 1, \dots, s-1$, an O-sequence (see [**BH98**, page 161] for the notation $h_i^{(i)}$). Since then there has been a great interest in commutative algebra in determining the Hilbert function of A with additional properties, as complete intersection, Gorenstein, level, etc. Recall that an Artinian local algebra $(A, \mathfrak{m})$ is said to be *level* of type τ if $(0 :_A \mathfrak{m}) = \mathfrak{m}^s$ and $\dim_K \mathfrak{m}^s = \tau$. In particular, if A has type 1, then A is *Gorenstein.* The Hilbert functions of graded Gorenstein algebras have been studied extensively in the literature. Relatively little is known in codimension three or larger about the Hilbert functions of Artinian local algebras that are not graded.

Recall that a graded K-algebra $B := \oplus_{i=0}^s B_i$, where $s := \max\{i : B_i \neq 0\}$, with the unique maximal ideal $\mathcal{M} = \oplus_{i>0} B_i$ is said to be level if $(0 :_B \mathcal{M}) = B_s$. The Hilbert functions of *graded* Artinian level algebras have been investigated in [**AS12, CI01, Iar05, Zan06, GHMS07**]. We refer to [**Iar05**] and references therein for a brief survey on the Hilbert functions of graded level algebras. The geometric properties of the family of Artinian graded level algebras of codimension two have been studied in [**Iar04, CG03**].

Since $gr_{\mathfrak{m}}(A)$ need not be level even if A is level, it is difficult to determine the structure of the Hilbert functions of *local* level algebras even if the corresponding result is known in the graded case. The Hilbert functions of Artinian level algebras of codimension two are known classically due to Macaulay [**Mac1904**]. However, the problem is widely open if A has codimension greater than two. In [**Ste14**] (resp. [**MR18**]) the author (resp. authors) classified the Hilbert functions of Artinian level algebras of socle degree at most 3 (resp. socle degree 4 and codimension 3). We say that a sequence $h = (1, h_1, \dots, h_s)$ is a *level (resp. Gorenstein) sequence of type* τ if h is the Hilbert function of some Artinian level (resp. Gorenstein) K-algebra A of type τ. By a *graded level sequence of type* τ we mean that h is the Hilbert function of some graded level K-algebra of type τ.

This paper deals with the Hilbert functions of Artinian level algebras of codimension three and type two. In our setting we assume that $A = R/I$ is an Artinian level ring of type two where $R = \mathrm{K}[\![x, y, z]\!]$ and $I \subseteq (x, y, z)^2$ is an ideal of R such that the Hilbert function h of A satisfies $h_1 = h_2 = 3$. Write h as a vector $h = (1, 3, 3, h_3, h_4, \dots, h_s)$. Since A is level of type 2, $h_s = 2$. For simplicity we call an O-sequence of the form $h = (1, 3, 3, h_3, h_4, \dots, h_s)$ a $(1, 3, 3)$ *sequence.* The $(1, 3, 3)$ Gorenstein sequences have been classified by the authors in [**JMR23**]. In this paper we generalize their result and classify $(1, 3, 3)$ level sequences of type two, provided $\Delta(h) \leq 2$ where

$$\Delta(h) := \max\{|h_i - h_{i-1}| : i = 3, \dots, s\}. \tag{1.2}$$

For an O-sequence h, we set $\max h := \max\{h_i : i \geq 0\}$. We say that h has a *fall* by m at i (resp. a *fall* m) if $h_{i+1} = h_i - m$ (resp. h has a fall by m at i for some i). By (1.1), if h is a $(1, 3, 3)$ sequence, then $h_3 \leq 4$, and h is unimodal. The smallest t such that $h_{t+1} < h_t$ is called the *peak position* of h.

The following is the main result of the paper.

THEOREM 1.1. *1 Let h be a $(1, 3, 3)$ sequence. If h is a level sequence of type 2, then $\Delta(h) \leq 4$. Moreover,*

(a) *If $\Delta(h) = 3$, then h has at most two positions with a fall by three;*

(b) *If $\Delta(h) = 4$, then h has a* ***unique*** *fall by* ***four*** *at the* ***peak*** *position and* ***no fall by three****. In this case, h is of the form:*

$$h_i = \begin{cases} i+1 & \text{for } i \leq d-2 \\ d & \text{for } i = d-1, d, \ldots, d+r-1 \\ d-4 & \text{for } i = d+r \\ h_{i-1} \text{ or } h_{i-1}-1 \text{ or } h_{i-1}-2 & \text{for } i \geq d+r+1. \end{cases}$$

for some integers $d = \max h \geq 4$ and $r \geq 0$.

Conversely, a $(1,3,3)$ sequence h with $\Delta(h) \leq 2$ is a level sequence of type 2.

As a consequence of this result and the list of graded level sequence given in [**GHMS07**, Part 2], we give a complete list of $(1,3,3)$ level sequences of type 2 and socle degree at most 6 in Section 5. In fact, for a given $(1,3,3)$ sequence with $s \leq 6$ and $h_s = 2$, we provide the inverse system $I^{\perp}$ of an ideal I such that A/I is level of type 2 and has the Hilbert function h.

In order to prove Theorem 1.1 we analyse the Macaulay inverse system of level ideals I in R. We obtain the Macaulay inverse system of $J := I \cap K[\![x,y]\!]$. This reduces the problem of classification of $(1,3,3)$ level sequences to the codimension two case. We remark that the Hilbert functions of Artinian codimension two algebras of any socle type are known, see [**Iar84**, Theorem 4.6C]. See also [**Mac1904, Ber09, MR15**] for a construction of Artinian codimension two algebras with a given number of generators. We emphasise that we heavily use the structure of the Hilbert functions of codimension two Artinian Algebras (not just level algebras). In fact, in the Step 2 of the algorithm presented in Section 4, we need to construct a level algebra of codimension two and type 2. A similar approach is considered for Gorenstein ideals in [**JMR23**]. However, in this paper we give a different method for constructing the level algebras that we require.

A reason for restricting the converse to $\Delta(h) \leq 2$ in Theorem 1.1 is that we are unable to construct a $(1,3,3)$ level sequence of type 2 with $\Delta(h) > 2$. Also, if $\Delta(h) = 3$, then we do not know the positions at which the falls by three occur.

We compare graded and local $(1,3,3)$ level sequences of type 2 in Proposition 4.5. We show that the only graded level $(1,3,3)$ sequence of type 2 is $(1,3,3,3,\ldots,3,3,2)$. Observe that there are more than one $(1,3,3)$ level sequences of socle degree 5 (see Table 1) and of socle degree 6 (see Table 2). This exhibits a difference between $(1,3,3)$ graded and level O-sequences.

The paper is organized as follows. We gather a few basic results on Macaulay inverse systems in Section 2. In Section 3 we provide necessary conditions for a $(1,3,3)$ sequence to be level of type 2. The Section 4 is devoted to constructing a level sequence of type 2 with the Hilbert function h and $\Delta(h) \leq 2$.

We have used the computer algebra systems [**Macaulay2, Singular**] and the library [**Eli15**] for various computations in this paper.

2. Macaulay Inverse System

In this section we recall some basic results about the Macaulay inverse system, which is an important tool used in this paper to determine the $(1,3,3)$ level sequences of type 2.

Let $R = K[\![x_1,\ldots,x_m]\!]$ and let R_i denote the K-vector space spanned by homogeneous polynomials of degree i. It is known that the injective hull of K as an R-module is isomorphic to the following ring (see [**Eis95**, Exercise A3.4(b)] and [**Gab59**]):

$$E_R(K) \cong \mathcal{D} := K_{DP}[X_1, \ldots, X_m] = \bigoplus_{i \geq 0} \operatorname{Hom}_K(R_i, K),$$

called the *divided power ring.* The ring $\mathcal{D}$ has a structure of R-module via the following contraction action: For $\mathbf{n} = (n_1, \ldots, n_m), \mathbf{n}' = (n_1', \ldots, n_m') \in \mathbb{N}^m$, $x^{\mathbf{n}} = x_1^{n_1} \cdots x_m^{n_m} \in R$ and $X^{\mathbf{n}'} = X_1^{n_1'} \cdots X_m^{n_m'} \in \mathcal{D}$, we define

$$x^{\mathbf{n}} \circ X^{\mathbf{n}'} := \begin{cases} X_1^{n_1'-n_1} \cdots X_m^{n_m'-n_m} & \text{if } \mathbf{n}' \geq \mathbf{n} \\ 0 & \text{otherwise,} \end{cases} \tag{2.1}$$

where by $\mathbf{n}' \geq \mathbf{n}$ we mean that $n_i' \geq n_i$ for all $i = 1, \ldots, m$. The contraction action is extended K-linearly to R and $\mathcal{D}$.

If $I \subseteq R$ is an ideal of R, then $(R/I)^\vee := \operatorname{Hom}_R(R/I, E_R(K))$ can be identified with the R-submodule $I^\perp$ of $\mathcal{D}$ which is defined as

$$I^\perp := \{F \in \mathcal{D} : f \circ F = 0 \text{ for all } f \in I\}.$$

This submodule is called the *Macaulay inverse system of I.* On the other hand, given an R-submodule W of $\mathcal{D}$, $W^\vee := \operatorname{Hom}_R(W, E_R(K))$ is the ring $R/\operatorname{Ann}_R(W)$ where

$$\operatorname{Ann}_R(W) := \{f \in R : f \circ F = 0 \text{ for all } F \in W\}$$

is an ideal of R. If $\{F_1, \ldots, F_t\} \subseteq \mathcal{D}$ is a set of polynomials, we will denote by $\langle F_1, \ldots, F_t\rangle$ the R-submodule of $\mathcal{D}$ generated by $F_1, \ldots, F_t$, i.e., the K-vector space generated by $F_1, \ldots, F_t$ and the corresponding elements of $\mathcal{D}$ generated by the contraction action. If M is an R-submodule of $\mathcal{D}$ generated by polynomials $\underline{F} := F_1, \ldots, F_t$, then we will write $\operatorname{Ann}_R(M) = \operatorname{Ann}_R(\underline{F})$ and $A_{\underline{F}} = R/\operatorname{Ann}_R(\underline{F})$. For the experts we note that $\mathcal{D}$ has another, more canonical, ring structure of a divided power ring, see [**Nor74**, Theorem 3] or [**IK99**, Appendix A]. To make the paper more accessible, we ignore this ring structure as we will not need it.

By Matlis duality, R/I is Artinian (resp. an R-submodule W of $\mathcal{D}$ is finitely generated) if and only if $(R/I)^\vee \cong I^\perp$ is finitely generated (resp. $W^\vee \cong R/\operatorname{Ann}_R(W)$ is Artinian) (see [**BH98**, Section 3.2]). Moreover, $\operatorname{type}(R/I) = \mu((R/I)^\vee)$ where $\mu(-)$ denotes the number of minimal generators as R-module. Emsalem in [**Ems78**, Proposition 2] and Iarrobino in [**Iar94**], based on the work of Macaulay in [**Mac1916**], gave a more precise description of the inverse system of Artinian level K-algebras:

PROPOSITION 2.1. *There is a one-to-one correspondence between ideals I such that R/I is an Artinian level local K-algebra of socle degree s and type τ and R-submodules of $\mathcal{D}$ generated by τ polynomials of degree s having linearly independent leading forms of degree s. The correspondence is defined as follows:*

$$\left\{ \begin{array}{c} I \subseteq R \textit{ such that } R/I \textit{ is an Artinian} \\ \textit{level (resp. graded) local } K\textit{-algebra} \\ \textit{of socle degree } s \textit{ and type } \tau \end{array} \right\} \overset{1-1}{\longleftrightarrow} \left\{ \begin{array}{c} W \subseteq \mathcal{D} \textit{ submodule generated by } \tau \\ \textit{(resp. homogeneous) polynomials} \\ \textit{of degree } s \textit{ with l.i. leading forms} \end{array} \right\}$$

$$\begin{array}{rcl} R/I & \longrightarrow & I^\perp \\ R/\operatorname{Ann}_R(W) & \longleftarrow & W \end{array}$$

Here by the leading form of an element of $\mathcal{D}$ we mean its top-degree form.

It is possible to compute the Hilbert function of $A = R/I$ via the inverse system. Namely, with the previous notation,

$$\mathrm{HF}_{R/I}(i) = \dim_K (I^\perp)_i \tag{2.2}$$

where

$$(I^\perp)_i := \frac{I^\perp \cap \mathcal{D}_{\leq i} + \mathcal{D}_{<i}}{\mathcal{D}_{<i}}.$$

3. Necessary conditions for $(1,3,3)$ level sequences of type 2

In this section we provide necessary conditions for $(1,3,3)$ level sequences of type 2. By (1.1) we know that if h is a $(1,3,3)$ sequence, then $h_3 \leq 4$. If $h_3 \leq 3$, then by (1.1) h is of the form $(1,3,3,3,\ldots,3,2,\ldots,2,h_s = 2)$ and hence $\Delta(h) \leq 1$. Therefore we assume $h_3 = 4$ from now onwards in this section. Such a sequence h we call a $(1,3,3,4)$ sequence. First we recall the following results from [**JMR23**] that we will need.

LEMMA 3.1. [**JMR23**, Lemma 3.1] *Let h be a $(1,3,3,4)$ sequence and let $A = R/I$ be an Artinian algebra with the Hilbert function h. Then there exist $f, g, p \in I$ such that up to a change of coordinates the quadratic parts of f, g, p are xz, yz, z^2, respectively.*

Recall that if $f = \sum_{j\geq v} f_j$ is a power series in R with f_v nonzero, then we say that f has the *initial degree* v and the *initial form* f_v. We denote the initial form of f by f^*. If I is an ideal in R, then the *initial ideal* I^* of I is the homogeneous ideal of $P = K[x,y,z]$ generated by the initial forms of the elements of I. It is well-known that if $A = R/I$, then $gr_{\mathfrak{m}}(A) \simeq P/I^*$.

Let T denote the set of monomials of P. For a fixed term order ν on the set of monomials of P, let $\overline{\nu}$ denotes the total order on T defined in the following way: given $m_1, m_2 \in T$, $m_1 >_{\overline{\nu}} m_2$ if and only if $\deg(m_1) < \deg(m_2)$ or $\deg(m_1) = \deg(m_2)$ and $m_1 >_\nu m_2$. For $f \in R$, let $\mathrm{LT}_{\overline{\nu}}(f)$ denotes the *leading term* of f with respect to $\overline{\nu}$, which is equal to the leading term of f^* with respect to ν.

PROPOSITION 3.2. [**JMR23**, Proposition 3.3] *Let $A = R/I$ be an Artinian algebra and ν a fixed term order on the set of monomials of P such that $z > y > x$. Assume that the ideal I contains elements f, g, p with $\mathrm{LT}_{\overline{\nu}}(f) = xz$, $\mathrm{LT}_{\overline{\nu}}(g) = yz$, $\mathrm{LT}_{\overline{\nu}}(p) = z^2$. Set $S := K[\![x,y]\!]$ and $J = I \cap S$. Then S/J has the Hilbert function*

$$\mathrm{HF}_{S/J}(i) = \begin{cases} 2 & \text{if } i = 1 \\ \mathrm{HF}_{R/I}(i) & \text{if } i \neq 1. \end{cases}$$

We recall the following result from [**Iar84**, Theorem 4.6B] on Hilbert functions of codimension two level algebras that will be used frequently in this paper. This result was first proved by Macaulay in [**Mac1904**]. See also [**MR15, Ber09**]. An O-sequence $h = (1,2,\ldots,h_i,\ldots,h_s)$ a level sequence of type τ if and only if

$$\max\{|h_i - h_{i-1}| : i \geq 1\} \leq \tau = h_s. \tag{3.1}$$

The following proposition provides necessary conditions for a $(1,3,3,4)$ sequence to be a level sequence of type 2.

PROPOSITION 3.3. *Let R/I be an Artinian level algebra of type 2 with Hilbert function of the form $(1,3,3,4)$. Set $S := K[\![x,y]\!]$ and $J = I \cap S$. Then there exist $F_1, F_2, G_1, G_2 \in K_{DP}[X,Y]$ such that*

(a) $\deg F_1 = \deg F_2 = s$ *and the leading forms of F_1, F_2 are linearly independent;*
(b) $x \circ G_i,\ y \circ G_i \in \langle F_i \rangle$ *for $i = 1,2$;*
(c) $J^\perp = \langle F_1, F_2, G_1, G_2 \rangle$.

In this case, $\mathrm{HF}_{S/J}(i) = \mathrm{HF}_{R/I}(i)$ for all $i \neq 1$.

In particular, if h is a level sequence of the form $(1,3,3,4)$, then $\Delta(h) \leq 4$, and the following holds true.
(1) If $\Delta(h) = 3$, then h has at most two positions with a fall by three.
(2) If $\Delta(h) = 4$, then h has exactly one position with a fall by four and it has no fall by three.

PROOF. By Lemma 3.1 the ideal I contains elements of the form $f = xz - U$, $g = yz - V$, $p = z^2 - W$, where $U, V, W \in R_{\geq 3}$. We may assume that $U, V, W \in S_{\geq 3}$ by taking into account the particular form of f^*, g^*, p^* and the fact that $(x,y,z)^k \subseteq I$ for $k \gg 0$.

Since R/I is a level algebra of type 2, there exist $T_1, T_2 \in K_{DP}[X,Y,Z]$ of degree s whose leading forms are linearly independent such that $I^\perp = \langle T_1, T_2 \rangle$. Write

$$T_1 = T_{1,0} + Z \cdot T_{1,1} + Z^2 \cdot T_{1,2} + \ldots + Z^n \cdot T_{1,n} + \ldots, \tag{3.2}$$

$$T_2 = T_{2,0} + Z \cdot T_{2,1} + Z^2 \cdot T_{2,2} + \ldots + Z^n \cdot T_{2,n} + \ldots \tag{3.3}$$

where $T_{1,i}, T_{2,i} \in K_{DP}[X,Y]$ for all i. Since $p \circ T_i = 0$ for $i = 1,2$, we get $z^2 \circ T_i = W \circ T_i$ for $i = 1,2$. This gives

$$T_{i,n+2} = W \circ T_{i,n} \quad \text{for all } n \geq 0 \text{ and } i = 1,2. \tag{3.4}$$

Therefore, $\langle T_{i,n+2} \rangle \subseteq \langle T_{i,n} \rangle$ for all $n \geq 0$ and $i = 1,2$. In particular, $\langle T_{i,2n} \rangle \subseteq \langle T_{i,2n-2} \rangle \subseteq \ldots \subseteq \langle T_{i,0} \rangle$ and similarly $\langle T_{i,2n+1} \rangle \subseteq \langle T_{i,1} \rangle$ for all $n \geq 0$ and $i = 1,2$. Moreover, $f \circ T_i = 0$ and $g \circ T_i = 0$ translate into the conditions

$$x \circ T_{i,n+1} = U \circ T_{i,n}, \quad y \circ T_{i,n+1} = V \circ T_{i,n} \quad \text{for all } n \geq 0 \text{ and } i = 1,2. \tag{3.5}$$

Let $F_i := T_{i,0}$ and $G_i := T_{i,1}$ for $i = 1,2$. Then by (3.5)

$$x \circ G_i = U \circ F_i \in \langle F_i \rangle \text{ and } y \circ G_i = V \circ F_i \in \langle F_i \rangle \text{ for } i = 1,2.$$

Since $J = I \cap S$, $J = \mathrm{Ann}_R(T_1, T_2) \cap S = \mathrm{Ann}_S(T_1, T_2) = \bigcap_{n \geq 0} \mathrm{Ann}_S(T_{1,n}, T_{2,n})$. As $\langle T_{i,2n} \rangle \subseteq \langle F_i \rangle$ and $\langle T_{i,2n+1} \rangle \subseteq \langle G_i \rangle$ for $i = 1,2$, we have $J = \mathrm{Ann}_S(F_1, F_2) \cap \mathrm{Ann}_S(G_1, G_2)$, hence $J^\perp = \langle F_1, F_2, G_1, G_2 \rangle$ as required. By Proposition 3.2 S/J and R/I have the same Hilbert function except at position 1. In particular, S/J has the socle degree s. By (3.2) and (3.3) degree $G_1, G_2 \leq s-1$. Hence we conclude that F_1 and F_2 are of degree s. Since the leading forms of T_1 and T_2 are linearly independent, the leading forms of F_1 and F_2 are also linearly independent.

Let $\omega_{S/J}$ be the canonical module of S/J. Then $\omega_{S/J} \cong J^\perp$. Hence $\mu(J) \leq 5$. By [**Ber09**, Theorem 2.3] we have $\Delta(\mathrm{HF}_{S/J}) \leq 4$. By (b)–(c) it follows that $\mathrm{HF}_{S/J}(i) = \mathrm{HF}_{S/\mathrm{Ann}_S(F_1,F_2)}(i)$ for all i except possibly $i = \deg G_1, \deg G_2$.

CASE 1 ($\Delta(h) = 3$). Since $\Delta(\mathrm{HF}_{S/\mathrm{Ann}_S(F_1,F_2)}) \leq 2$ by (3.1), it follows that if $\Delta(h) = 3$, then h has at most two falls by 3, which can occur at positions $\deg G_1$ and $\deg G_1$.

CASE 2 ($\Delta(h) = 4$). Since $\Delta(\mathrm{HF}_{S/\operatorname{Ann}_S(F_1,F_2)}) \leq 2$ by (3.1), it is clear that if $\Delta(h) = 4$, then h has a unique fall by 4 which is at the position $\deg G_1 = \deg G_2$. Also, h has no fall by 3.

□

The following example shows that h need not be a level sequence of type 2 even if h has exactly one fall by four and no fall by three.

EXAMPLE 3.4. Consider the O-sequence $(1, 3, 3, 4, 5, 6, 7, 6, 2)$. We claim that h is not a level sequence. Suppose h is a level sequence. Then by Proposition 3.3 there exist F_1, F_2, G_1, G_2 such that $J := \operatorname{Ann}_S(F_1, F_2, G_1, G_2)$ has the Hilbert function $(1, 2, 3, 4, 5, 6, 7, 6, 2)$. Since by (b)–(c) of Proposition 3.3 $\mathrm{HF}_{S/J}(i) = \mathrm{HF}_{S/\operatorname{Ann}_S(F_1,F_2)}(i)$ for all i except possibly $i = \deg G_1, \deg G_2$, the possibility for the Hilbert function of $S/\operatorname{Ann}_S(F_1, F_2)$ is $h_1 = (1, 2, 3, 4, 5, 6, 7, 4, 2)$ or $h_2 = (1, 2, 3, 4, 5, 6, 6, 5, 2)$. But neither h_1 nor h_2 are level sequences as $\Delta(h_1) = \Delta(h_2) = 3$ whereas $\Delta(\mathrm{HF}_{S/\operatorname{Ann}_S(F_1,F_2)}) \leq 2$ by (3.1).

In the rest of this section we investigate the position at which a jump by 4 can occur. First we would like to determine the position at which $\mathrm{HF}_{S/J}$ and $\mathrm{HF}_{S/\operatorname{Ann}_S(F_1,F_2)}$ in Proposition 3.3 possibly differ by 2.

LEMMA 3.5. *Let $k \geq 3$ and let $J \subseteq I$ be ideals in $S := K[\![x, y]\!]$ such that $\mathfrak{m}_S I \subseteq J$ where $\mathfrak{m}_S = (x, y)S$, I is a level ideal of type 2 and*

$$\mathrm{HF}_{S/J}(i) = \begin{cases} \mathrm{HF}_{S/I}(i) & \text{for } i \neq k \\ \mathrm{HF}_{S/I}(i) + 2 & \text{for } i = k. \end{cases}$$

Then I has two minimal generators of initial degree k and k is the peak position of $\mathrm{HF}_{S/J}$.

PROOF. We have $\dim_K I/J = \dim_K S/J - \dim_K S/I = 2$. Since I is a level ideal of type 2, $\mu(I) = 3$ and hence $\dim_K I/\mathfrak{m}_S I = 3$. Therefore

$$\dim_K J/\mathfrak{m}_S I = \dim_K I/\mathfrak{m}_S I - \dim_K I/J = 3 - 2 = 1.$$

Choose $g_1 \in J \setminus \mathfrak{m}_S I$. For $g \in J$, let $\overline{g}$ denote the image of g in $J/\mathfrak{m}_S I$. Then $K\overline{g_1} = J/\mathfrak{m}_S I$. Therefore $J = (g_1)S + \mathfrak{m}_S I$. Choose $g_2, g_3 \in I \setminus \mathfrak{m}_S I$ such that $\overline{g_1}, \overline{g_2}$ and $\overline{g_3}$ are K-linearly independent elements of $I/\mathfrak{m}_S I$. Then by Nakayama lemma $I = (g_1, g_2, g_3)$. Also,

$$J = (g_1) + \mathfrak{m}_S I = (g_1) + (x, y)(g_1, g_2, g_3) = (g_1) + (x, y)(g_2, g_3).$$

We know that the Hilbert function of S/J is the Hilbert function of $K[x, y]/J^*$. Without loss of generality we may assume that the initial forms of g_1, g_2 and g_3 are K-linearly independent. Then the associated graded ideal $I^* = (g_2^*, g_3^*) + J^*$. Since the Hilbert functions of $K[x, y]/I^*$ and $K[x, y]/J^*$ are the same except at the position k, the initial forms g_2^* and g_3^* are in I^* but are not in J^*. Thus, g_2 and g_3 have initial degree k.

Let $h_I := \mathrm{HF}_{S/I}$, $h_J = \mathrm{HF}_{S/J}$ and $d := \max h_J$. If $\max h_I = d - 1$, then $k = d - 1$, and is the peak position of h_J. Suppose $\max h_I = d$. Then there exist elements g_1 of initial degree d, g_2 of initial degree $t + 1$ and g_3 of initial degree $\geq t + 1$ in I such that $I = (g_1, g_2, g_3)$ where t is the peak position of h_I. Since I has two minimal generators of initial degree k, it follows that g_3 has initial degree $t + 1$ and $k = t + 1$. Since t is the peak position of h_I and $\Delta(h_I) \leq 2$ by [**Ber09**, Theorem

2.3], $h_I(t+1) = d-1$ or $h_I(t+1) = d-2$. If $h_I(t+1) = d-1$, then $h_J(t+1) = d+1$, a contradiction since $\max h_J = d$. Hence $h_I(t+1) = d-2$, and so $h_I(t+2) \leq d-2$. This implies that $h_J(k+1) \leq d-2$ while $h_J(k) = h_I(k) + 2 = d$. Hence k is the peak position of h_J. □

We know that a $(1,3,3,4)$ level sequence of type 2 is of the form $h = (1,3,3,4,5, \ldots, d, d, h_t, \ldots, 2)$ where $d := \max h$ and $h_t < d$. In the following proposition we observe that such a level sequence with $\Delta(h) = 4$ has a unique fall by 4 at $t-1$.

PROPOSITION 3.6. *Let h be a $(1,3,3,4)$ level sequence of type 2 with $\Delta(h) = 4$. Let $d := \max h$ and let r be a nonnegative integer such that d is repeated $r+1$ times in h. Then*

$$h_i = \begin{cases} i+1 & \text{for } i \leq d-2 \\ d & \text{for } i = d-1, d, \ldots, d+r-1 \\ d-4 & \text{for } i = d+r \\ h_{i-1} \text{ or } h_{i-1}-1 \text{ or } h_{i-1}-2 & \text{for } i \geq d+r+1. \end{cases}$$

PROOF. Let h be a $(1,3,3,4)$ level sequence of type 2 with $\Delta(h) = 4$. By Proposition 3.3 h has a unique fall by 4, say at position m. By Proposition 3.3 there exist $F_1, F_2, G_1, G_2 \in K_{DP}[X,Y]$ such that $\mathrm{HF}_{S/\operatorname{Ann}_S(F_1,F_2)}$ and $\mathrm{HF}_{S/\operatorname{Ann}_S(F_1,F_2,G_1,G_2)}$ differ, possibly, at two positions, namely $\deg G_1$ and $\deg G_2$. Let $J = \operatorname{Ann}_S(F_1, F_2, G_1, G_2)$. By Proposition 3.2 $\mathrm{HF}_{S/J}$ and h differ only at position 1. Hence

$$\mathrm{HF}_{S/J} = (1,2,3,4,\ldots,h_{m-1},h_m,h_{m+1} = h_m - 4,\ldots,2).$$

Since $\Delta(h) = 4$ and $\Delta(\mathrm{HF}_{S/\operatorname{Ann}_S(F_1,F_2)}) \leq 2$ by (3.1), $\deg G_1 = \deg G_2 = m$, and $\mathrm{HF}_{S/J}$ and $\mathrm{HF}_{S/\operatorname{Ann}_S(F_1,F_2)}$ differ exactly at one position, namely $m = \deg G_1 = \deg G_2$. Therefore

$$\mathrm{HF}_{S/\operatorname{Ann}_S(F_1,F_2)} = (1,2,3,4,\ldots,h_{m-1},h_m - 2,h_{m+1} = h_m - 4,\ldots,2).$$

Since $x \circ G_i \in \langle F_i \rangle$ for $i = 1,2$, $\mathfrak{m}_S I' \subseteq J$ where $I' := \operatorname{Ann}_S(F_1,F_2)$. By Proposition 3.3 I' is a level ideal of type 2. Hence by Lemma 3.5 the position m is the peak position of $\mathrm{HF}_{S/J}$, and so m is the peak position of h. Since h has no fall by 3 by Proposition 3.3, it is easy to check that h has the required form. □

EXAMPLE 3.7. Consider the O-sequence $h = (1,3,3,4,5,6,7,6,2)$. Then h has a fall by 4 at position 7 which is not the peak position. Hence h is not a level sequence.

4. Construction of $(1,3,3)$ level sequences of type 2

In this section we construct a level algebra of type 2 with the Hilbert function h where h is a $(1,3,3)$ sequence with $\Delta(h) \leq 2$. Note that if h is a $(1,3,3)$ sequence, then $h_3 \leq 4$. First we construct a level algebra of type 2 when $h_3 \leq 3$. In this case, we provide explicit $F_1, F_2 \in K_{DP}[X,Y]$ such that A_{F_1,F_2} is a level K-algebra of type 2 with the Hilbert function h. If $h_3 = 4$, then we reduce to the codimension two case and take advantage of the structure of codimension two level ideals. Then we modify the inverse system of a codimension two level ideal and produce a codimension three level ideal with the Hilbert function h.

PROPOSITION 4.1. *Let*

$$h := \left(1, 3, 3, \underbrace{3, \ldots, 3}_{u}, \underbrace{2, \ldots, 2}_{v}, 2\right)$$

be an O-sequence. Then there exists a level K-algebra of type 2 with the Hilbert function h. More precisely, A_{F_1,F_2} where $F_1 := X^{u+v+3} + Z^{u+3}$ and $F_2 = Y^{u+v+3}$ is a level K-algebra of type 2 and has the Hilbert function h.

PROOF. It is easy to verify that A_{F_1,F_2} has the Hilbert function h. Since the leading forms of F_1 and F_2 are linearly independent and $\deg F_1 = \deg F_2$, A_{F_1,F_2} is a level algebra of type 2. $\square$

In the next result we show that if h is a $(1, 3, 3)$ sequence with $h_3 = 4$ and $\Delta(h) \leq 2$, then h is a level sequence of type 2.

PROPOSITION 4.2. *Let $h = (1, 3, 3, 4, \ldots, h_s = 2)$ be an O-sequence with $\Delta(h) \leq 2$. Then h is a level O-sequence of type 2.*

PROOF. Define a numerical sequence h' as $h'_i = \begin{cases} h_i & \text{if } i \neq 1 \\ 2 & \text{if } i = 1. \end{cases}$ Then $\Delta(h') \leq 2$. Hence by (3.1) there exists an ideal I' in $S = K[\![x, y]\!]$ such that S/I' is level of type 2 and has the Hilbert function h'. Therefore there exist $F_1, F_2 \in K_{DP}[X, Y]$ of degree s whose leading forms are linearly independent such that $I' = \operatorname{Ann}_S(F_1, F_2)$. Define $T_1 = F_1 + Z^2$ and $T_2 = F_2$. Then the leading forms of T_1 and T_2 are linearly independent. Let $I := \operatorname{Ann}_R(T_1, T_2)$. Then R/I is a level algebra of type 2 with the Hilbert function h. $\square$

PROOF OF THEOREM 1.1: Follows from Propositions 3.3, 3.6, 4.1 and 4.2. $\square$

The proof of Proposition 4.2 is constructive. In fact, the following is an algorithm to construct a level K-algebra of type 2 with the Hilbert function as $(1, 3, 3, 4)$ sequence h and $\Delta(h) \leq 2$.

Algorithm:
Given: a $(1, 3, 3, 4)$ sequence h with $\Delta(h) \leq 2$.
Aim: To construct a level K-algebra of type 2 with the Hilbert function h.
Step 1: Define

$$h'_i = \begin{cases} h_i & \text{if } i \neq 1 \\ 2 & \text{if } i = 1. \end{cases}$$

Step 2: Construct $F_1, F_2 \in K_{DP}[X, Y]$ such that $S/\operatorname{Ann}_S(F_1, F_2)$ is a level K-algebra of type 2 and has the Hilbert function h'.

Step 3: Define $T_1 = F_1 + Z^2$ and $T_2 = F_2$. Then $R/\operatorname{Ann}_R(T_1, T_2)$ is a level K-algebra of type 2 with the Hilbert function h.

We illustrate the above algorithm by the following examples. First we construct a level K-algebra with the Hilbert function h where $\Delta(h) = 1$.

EXAMPLE 4.3. Let $h = (1,3,3,4,3,2)$. Define $h' = (1,2,3,4,3,2)$. Consider $F_1 = X^2Y^3$, $F_2 = X^3Y^2 \in \mathbb{Q}_{DP}[X,Y]$. Then A_{F_1,F_2} has the Hilbert function h'. Now, let $T_1 = F_1 + Z^2$, $T_2 = F_2$. Then $R/\operatorname{Ann}_R(T_1,T_2)$ has the Hilbert function h. Here $\operatorname{Ann}_R(T_1,T_2) = (x^4, y^4, z^2 - x^2y^3, xz, yz)$.

We now construct a level K-algebra having the Hilbert function h where $\Delta(h) = 2$.

EXAMPLE 4.4. Consider $h = (1,3,3,4,4,2)$. Then $h' = (1,2,3,4,4,2)$. Let $F_1 = X^5 + Y^5$, $F_2 = X^3Y^2 \in \mathbb{Q}_{DP}[X,Y]$. Then $S/\operatorname{Ann}_S(F_1,F_2)$ has the Hilbert function h'. Define $T_1 = F_1 + Z^2$, $T_2 = F_2$. Then $R/\operatorname{Ann}_R(T_1,T_2)$ has the Hilbert function h. We have $\operatorname{Ann}_R(T_1,T_2) = (x^4y, x^5 - y^5, xy^3, z^2 - y^5, xz, yz)$.

We now compare the $(1,3,3)$ level sequences of type 2 and the $(1,3,3)$ graded level sequences of type 2. We remark that by Stanley's result [**Sta78**, Theorem 4.2] an O-sequence with $h_1 = h_2 = 3$ and $h_s = 1$ is a graded Gorenstein sequence if and only if $h = (1,3,3,3,\ldots,3,3,1)$. In the following proposition we prove a similar result for graded level O-sequences.

PROPOSITION 4.5. *Let h be a $(1,3,3)$ O-sequence. Then h is a graded level O-sequence of type 2 if and only if $h = (1,3,3,3,\ldots,3,3,2)$, that is,* $h_i = \begin{cases} 1 & \textit{if } i = 0 \\ 3 & \textit{if } i \neq 0, s \\ 2 & \textit{if } i = s. \end{cases}$

PROOF. Suppose h is a graded level O-sequence of type 2. Then $h_3 \leq 4$. Let $h_3 = 4$. Taking $d = 3$ and $h'_1 = 2, h'_2 = 3, h'_3 = 4$ in [**GHMS07**, Corollary 3.5], we get that h is not a graded level O-sequence of type 2. Hence assume that $h_3 \leq 3$. Then

$$h := \left(1,3,3,\underbrace{3,\ldots,3}_{u},\underbrace{2,\ldots,2}_{v},2\right).$$

If $v \neq 0$, then h is not a graded level sequence by [**GHMS07**, Remark 2.12(b)]. Therefore

$$h = (1,3,3,3,\ldots,3,3,2).$$

Conversely, suppose $h = (1,3,3,3,\ldots,3,3,2)$. Then taking $v = 0$ in Proposition 4.1 we get a graded level O-sequence. □

We end this section with a few questions. In [**Jel17**] Jelisiejew classified Artinian Gorenstein K-algebras with the Hilbert function $(1,3,3,3,1)$. Inspired by this we ask:

PROBLEM 4.1. Classify level K-algebras, up to isomorphism, with the Hilbert function $(1,3,3,3,2)$.

In [**Iar94**] Iarrobino developed a beautiful theory of symmetric decomposition of the Hilbert function of Artinian Gorenstein K-algebra. This theory has a far reaching consequence on the structure of the Hilbert function, which in our opinion is yet to be explored. This decomposition is also useful to construct an Artinian Gorenstein K-algebra with the required Hilbert function, see for instance [**IM21**].

We hope that if there is a "level" decomposition of the Hilbert function of an Artinian level K-algebra, then we might use it to construct a level K-algebra having a required Hilbert function. Notice that a level K-algebra of type 2 is the intersection of two Gorenstein ideals, but still we do not know about the decomposition of the Hilbert function of Artinian level K-algebra of type 2. Hence we ask:

PROBLEM 4.2. Is there a "level" decomposition of the Hilbert function of Artinian level K-algebra of type 2?

5. Classification of $(1,3,3)$ level sequences of type 2 and socle degree ≤ 6

In this section we give a complete list of $(1,3,3)$ level sequences of type 2 and socle degree at most 6. For this we take advantage of classification of graded level sequences of socle degree at most 6 given in the Appendix of [**GHMS07**].

5.1. $(1,3,3)$ **level sequences of type** 2 **and** $s \leq 4$. Observe that if h is a $(1,3,3)$ level sequence of type 2, then $s \geq 3$. If $s = 3$, then $h = (1,3,3,2)$. By Proposition 4.1, h is a level sequence of type 2. In fact, h is realisable by a graded level K-algebra A_{F_1,F_2} where $F_1 = X^3 + Z^3$ and $F_2 = Y^3$.

Suppose $s = 4$. Then $(1,3,3)$ O-sequence with $h_4 = 2$ is either $(1,3,3,2,2)$, $(1,3,3,3,2)$ or $(1,3,3,4,2)$. All these sequences are level sequences by [**MR15**, Theorem 3.8]. In fact, by Proposition 4.1 the sequences $(1,3,3,2,2)$ and $(1,3,3,3,2)$ are realized by A_{F_1,F_2} where $F_1 = X^4 + Z^3$, $F_2 = Y^4$ and $F_1 = X^4 + Z^4, F_2 = Y^4$, respectively. The sequence $(1,3,3,4,2)$ is realized by A_{F_1,F_2} where $F_1 = X^4 + X^2Y^2 + Z^2$, $F_2 = Y^4 + X^2Y^2$ (see the proof of [**MR15**, Theorem 3.8]).

5.2. $(1,3,3)$ **level sequences of type** 2 **and** $s = 5$. Let $h = (1,3,3,h_3,h_4,2)$ be an O-sequence. If $h_3 \leq 3$, then h is a level sequence of type 2 by Proposition 4.1. So, we assume $h_3 = 4$. Then $2 \leq h_4 \leq 5$. If $h_4 \leq 4$, then h is a level sequence of type 2 by Proposition 4.2. So, it remains to check whether $(1,3,3,4,5,2)$ is a level sequence. In the following proposition we show that $(1,3,3,4,5,2)$ is not a level sequence.

Recall that if $A = R/I$ is an Artinian level K-algebra with the Hilbert function h of socle degree s and type 2, then for all $i \geq 0$

$$h_i \leq \min\{\dim_K R_i, 2\dim_K R_{s-i}\} \tag{5.1}$$

(see [**Iar84**]).

PROPOSITION 5.1. *The sequence* $h = (1,3,3,4,5,2)$ *is not a level sequence of type* 2.

PROOF. Suppose $h = (1,3,3,4,5,2)$ is a level sequence, and $A = R/I$ is a level K-algebra with the Hilbert function h. Then there exist polynomials $T_1, T_2 \in K_{DP}[X,Y,Z]$ of degree 5 whose leading forms are linearly independent such that $I^{\perp} = \langle T_1, T_2\rangle$. Write

$$T_1 = T_1[5] + \text{ terms of degree } \leq 4 \tag{5.2}$$

$$T_2 = T_2[5] + \text{ terms of degree } \leq 4 \tag{5.3}$$

where $T_1[5]$ and $T_2[5]$ are the leading forms of degree 5 in T_1 and T_2, respectively. Since $T_1[5]$ and $T_2[5]$ are linearly independent, $B := K[x,y,z]/\mathrm{Ann}_{K[x,y,z]}(T_1[5], T_2[5])$ is a graded level K-algebra of type 2. Let $h_B = (1, h'_1, h'_2, h'_3, h'_4, 2)$ be the Hilbert function of B. Since $h_4 = 5, h_4 = \dim_K\langle T_1, T_2\rangle_4 = 5$. This implies that

$h'_4 = \dim_K\langle T_1[5], T_2[5]\rangle_4 = 5$. By (5.1), $h'_4 \leq 2h'_1$. Hence $h'_1 = 3$. Thus $h_B = (1, 3, h'_2, h'_3, 5, 2)$ is a graded level sequence of type 2. Now, by the list given [**GHMS07**, page 90, Table 5.2], we conclude that $h'_2 \geq 4$. Hence $h_2 \geq 4$, a contradiction. Therefore h is not a level sequence. □

The Table 1 below gives a complete list of $(1, 3, 3)$ level sequences h of type 2 and socle degree 5. For each h, we give $F_1, F_2 \in K_{DP[X,Y,Z]}$ such that A_{F_1,F_2} is a level K-algebra of type 2 and has the Hilbert function h, see also Examples 4.3 and 4.4.

TABLE 1. Level sequences of $s = 5$ and type 2

level sequence h of $s = 5$ and type 2	$F_1, F_2 \in K_{DP}[X, Y, Z]$ **such that A_{F_1,F_2} has the HF h**
$(1, 3, 3, 2, 2, 2)$	$F_1 = X^5 + Z^3,\ F_2 = Y^5$
$(1, 3, 3, 3, 2, 2)$	$F_1 = X^5 + Z^4,\ F_2 = Y^5$
$(1, 3, 3, 3, 3, 2)$	$F_1 = X^5 + Z^5,\ F_2 = Y^5$
$(1, 3, 3, 4, 2, 2)$	$F_1 = X^5 + X^2Y^2 + Z^2,\ F_2 = Y^5 + X^2Y^2$
$(1, 3, 3, 4, 3, 2)$	$F_1 = X^2Y^3 + Z^2,\ F_2 = X^3Y^2$
$(1, 3, 3, 4, 4, 2)$	$F_1 = X^5 + Y^5 + Z^2,\ F_2 = X^3Y^2$

5.3. $(1, 3, 3)$ **level sequences of type** 2 **and** $s = 6$. Let $h = (1, 3, 3, h_3, h_4, h_5, 2)$ be on O-sequence. Observe that among these O-sequences the only sequences with $\Delta(h) > 2$ are $(1, 3, 3, 4, 5, 2, 2)$, $(1, 3, 3, 4, 5, 5, 2)$, and $(1, 3, 3, 4, 5, 6, 2)$. In the following proposition we show that none of these are level sequences.

PROPOSITION 5.2. *Suppose $h = (1, 3, 3, h_3, h_4, h_5, 2)$ is a level sequence of type 2. Then $\Delta(h) \leq 2$.*

PROOF. Suppose $h = (1, 3, 3, h_3, h_4, h_5, 2)$ is a level sequence of type 2 and $\Delta(h) > 2$. Then $h = (1, 3, 3, 4, 5, 2, 2)$, $(1, 3, 3, 4, 5, 5, 2)$, or $(1, 3, 3, 4, 5, 6, 2)$. We show that these sequences are not level sequences.

CASE 1 ($h = (1, 3, 3, 4, 5, 2, 2)$). Suppose h is a level sequence of type 2. Let $F_1, F_2 \in K_{DP}[X, Y, Z]$ be polynomials of degree 6 such that their leading forms are linearly independent and A_{F_1,F_2} has the Hilbert function h. Then by (2.2) $\dim_K\langle F_1, F_2\rangle_5 = 2$. Hence there exist polynomials $G_1, G_2 \in \langle F_1, F_2\rangle$ of degree 5 whose leading forms are linearly independent and $\langle F_1, F_2\rangle_5 = \mathrm{Span}_K\{G_1, G_2\}$. Then A_{G_1,G_2} is a level K-algebra of type 2 with the Hilbert function $(1, 3, 3, 4, 5, 2)$. This contradicts Proposition 5.1. Hence h is not a level sequence.

CASE 2 ($h = (1, 3, 3, 4, 5, 5, 2)$). Let $F_1, F_2 \in K_{DP}[X, Y, Z]$ be polynomials of degree 6 whose leading forms are linearly independent such that A_{F_1,F_2} has the Hilbert function h. Write

$$T_1 = T_1[6] + \text{ terms of degree} \leq 5 \tag{5.4}$$

$$T_2 = T_2[6] + \text{ terms of degree} \leq 5 \tag{5.5}$$

where $T_1[6]$ and $T_2[6]$ are the leading forms of degree 6 in T_1 and T_2, respectively. Then

$$B := K[x, y, z]/\operatorname{Ann}_{K[x,y,z]}(T_1[6], T_2[6])$$

is a graded level K-algebra of type 2 with the Hilbert function $h_B = (1, h'_1, h'_2, h'_3, h'_4, h'_5, 2)$. Since $h_5 = \dim_K\langle F_1, F_2\rangle_5 = 5$, $h'_5 = \dim_K\langle F_1[6], F_2[6]]_5\rangle = 5$. Hence by (5.1) $h'_1 \geq 3$. But as $h_1 = 3$, $h'_1 = 3$. Thus h is of the form $(1, 3, h'_2, h'_3, h'_4, 5, 2)$. Now by [**GHMS07**, Page 77, Table 6.2] $h'_2 \geq 4$. This gives that $h_2 \geq 4$, a contradiction. Therefore $h = (1, 3, 3, 4, 5, 5, 2)$ is not a level sequence.

CASE 3 ($h = (1, 3, 3, 4, 5, 6, 2)$). By a similar argument in Case 2, we can show that $h = (1, 3, 3, 4, 5, 6, 2)$ is not a level sequence.

□

We remark that $(1, 3, 3, 4, 5, 6, 2)$ is a $(1, 3, 3, 4)$ sequence with a unique fall by 4 at the peak position, but is not a level sequence. This shows that the converse of Theorem 1.1 for $\Delta(h) = 4$ need not be true.

The Table 2 is a complete list of level sequences of socle degree 6 and type 2. In each case we give an explicit $F_1, F_2 \in K_{DP}[X, Y, Z]$ such that A_{F_1,F_2} is a level K-algebra of type 2 with the Hilbert function h.

TABLE 2. Level sequences of $s = 6$ and type 2

level sequence h **of** $s = 6$ **and type** 2	$F_1, F_2 \in K_{DP}[X, Y, Z]$ **such that** A_{F_1,F_2} **has the HF** h
$(1, 3, 3, 2, 2, 2, 2)$	$F_1 = X^6 + Z^3,\ F_2 = Y^6$
$(1, 3, 3, 3, 2, 2, 2)$	$F_1 = X^6 + Z^4,\ F_2 = Y^6$
$(1, 3, 3, 3, 3, 2, 2)$	$F_1 = X^6 + Z^5,\ F_2 = Y^6$
$(1, 3, 3, 3, 3, 3, 2)$	$F_1 = X^6 + Z^6,\ F_2 = Y^6$
$(1, 3, 3, 4, 2, 2, 2)$	$F_1 = X^6 + X^2Y^2 + Z^2,\ F_2 = Y^6 + X^2Y^2$
$(1, 3, 3, 4, 3, 2, 2)$	$F_1 = X^6 + X^3Y^2 + Z^2,\ F_2 = Y^6 + X^2Y^2$
$(1, 3, 3, 4, 3, 3, 2)$	$F_1 = X^5Y + Z^2,\ F_2 = Y^6 + X^2Y^2$
$(1, 3, 3, 4, 4, 2, 2)$	$F_1 = X^6 + XY^4 + Z^2,\ F_2 = Y^6 + X^4Y$
$(1, 3, 3, 4, 4, 3, 2)$	$F_1 = X^5Y + Z^2,\ F_2 = Y^6 + X^2Y^3$
$(1, 3, 3, 4, 4, 4, 2)$	$F_1 = X^5Y + Z^2,\ F_2 = XY^5$
$(1, 3, 3, 4, 5, 3, 2)$	$F_1 = X^4Y^2 + Z^2,\ F_2 = Y^6 + X^2Y^3$
$(1, 3, 3, 4, 5, 4, 2)$	$F_1 = X^4Y^2 + Z^2,\ F_2 = X^2Y^4$

Acknowledgments

We thank Prof. M. E. Rossi for comments on an earlier version of the draft. The author is also thankful to the referees for several suggestions which have greatly improved the presentation of the paper.

References

[AS12] J. Ahn and Y. S. Shin, *Artinian level algebras of codimension 3*, J. Pure Appl. Algebra **216** (2012), no. 1, 95–107, DOI 10.1016/j.jpaa.2011.05.006. MR2826422

[Boi94] M. Boij, *Artin level algebras*, Ph. D. thesis, Royal Institute of Technology, Stockholm, (1994).

[Boi99] M. Boij, *Betti numbers of compressed level algebras*, J. Pure Appl. Algebra **134** (1999), no. 2, 111–131, DOI 10.1016/S0022-4049(97)90163-8. MR1663785

[Boi00] M. Boij, *Artin level modules*, J. Algebra **226** (2000), no. 1, 361–374, DOI 10.1006/jabr.1999.8185. MR1749894

[BI09] M. Boij and A. Iarrobino, *Reducible family of height three level algebras*, J. Algebra **321** (2009), no. 1, 86–104, DOI 10.1016/j.jalgebra.2008.10.001. MR2469350

[Ber09] V. Bertella, *Hilbert function of local Artinian level rings in codimension two*, J. Algebra **321** (2009), no. 5, 1429–1442, DOI 10.1016/j.jalgebra.2008.11.023. MR2494398

[BH98] W. Bruns and J. Herzog, *Cohen-Macaulay rings*, Revised Edition, Cambridge University Press, 1998.

[CEVV09] D. A. Cartwright, D. Erman, M. Velasco, and B. Viray, *Hilbert schemes of 8 points*, Algebra Number Theory **3** (2009), no. 7, 763–795, DOI 10.2140/ant.2009.3.763. MR2579394

[CENR13] G. Casnati, J. Elias, R. Notari, and M. E. Rossi, *Poincaré series and deformations of Gorenstein local algebras*, Comm. Algebra **41** (2013), no. 3, 1049–1059, DOI 10.1080/00927872.2011.636643. MR3037178

[CJN15] G. Casnati, J. Jelisiejew, and R. Notari, *Irreducibility of the Gorenstein loci of Hilbert schemes via ray families*, Algebra Number Theory **9** (2015), no. 7, 1525–1570, DOI 10.2140/ant.2015.9.1525. MR3404648

[CG03] J. V. Chipalkatti and A. V. Geramita, *On parameter spaces for Artin level algebras*, Michigan Math. J. **51** (2003), no. 1, 187–207, DOI 10.1307/mmj/1049832900. MR1960928

[CI01] Y. H. Cho and A. Iarrobino, *Hilbert functions and level algebras*, J. Algebra **241** (2001), no. 2, 745–758, DOI 10.1006/jabr.2001.8787. MR1843323

[Ste14] A. De Stefani, *Artinian level algebras of low socle degree*, Comm. Algebra **42** (2014), no. 2, 729–754, DOI 10.1080/00927872.2012.722736. MR3169599

[Singular] W. Decker, G.-M. Greuel, G. Pfister and H. Schönemann, SINGULAR 4-2-1 — A computer algebra system for polynomial computations available at `http://www.singular.uni-kl.de`, (2021).

[Eis95] D. Eisenbud, *Commutative algebra*, Graduate Texts in Mathematics, vol. 150, Springer-Verlag, New York, 1995. With a view toward algebraic geometry, DOI 10.1007/978-1-4612-5350-1. MR1322960

[Eli15] J. Elias, INVERSE-SYST.LIB–*Singular library for computing Macaulay's inverse systems*, `http://www.ub.edu/C3A/elias/inverse-syst-v.5.2.lib`, 2015.

[EV08] J. Elias and G. Valla, *Structure theorems for certain Gorenstein ideals*, Michigan Math. J. **57** (2008), 269–292, DOI 10.1307/mmj/1220879409. Special volume in honor of Melvin Hochster. MR2492453

[Ems78] J. Emsalem, *Géométrie des points épais* (French, with English summary), Bull. Soc. Math. France **106** (1978), no. 4, 399–416. MR518046

[Gab59] P. Gabriel, *Objets injectifs dans les catégories abéliennes*, Séminaire P. Dubriel 1958-1959 no. 17, 1-32, (1959).

[GHMS07] A. V. Geramita, T. Harima, J. C. Migliore, and Y. S. Shin, *The Hilbert function of a level algebra*, Mem. Amer. Math. Soc. **186** (2007), no. 872, vi+139, DOI 10.1090/memo/0872. MR2292384

[Macaulay2] D. Grayson and M. Stillman, *Macaulay2, a software system for research in algebraic geometry*, available at `http://www2.macaulay2.com`.

[Iar84] A. Iarrobino, *Compressed algebras: Artin algebras having given socle degrees and maximal length*, Trans. Amer. Math. Soc. **285** (1984), no. 1, 337–378, DOI 10.2307/1999485. MR748843

[Iar94] A. A. Iarrobino, *Associated graded algebra of a Gorenstein Artin algebra*, Mem. Amer. Math. Soc. **107** (1994), no. 514, viii+115, DOI 10.1090/memo/0514. MR1184062

[Iar04] A. Iarrobino, *Ancestor ideals of vector spaces of forms, and level algebras*, J. Algebra **272** (2004), no. 2, 530–580, DOI 10.1016/S0021-8693(03)00425-3. MR2028071

[Iar05] A. Iarrobino, *Hilbert functions of Gorenstein algebras associated to a pencil of forms*, Projective varieties with unexpected properties, Walter de Gruyter, Berlin, 2005, pp. 273–286. MR2202259

[IM21] A. Iarrobino and P. Macias Marques, *Symmetric decomposition of the associated graded algebra of an Artinian Gorenstein algebra*, J. Pure Appl. Algebra **225** (2021), no. 3, Paper No. 106496, 49, DOI 10.1016/j.jpaa.2020.106496. MR4134067

[IK99] A. Iarrobino and V. Kanev, *Power sums, Gorenstein algebras, and determinantal loci*, Lecture Notes in Mathematics, vol. 1721, Springer-Verlag, Berlin, 1999. Appendix C by Iarrobino and Steven L. Kleiman, DOI 10.1007/BFb0093426. MR1735271

[Jel17] J. Jelisiejew, *Classifying local Artinian Gorenstein algebras*, Collect. Math. **68** (2017), no. 1, 101–127, DOI 10.1007/s13348-016-0183-1. MR3591467

[JMR23] J. Jelisiejew, S. K. Masuti, and M. E. Rossi, *On the Hilbert function of Artinian local complete intersections of codimension three*, J. Pure Appl. Algebra **227** (2023), no. 7, Paper No. 107326, 21, DOI 10.1016/j.jpaa.2023.107326. MR4541894

[Mac1904] F. S. Macaulay, *On a method of dealing with the intersections of plane curves*, Trans. Amer. Math. Soc. **5** (1904), no. 4, 385–410, DOI 10.2307/1986270. MR1500679

[Mac1916] F. S. Macaulay, *The algebraic theory of modular systems*, Cambridge Mathematical Library, Cambridge University Press, Cambridge, 1994. Revised reprint of the 1916 original; With an introduction by Paul Roberts. MR1281612

[Mac27] F. S. Macaulay, *Some Properties of Enumeration in the Theory of Modular Systems*, Proc. London Math. Soc. (2) **26** (1927), 531–555, DOI 10.1112/plms/s2-26.1.531. MR1576950

[MR15] M. Mandal and M. E. Rossi, *The tangent cone of a local ring of codimension 2*, Acta Math. Vietnam. **40** (2015), no. 1, 85–100, DOI 10.1007/s40306-014-0102-z. MR3331934

[MR18] S. K. Masuti and M. E. Rossi, *Artinian level algebras of socle degree 4*, J. Algebra **507** (2018), 525–546, DOI 10.1016/j.jalgebra.2018.03.043. MR3807059

[Nor74] D. G. Northcott, *Injective envelopes and inverse polynomials*, J. London Math. Soc. (2) **8** (1974), 290–296, DOI 10.1112/jlms/s2-8.2.290. MR360555

[Poo08] B. Poonen, *Isomorphism types of commutative algebras of finite rank over an algebraically closed field*, Computational arithmetic geometry, Contemp. Math., vol. 463, Amer. Math. Soc., Providence, RI, 2008, pp. 111–120, DOI 10.1090/conm/463/09050. MR2459993

[Sta78] R. P. Stanley, *Hilbert functions of graded algebras*, Advances in Math. **28** (1978), no. 1, 57–83, DOI 10.1016/0001-8708(78)90045-2. MR485835

[Zan06] F. Zanello, *Level algebras of type 2*, Comm. Algebra **34** (2006), no. 2, 691–714, DOI 10.1080/00927870500387986. MR2211949

Department of Mathematics, Indian Institute of Technology Dharwad, WALMI Campus, PB Road, Dharwad - 580011, Karnataka, India

Email address: `shreedevi@iitdh.ac.in`

Contemporary Mathematics
Volume **805**, 2024
https://doi.org/10.1090/conm/805/16135

Isolated singularities, inverse systems and the punctual Hilbert scheme

Bernard Mourrain

To the memory of Agnes Szanto

ABSTRACT. In this paper, we present an effective study of the punctual Hilbert scheme. First, we recall algorithms to compute the inverse system of an isolated singular point. These inverse systems are points of the punctual Hilbert scheme, that we define as a subvariety of a Grassmannian variety, providing explicit defining equations. We localise our study to the algebraic variety Hilb_B of inverse systems, which admits a given dual (monomial) basis B. Building on the algorithm to compute inverse systems and exploiting combinatorial properties of the monomial basis B, we derive a set of equations defining Hilb_B. Using this effective presentation, we prove new properties of transversality of Hilb_B, give new dimension formula for its irreducible components and prove that they are birational.

1. Introduction

Solving a system of equations $f_1 = 0, \ldots, f_s = 0$ with multivariate polynomials $f_i \in \mathbb{K}[x_1, \ldots, x_n]$ is a task which appears in many contexts and applications. One of the most popular method used to find *a* solution of such non-linear equations is Newton method (or variants like Gauss-Newton method). However, the methods fail when the solution is not a simple isolated point.

To address this problem and solve more efficiently non-linear equations with isolated singular solutions, one can analyse the multiplicity structure and then exploit the properties of this structure to develop efficient algorithms. This leads to the conceptualisation of families of multiplicity structures of a given length as an algebraic variety, whose geometric properties give structural insights into the multiplicity classes themselves.

This approach, which underlies the construction of (punctual Hilbert) schemes by A. Grothendieck [**17**], has opened up a very rich field of mathematical investigation. The study of Hilbert schemes is an active domain of algebraic geometry since several decades. In [**7**] an analysis of the punctual Hilbert scheme in two variables and of the generic staircase strata is developed. Bounds on the dimension of the punctual Hilbert scheme are provided in [**8**]. A review of geometric properties of the punctual Hilbert scheme can be found in [**23**]. The study of the punctual Hilbert schemes is further developed in [**16**]. The subject also progresses

2020 *Mathematics Subject Classification.* Primary 14B05, 14B12, 13H15, 13P10, 13P15, 68W30.

on the analysis of the *global* Hilbert scheme and the functorial properties. See e.g. [**3**] for a general overview. In a more effective perspective, works like [**19**] aimed to provide an explicit description of the global Hilbert scheme (in two variables). In [**35**], graphs of irreducible components of the Hilbert scheme associated to connections of Borel initial ideals are analyzed. In [**34**], a stratification of the Hilbert scheme in terms of initial ideals is further investigated. In [**20**], Hilbert schemes in multigraded algebra are considered. A case-study of the Hilbert schemes of 8 points is developed in [**9**]. In [**22**], charts of the Hilbert scheme in two variables are analysed. Explicit coordinate rings and relations related to corner monomials are described. This approach is further developed in [**27, 36**], with the study of so-called Gröbner strata. Border basis schemes are studied in [**25, 26**], exploiting the characterization of border bases by commutation of the multiplication operators [**32**]. These commuting endomorphisms are also used in [**18**] to provide a representation of the Hilbert functor and equations in local coordinates. As shown in [**1**], the homogeneization of the commutation relations define projectively the Hilbert scheme of points. As we will see, commutation relations also play an important role in the effective description of the punctual Hilbert scheme. In [**10**], strata of the Hilbert scheme of points are analysed via initial ideals, in particular those associated with Borel initial ideals. This approach is further investigated in [**5**]. Charts of the Hilbert scheme with marked bases are studied in [**4**]. In [**6**], explicit equations of low degree in the Plücker coordinates defining general Hilbert schemes are described. In [**2**], the tangent space of the punctual Hilbert scheme in two variables is analysed. Bounds on its dimension are provided in terms of Young diagrams associated with monomial bases. This list, which is far from being exhaustive, shows that many works study the properties of the Hilbert scheme, via explicit descriptions coming from the algebraic calculus on ideals.

Despite these numerous developments, a complete understanding of the geometry of Hilbert schemes is not yet achieved and many questions remain open.

In this paper, we approach the study of the punctual Hilbert scheme from an algorithmic point of view. We first present algorithms, which allow to compute the inverse system or multiplicity structure associated to an isolated point. The main steps of the algorithm are the solution of linear systems, which are built by formal integration. We define the punctual Hilbert scheme as a subvariety of a Grassmannian variety and provide explicit equations defining it. Then we localised our study to the varieties Hilb_B of inverse systems, which admits a given dual (monomial) basis B. Building on the algorithm to compute the inverse system, we derive a set of integration and commutation relations, which define Hilb_B. Exploiting combinatorial properties of the monomial basis B, we provide a reduced set of defining equations. Using this effective presentation of Hilb_B, we prove new properties of transversality of Hilb_B, give new dimension formula for its irreducible components and prove that they are birational.

2. The inverse system of a singular point

In this section, we provide the notation and basic material needed for the analysis of the punctual Hilbert scheme. We refer to [**14, 15, 28, 33**] for more details. Without loss of generality, we will assume that the singular point that we analyse is the origin $\mathbf{0} = (0, \dots, 0)$.

2.1. Preliminaries. Let $\mathbb{K}$ be an algebraic closed field (not necessarily of characteristic 0). Throughout the paper we use the notation $\mathbf{e}_1, \ldots, \mathbf{e}_n$ for the standard basis of $\mathbb{K}^n$ or for a *canonical* basis of a vector space V of dimension n.

The algebra of polynomials in the variables $\mathbf{x} = (x_1, \ldots, x_n)$ with coefficients in $\mathbb{K}$ is denoted $\mathbb{K}[\mathbf{x}]$. The degree of a polynomial $p = \sum_{\alpha \in \mathbb{N}^n} p_\alpha \mathbf{x}^\alpha \in \mathbb{K}[\mathbf{x}]$ with $\mathbf{x}^\alpha = x_1^{\alpha_1} \cdots x_n^{\alpha_n}$ is $\deg(p) = \max_{\alpha \text{ s.t. } p_\alpha \neq 0}\{|\alpha|\}$ where $|\alpha| = |(\alpha_1, \ldots, \alpha_n)| = \sum_i \alpha_i$. Its valuation is $\mathrm{val}(p) = \min_{\alpha \text{ s.t. } p_\alpha \neq 0}\{|\alpha|\}$. For $t \in \mathbb{N}$, the space of polynomials of degree $\leq t$ is $\mathbb{K}[\mathbf{x}]_t$. Let $s_{n,t} = \binom{n+t}{n} = \dim \mathbb{K}[\mathbf{x}]_t$ be the number of monomials of degree $\leq t$ in n variables.

For a set B of monomials in $\mathbb{K}[\mathbf{x}]$, $B_t = B \cap \mathbb{K}[\mathbf{x}]_t$ and $B_{[t]}$ is the subset of elements in B of degree exactly t. For $B \subset \mathbb{K}[\mathbf{x}]$ let $B^+ = B \cup x_1 B \cup \cdots \cup x_n B$ be the *prolongation* of B and let $\partial B = B^+ \setminus B$ be the *border* of B. We say that B is closed by division if for any $b \in B$ of the form $b = x_j b'$ for some $1 \leq j \leq n$, we have $b' \in B$. In particular, if B is a set of monomials closed by division, it must contains 1. In the following, we will identify a monomial $\mathbf{x}^\beta$ with its exponent $\beta = (\beta_1, \ldots, \beta_n) \in \mathbb{N}^n$.

Let $\mathbf{m} = (x_1, \ldots, x_n)$ be the maximal ideal defining the origin $\mathbf{0}$. For a $\mathbf{m}$-primary ideal Q (with $\sqrt{Q} = \mathbf{m}$), $\mathcal{A} = \mathbb{K}[\mathbf{x}]/Q$ is a local Artinian algebra. Its dimension δ over $\mathbb{K}$ is the *multiplicity* of $\mathbf{0}$. More generally, the multiplicity of a point $\zeta \in \mathbb{K}^n$ is the dimension of $\mathbb{K}[\mathbf{x}]/Q_\zeta$ where Q_ζ is a $\mathbf{m}_\zeta$-primary ideal and $\mathbf{m}_\zeta$ is the maximal ideal defining ζ.

The dual $\mathbb{K}[\mathbf{x}]^* = \mathrm{Hom}_{\mathbb{K}}(\mathbb{K}[\mathbf{x}], \mathbb{K})$ is the space of linear functionals $\lambda : p \in \mathbb{K}[\mathbf{x}] \mapsto \lambda(p) \in \mathbb{K}$ on $\mathbb{K}[\mathbf{x}]$. An element $\lambda \in \mathbb{K}[\mathbf{x}]^*$ can be represented by an element of the ring of formal power series $\mathbb{K}[[\mathbf{y}]] := \mathbb{K}[[y_1, \ldots, y_n]]$ as

$$\lambda = \sum_{\alpha \in \mathbb{N}^n} \lambda(\mathbf{x}^\alpha)\, \mathbf{y}^\alpha$$

where $\mathbf{y}^\alpha = y_1^{\alpha_1} \cdots y_n^{\alpha_n}$ is the basis dual to the monomial basis $(\mathbf{x}^\alpha)_{\alpha \in \mathbb{N}^n}$ of $\mathbb{K}[\mathbf{x}]$. Notice that $1 \in \mathbb{K}[\mathbf{y}]$ represents the evaluation $p \in \mathbb{K}[\mathbf{x}] \mapsto p(\mathbf{0})$ at $\mathbf{0} \in \mathbb{K}^n$. This defines a pairing between $\mathbb{K}[\mathbf{x}]^*$ and $\mathbb{K}[\mathbf{x}]$: For $\lambda = \sum_{\alpha \in \mathbb{N}^n} \lambda_\alpha \mathbf{y}^\alpha \in \mathbb{K}[\mathbf{x}]^*$ and $p = \sum_{\beta, |\beta| \leq d} p_\beta \mathbf{x}^\beta$ where $d = \deg(p)$,

$$\langle \lambda | p \rangle = \lambda(p) = \sum_\alpha \lambda_\alpha p_\alpha. \tag{1}$$

As $\mathbb{K}[\mathbf{x}]$ acts on $\mathbb{K}[\mathbf{x}]$ by multiplication, it also acts on its dual $\mathbb{K}[\mathbf{x}]^*$, by the transposed multiplication defined as follows. For $p \in \mathbb{K}[\mathbf{x}], \lambda \in \mathbb{K}[\mathbf{x}]^*$, $p \star \lambda : q \mapsto \lambda(p\, q)$. We check that $p = x_i$ acts as a divided difference on $\mathbb{K}[[\mathbf{y}]]$ for this transposed product:

$$x_i \star \mathbf{y}^\alpha = \begin{cases} y_1^{\alpha_1} \cdots y_{i-1}^{\alpha_{i-1}}\, y_i^{\alpha_i - 1}\, y_{i+1}^{\alpha_{i+1}} \cdots y_n^{\alpha_n} & \text{if } \alpha_i > 0 \\ 0 & \text{otherwise} \end{cases}$$

We also denote this transpose product by x_i or *divided difference* $\partial_i : \lambda \in \mathbb{K}[\mathbf{x}]^* \mapsto \partial_i(\lambda) := x_i \star \lambda$. For this product x_i acts as the *inverse* of y_i. This explains why, in the (old) literature, the elements in $\mathbb{K}[\mathbf{x}]^*$ generate so called *inverse systems.* In characteristic 0, using a scaled dual basis, one can see the transpose product as a derivation [**33**].

Considering an integration operation reverse to the divided difference, we introduce the following integral notation to denote the linear operator of $\mathbb{K}[[\mathbf{y}]]$ defined for $\alpha \in \mathbb{N}^n$ on the dual basis $(\mathbf{y}^\alpha)$ by

$$\oint_k \mathbf{y}^\alpha := \begin{cases} y_1^{\alpha_1} \cdots y_{k-1}^{\alpha_{k-1}} y_k^{\alpha_k+1} & \text{if } \alpha_{k+1} = \cdots = \alpha_n = 0 \\ 0 & \text{otherwise} \end{cases} \tag{2}$$

2.2. The orthogonal to an ideal. For an ideal $I \subset \mathbb{K}[\mathbf{x}]$, let $I^\perp = \{\lambda \in \mathbb{K}[[\mathbf{y}]] \mid \forall p \in I, \langle \lambda | p \rangle = 0\}$. For a vector space $\mathcal{L} \subset \mathbb{K}[[\mathbf{y}]]$, we denote $\mathcal{L}^\perp = \{p \in \mathbb{K}[\mathbf{x}] \mid \forall \lambda \in \mathcal{L}, \langle \lambda \,|\, p \rangle = 0\}$.

DEFINITION 2.1. The vector space $I^\perp \subset \mathbb{K}[[\mathbf{y}]]$ is called the *inverse system* of I.

The vector space $I^\perp$ is naturally identified with the dual space of $\mathbb{K}[\mathbf{x}]/I$. We check that $\mathcal{L} = I^\perp$ is a vector subspace of $\mathbb{K}[[\mathbf{y}]]$, which is closed by divided difference ∂_i for $i = 1, \ldots, n$. Conversely, $\mathcal{L} \subset \mathbb{K}[[\mathbf{y}]]$ is closed by divided difference iff $Q = \mathcal{L}^\perp$ is an ideal.

The following classical property says that inverse systems of $\mathbf{m}$-primary ideals are vector spaces of $\mathbb{K}[\mathbf{y}]$ (see e.g. [**14**][Théorème 7.31]):

LEMMA 2.2. *If Q is a $\mathbf{m}$-primary isolated component of I, then $Q^\perp = I^\perp \cap \mathbb{K}[\mathbf{y}]$. Its dimension is denoted by $\delta_{\mathbf{m}}(I)$ or simply $\delta(I)$.*

Therefore, we can compute $Q^\perp$, as the space of all polynomials belonging to $\mathbb{K}[\mathbf{y}]$ which are in $I^\perp$, without knowing the primary decomposition of I.

Let us denote this set $\mathcal{L} = I^\perp \cap \mathbb{K}[\mathbf{y}]$. It is a vector space stable under the divided differences ∂_i. Its dimension is the dimension of $Q^\perp$ or $\mathbb{K}[\mathbf{x}]/Q$, that is the *multiplicity* of $\mathbf{0}$, denoted $\delta(I)$, or simply δ if I is clear from the context.

EXAMPLE 2.3. Let $Q = (x_1^2, x_2^2) \subset \mathbb{K}[x_1, x_2]$ defining $\zeta = (0,0) \in \mathbb{K}^2$ as an isolated point of multiplicity 4. Then its inverse system is $Q^\perp = \text{span}_{\mathbb{K}}(1, y_1, y_2, y_1 y_2)$.

For an element $\lambda = \sum_\beta \lambda_\beta \mathbf{y}^\beta \in \mathbb{K}[\mathbf{y}]$ we define the degree or *order* $\omega(\lambda)$ to be the maximal $|\beta|$ s.t. $\mathbf{y}^\beta$ appears in λ with a non-zero coefficient λ_β.

For $t \in \mathbb{N}$, let $\mathcal{L}_t$ be the elements of $\mathcal{L}$ of order $\leq t$. We directly check that $\mathcal{L}_t = (Q + \mathbf{m}^{t+1})^\perp$. As $\mathcal{L}$ is of dimension δ, there exists a smallest $t \geq 0$ s.t. $\mathcal{L}_{t+1} = \mathcal{L}_t$. Let us call this smallest t, the *nil-index* of $\mathcal{L}$ and denote it by $\omega(I)$, or simply by ω.

REMARK 2.4. As $\mathcal{L}$ is stable by the divided differences ∂_i, we easily check that $\omega < \delta$ so that $\mathcal{L}_{\delta-1} = \mathcal{L}$.

Let $B = \{\mathbf{x}^{\beta_1}, \ldots, \mathbf{x}^{\beta_\delta}\}$ be a basis of $\mathbb{K}[\mathbf{x}]/Q$. We can identify the elements of $\mathbb{K}[\mathbf{x}]/Q$ with the elements of the vector space $\text{span}_{\mathbb{K}}(B)$ spanned by B.

We define the normal form $N(p)$ of a polynomial p in $\mathbb{K}[\mathbf{x}]$ as the projection of p on $\text{span}_{\mathbb{K}}(B)$ along Q, that is, the unique element b of $\text{span}_{\mathbb{K}}(B)$ such that $p - b \in Q$. We easily verify that if $B = \{\mathbf{x}^{\beta_1}, \ldots, \mathbf{x}^{\beta_\delta}\}$, then $\forall p \in \mathbb{K}[\mathbf{x}]$

$$N(p) = \sum_{i=1}^{\delta} \lambda_i(p) \mathbf{x}^{\beta_i}$$

where $\lambda_i \in Q^\perp$ is a linear functional, which vanishes on Q. We have the following property:

LEMMA 2.5. *The coefficients $\Lambda = \{\lambda_1, \ldots, \lambda_\delta\}$ of the normal form N modulo Q in the basis B of $\mathbb{K}[\mathbf{x}]/Q$ form a basis of $Q^\perp$ dual to B.*

PROOF. Given any $\lambda \in Q^\perp$, let us consider

$$\tilde{\lambda} = \lambda - \sum_{i=1}^{\delta} \lambda(\mathbf{x}^{\beta_i})\, \lambda_i,$$

where $\lambda_i \in Q^\perp$ for $i = 1, \dots, \delta$ are the coefficients of N. They verify $\lambda_i(\mathbf{x}^{\beta_j}) = 1$ if $i = j$ and 0 otherwise. Thus $\tilde{\lambda}(p) = 0$ for $p \in Q$ and $p \in \mathrm{span}_{\mathbb{K}}(B)$, that is, $\tilde{\lambda} = 0$ or equivalently $\lambda = \sum_{i=1}^{\delta} \lambda(\mathbf{x}^{\beta_i})\, \lambda_i$. This shows that (λ_i) is a generating set of $Q^\perp$ and a basis, since (λ_i) is dual to $(\mathbf{x}^{\beta_i})$. □

EXAMPLE 2.6. Let $Q = (x_1^2, x_2^2) \subset \mathbb{K}[x_1, x_2]$ defining $\zeta = (0,0) \in \mathbb{K}^2$ with multiplicity 4. Let $B = \{1, x_1, x_2, x_1x_2\}$. It is a basis of $\mathbb{K}[x_1, x_2]/Q$ and for any $p = \sum_\alpha p_\alpha \mathbf{x}^\alpha$, we have

$$N(p) = p_{0,0} + p_{1,0}x_1 + p_{0,1}x_2 + p_{1,1}x_1x_2 = \langle 1 \,|\, p\rangle + \langle y_1 \,|\, p\rangle\, x_1 + \langle y_2 \,|\, p\rangle\, x_2 + \langle y_1y_2 \,|\, p\rangle\, x_1x_2$$

and the coefficients of the normal form in the basis $B = \{1, x_1, x_2, x_1x_2\}$ are the linear functionals $\{1, y_1, y_2, y_1y_2\}$, which form a basis of $Q^\perp$ dual to B.

2.3. Computing inverse systems. We describe now how the inverse system of the isolated point $\mathbf{0}$ of the variety defined by $\mathbf{f} = \{f_1, \dots, f_s\} \subset \mathbb{K}[\mathbf{x}]$ can be computed from the input polynomials $\mathbf{f}$. The way a basis of the inverse system is computed relies on structural properties, that will be useful to analyse the geometry of the Punctual Hilbert Scheme.

Macaulay's dialytic method. The first "algorithm" was proposed by F.S. Macaulay [**28**]. It is an iterative method, which degree by degree computes a basis of the kernel of matrices built from monomial multiples of $\mathbf{f}$.

As $f_i(\mathbf{0}) = 0$, the valuation $\mathrm{val}(f_i)$ is greater than 1 for $i = 1, \dots, s$. We denote by $\mathbf{m} = (x_1, \dots, x_n)$ the maximal ideal defining $\mathbf{0}$. For $t \in \mathbb{N}$ let $\mathbf{m}^{[t]}$ be the set of monomials of degree t. Then the ideal $\langle \mathbf{f} \rangle$ is the vector space generated by $\mathbf{f} \cdot \mathbf{m}^{[t]}$ for $t \in \mathbb{N}$. Notice that the polynomials $\mathbf{f} \cdot \mathbf{m}^{[t]}$ are of valuation $\geq t + 1$. Let $\mathbf{h}_t = \{\mathbf{f}, \mathbf{f} \cdot \mathbf{m}^{[1]}, \dots, \mathbf{f} \cdot \mathbf{m}^{[t-1]}\} \subset \langle \mathbf{f} \rangle$. The elements $p \in \langle \mathbf{f} \rangle$ of valuation $\leq t$ are in the space spanned by $\mathbf{h}_t$. We consider the linear map defined as follows:

$$\begin{aligned} H_t : \mathbb{K}[\mathbf{y}]_t &\to \mathbb{K}^{|\mathbf{h}_t|} \\ \lambda &\mapsto (\langle \lambda \,|\, h \rangle)_{h \in \mathbf{h}_t}. \end{aligned}$$

By Lemma 2.2, the elements of the inverse systems $\mathcal{L}_t$ of degree $\leq t$ are the elements of $\mathbb{K}[\mathbf{y}]_t$ orthogonal to $\langle \mathbf{f} \rangle$. Thus they are the elements of $\mathbb{K}[\mathbf{y}]_t$ orthogonal to the elements of $\langle \mathbf{f} \rangle$ of valuation $\leq t$. In other words, $\mathcal{L}_t = \ker H_t$.

We easily check that $\ker H_0 = \mathrm{span}_{\mathbb{K}}(1) \subset \mathbb{K}[\mathbf{y}]$. The kernel of H_t can be computed from the kernel H_{t-1} since H_{t-1} is a submap of H_t and its matrix in the monomial basis is of the form

$$\begin{bmatrix} H_{t-1} & H_t' \\ \mathbf{0} & N_t \end{bmatrix}.$$

Thus $\ker H_{t-1} \hookrightarrow \ker H_t$ and new basis elements of $\ker H_t$ are obtained from the basis of $\ker H_{t-1}$ by computing elements $\varphi \in \ker N_t$ and $\lambda \in \mathbb{K}[\mathbf{y}]_{t-1}$ such that $H_t(\lambda) + H_t'(\varphi) = 0$. This leads to Algorithm 1.

The iteration stops at a degree t which is the nil-index ω of the singular point $\zeta = \mathbf{0}$. This method has been used and improved in recent works. See e.g. [**11, 12**]. However, its complexity is in $\mathcal{O}(\ell_\omega s_{n,\omega}^2)$ where $\ell_\omega = |\mathbf{h}_\omega|$ and $s_{n,\omega} = \dim \mathbb{K}[\mathbf{y}]_\omega$.

Algorithm 1 INVERSE SYSTEM BY THE DIALYTIC METHOD

Starting with $\mathcal{D}_0 = \{1\}$ and $t = 0$
- $t := t + 1$
- Compute H_t, a basis of $\ker N_t$ and the new basis elements in $\mathcal{L}_t$.

until $\ker N_t = \{0\}$.

Integration method. Another approach to compute a basis of the inverse system $\mathcal{L}$ exploits the stability of $\mathcal{L}$ by divided difference. In particular, $\partial_k \mathcal{L}_t \subset \mathcal{L}_{t-1}$ for $k = 1, \ldots, n$. Therefore, a basis of $\mathcal{L}_t$ can be computed from a basis of $\mathcal{L}_{t-1}$ by *integration*, which consists in "inverting" the actions of the divided differences ∂_k.

The method computes a so-called *graded primal-dual pair* (B, Λ) with $B = \{\mathbf{x}^{\beta_1}, \ldots, \mathbf{x}^{\beta_\delta}\}$, $|\beta_1| \leq \cdots \leq |\beta_\delta|$, $\Lambda = \{\lambda_{\beta_1}, \ldots, \lambda_{\beta_\delta}\}$, $\omega(\lambda_{\beta_i}) \leq |\beta_i|$ and

$$\langle \lambda_{\beta_i} \,|\, \mathbf{x}^{\beta_j} \rangle = \begin{cases} 1 & \text{if } i = j \\ 0 & \text{otherwise} \end{cases} \tag{3}$$

It is based on the following characterisation of spaces stable by divided differences:

THEOREM 2.7 ([**29**–**31**]). *Let* $B = \{\mathbf{x}^{\beta_1}, \ldots, \mathbf{x}^{\beta_\delta}\} \subset \mathbb{K}[\mathbf{x}]$ *with* $|\beta_1| \leq \cdots \leq |\beta_\delta|$. *Let* $\mathcal{L} = \langle \lambda_{\beta_1}, \ldots, \lambda_{\beta_\delta} \rangle \subset \mathbb{K}[\mathbf{y}]$ *with* $\omega(\lambda_{\beta_i}) \leq |\beta_i|$.

Then $\mathcal{L} = \langle \lambda_1, \ldots, \lambda_r \rangle$ *is closed by divided difference with*[1]

$$\partial_k \lambda_{\beta_i} = \sum_{|\beta_j| < |\beta_i|} \mu_{\beta_i, \beta_j + \mathbf{e}_k} \lambda_{\beta_j}. \tag{4}$$

iff for $1 \leq i \leq \delta$ *we have the* integration relations

$$\lambda_{\beta_i} = \sum_{|\beta_j| < |\beta_i|} \sum_{k=1}^{n} \mu_{\beta_i, \beta_j + \mathbf{e}_k} \oint_k \lambda_{\beta_j} \tag{5}$$

and for all $i, s = 1, \ldots, r$, $|\beta_s| < |\beta_i|$, $k \neq l \in \{1, \ldots, n\}$ *we have the* commutation relations

$$\sum_{j : |\beta_h| < |\beta_j| < |\beta_i|} \mu_{\beta_i, \beta_j + \mathbf{e}_k} \mu_{\beta_j, \beta_h + \mathbf{e}_l} - \mu_{\beta_i, \beta_j + \mathbf{e}_l} \mu_{\beta_j, \beta_h + \mathbf{e}_k} = 0. \tag{6}$$

Furthermore, (B, Λ) *is a primal-dual basis pair iff they satisfy* (5), (6) *and* orthogonality relations

$$\mu_{\beta_i, \beta_j + \mathbf{e}_k} = \begin{cases} 1 & \text{for } \beta_i = \beta_j + \mathbf{e}_k \\ 0 & \text{for } \beta_j + \mathbf{e}_k = \beta_l \text{ for some } 1 \leq l \leq \delta,\ \beta_i \neq \beta_j + \mathbf{e}_k, \end{cases} \tag{7}$$

Moreover Λ *is a basis of the inverse system of* $\mathbf{f} = (f_1, \ldots, f_s)$ *at* $\zeta = \mathbf{0}$ *iff*

$$\langle \lambda_{\beta_i} \,|\, f_j \rangle = 0 \quad \text{for} \quad 1 \leq i \leq \delta, 1 \leq j \leq s. \tag{8}$$

This leads to Algorithm 2 to compute a graded primal-dual basis pair (B, Λ) for the inverse system $\mathcal{L}$ of $\mathbf{f}$ at $\mathbf{0}$.

The monomials of $B_{[t]}$ computed at step t are of degree t. They can be obtained by triangularization of the system 3. in Algorithm 2, so that $(B_{[t]}, \Lambda_{[t]})$ is a graded primal-dual basis pair. By construction, the algorithm yields a graded primal-dual basis pair (B, Λ) of $\mathcal{L}$. It can moreover compute a basis B, which is closed by

[1] Recall that $(\mathbf{e}_k)_{k=1,\ldots,n}$ denotes the canonical basis of $\mathbb{R}^n \supset \mathbb{N}^n$.

Algorithm 2 INVERSE SYSTEM BY INTEGRATION

Starting with $B_0 = \{1\} \subset \mathbb{K}[\mathbf{x}]$, $\Lambda_0 = \{1\} \subset \mathbb{K}[\mathbf{y}]$, $t = 0$

1. $t := t + 1$
2. For unknown coefficients $\boldsymbol{\nu} = (\nu_{\beta_j + \mathbf{e}_k})$, let

$$\lambda = \sum_{|\beta_j|<t} \sum_{k=1}^{n} \nu_{\beta_j+\mathbf{e}_k} \oint_k \lambda_{\beta_j}$$

3. Compute $B_{[t]} = \{\mathbf{x}^{\beta_{\delta_{t-1}+1}}, \ldots, \mathbf{x}^{\beta_{\delta_t}}\}$ and a basis $\Lambda_{[t]} = \{\lambda_{\beta_{\delta_{t-1}+1}}, \ldots, \lambda_{\beta_{\delta_t}}\}$ of the solutions of the linear system in $\boldsymbol{\nu}$:

$$\sum_{j:|\beta_h|<|\beta_j|<t} \nu_{\beta_j+\mathbf{e}_k} \mu_{\beta_j,\beta_h+\mathbf{e}_l} - \nu_{\beta_j+\mathbf{e}_l} \mu_{\beta_j,\beta_h+\mathbf{e}_k} = 0 \ \text{ for } 1 \le h \le \delta_{t-1}, 1 \le k < l \le n,$$

$$\langle \lambda \,|\, f_j \rangle = 0 \quad \text{for} \quad 1 \le j \le s,$$
$$\langle \lambda \,|\, \mathbf{x}^{\beta_h} \rangle = 0 \quad \text{for} \quad 1 \le h \le \delta_{t-1}$$

such that $\langle \lambda_{\beta_j} \,|\, \mathbf{x}^{\beta_i} \rangle = \begin{cases} 1 & \text{if } i = j \\ 0 & \text{otherwise} \end{cases}$ for $\delta_{t-1} + 1 \le i, j \le \delta_t$.

4. $B_t := B_{t-1} \cup B_{[t]}$, $\Lambda_t := \Lambda_{t-1} \cup \Lambda_{[t]}$, $\mu_{\beta_i,\beta_j+\mathbf{e}_k} = \langle \lambda_{\beta_i} \,|\, \mathbf{x}^{\beta_j+\mathbf{e}_k} \rangle$ for $\delta_{t-1} + 1 \le i \le \delta_t, 1 \le j \le \delta_{t-1}, 1 \le k \le n$.

until $B_{[t]} = \{\}$.

divided difference. Its complexity is in $\mathcal{O}(\delta^3)$. For more details on the algorithm, we refer to [**29**–**31**].

REMARK 2.8. We can notice the analogy between these relations and the characterisation of border bases by commutation properties [**32**]. Indeed, the relations (6) corresponds to the commutation of the transposed operators of multiplication by the variables, which corresponds to the derivations in the dual setting [**31**]. As mentioned in [**21**], a border basis can be directly extracted from the coefficients of the dual basis.

EXAMPLE 2.9. We consider the polynomials $f_1 = x_1 - x_2 + x_1^2$, $f_2 = x_1 - x_2 + x_2^2 \in \mathbb{C}[x_1, x_2]$, for which $\zeta = (0, 0)$ is an isolated root of multiplicity 3 as we will see by applying the integration algorithm:

- We start with $B_0 = \{1\}$, $\Lambda_0 = \{1\}$;
- At step 2. we have $\lambda = \nu_{1,0}\, y_1 + \nu_{0,1}\, y_2$;
- Solving the system 3. yields $\nu_{1,0} = \nu_{0,1} = 1$ and $B_{[1]} = \{x_1\}$, $\Lambda_{[1]} = \{y_1 + y_2\}$;
- At the iteration for $t = 2$, we have $\lambda = \nu_{1,0}\, y_1 + \nu_{0,1}\, y_2 + \nu_{2,0}\, y_1^2 + \nu_{1,1}\, (y_1 y_2 + y_2^2)$.
- In step 3., we obtain $B_{[2]} = \{x_1^2\}$, $\Lambda_{[2]} = \{y_1^2 + y_1 y_2 + y_2^2\}$;
- At $t = 3$, there is no solution and the iteration stops.

We obtain the primal-dual basis pair $B = \{1, x_1, x_1^2\}$, $\Lambda = \{1, y_1 + y_2, y_1^2 + y_1 y_2 + y_2^2\}$. This shows that $\zeta = (0, 0)$ is an isolated point of multiplicity $3 = |B| = |\Lambda|$. We verify that Λ is stable by the divided differences ∂_k for $k = 1, 2$.

3. The punctual Hilbert scheme Hilb_δ

In this section, we define the punctual Hilbert scheme, affine charts and describe their defining equations.

3.1. The projective variety Hilb_δ**.** For a vector space V, let $\mathscr{G}_\delta(V)$ be the projective variety of the δ dimensional linear subspaces of V, also known as the *Grassmannian* of δ-spaces of V. The points in $\mathscr{G}_\delta(V)$ are the projective points of $\mathbb{P}(\wedge^\delta V)$ of the form $\mathbf{v} = v_1 \wedge \cdots \wedge v_\delta$ for $v_i \in V$.

Fixing a basis $\mathbf{e}_1, \ldots, \mathbf{e}_s$ of V, the Plücker coordinates of $\mathbf{v} = v_1 \wedge \cdots \wedge v_\delta \in \wedge^\delta V$ are the coefficients of $\Delta_{i_1,\ldots,i_\delta}(\mathbf{v})$ of $\mathbf{v} = \sum_{i_1<\cdots<i_\delta} \Delta_{i_1,\ldots,i_\delta}(\mathbf{v})\, \mathbf{e}_{i_1} \wedge \cdots \wedge \mathbf{e}_{i_\delta}$.

To define the punctual Hilbert scheme Hilb_δ, we will consider the family of linear spaces $\mathcal{L}$, which are the orthogonal of an ideal Q defining $\mathbf{0}$ with multiplicity δ. By Lemma 2.2 and Remark 2.4, $\mathcal{L}$ is a linear subspace of $\mathbb{K}[\mathbf{y}]_{\delta-1}$, that is a point of the Grassmannian $\mathscr{G}_\delta(\mathbb{K}[\mathbf{y}]_{\delta-1})$.

A natural basis of $\mathbb{K}[\mathbf{y}]_{\delta-1}$ is the dual monomial basis $(\mathbf{y}^\alpha)_{|\alpha|<\delta}$. The Plücker coordinates of an element $\mathcal{L} \in \mathscr{G}_\delta(\mathbb{K}[\mathbf{y}]_{\delta-1})$ for this canonical basis are denoted $\Delta_{\alpha_1,\ldots,\alpha_\delta}(\mathcal{L})$ where $\alpha_i \in \mathbb{N}^n$, $|\alpha_i| < \delta$. If $\Lambda = \{\lambda_1, \ldots, \lambda_\delta\}$ is a basis of $\mathcal{L}$ in $\mathbb{K}[\mathbf{y}]_{\delta-1}$ with $\lambda_i = \sum_{|\alpha|<\delta} \mu_{i,\alpha}\mathbf{y}^\alpha$, its coefficient matrix in the monomial basis is $[\mu_{i,\alpha}]_{1\le i\le n, |\alpha|<\delta}$. The Plücker coordinates of $\mathcal{L} = \mathrm{span}_{\mathbb{K}}(\Lambda)$ are, up to a scalar,

$$\Delta_{\alpha_1,\ldots,\alpha_\delta}(\mathcal{L}) = \det\left[\mu_{i,\alpha_j}\right]_{1\le i,j\le\delta} = \det\left[\langle \lambda_i \,|\, \mathbf{x}^{\alpha_j}\rangle\right]_{1\le i,j\le\delta}.$$

In relation with the construction of Hilbert Schemes of points in terms of dual spaces described in [**24**], we define the punctual Hilbert scheme as follows:

DEFINITION 3.1. Let $\mathrm{Hilb}_\delta \subset \mathscr{G}_\delta(\mathbb{K}[\mathbf{y}]_{\delta-1})$ be the set of linear spaces $\mathcal{L}$ of dimension δ in $\mathbb{K}[\mathbf{y}]_{\delta-1}$ which are stable by the divided differences ∂_i with respect to the variables $\mathbf{y}$ (i.e. $\partial_i\mathcal{L} \subset \mathcal{L}$ for $i = 1, \ldots, n$). We called Hilb_δ the *punctual Hilbert scheme* of points of multiplicity δ.

If $\mathcal{L} \subset \mathbb{K}[\mathbf{y}]$ is stable by the divided differences ∂_i, then by duality $Q = \mathcal{L}^\perp := \{p \in \mathbb{K}[\mathbf{x}] \mid \forall\lambda \in \mathcal{L}, \langle \lambda \,|\, p\rangle = 0\}$ is a vector space of $\mathbb{K}[\mathbf{x}]$ stable by multiplication by x_i, i.e. an ideal of $\mathbb{K}[\mathbf{x}]$. Consequently, $\mathcal{L} \in \mathrm{Hilb}_\delta$ iff $\mathcal{L}^\perp = Q$ is a $\mathbf{m}$-primary ideal such that $\dim \mathbb{K}[\mathbf{x}]/Q = \delta$ (see [**30**]).

To describe equations defining Hilb_δ, we recall that ∂_k ($1 \le k \le n$) is a linear operator in $\mathbb{K}[\mathbf{y}]_{\delta-1}$ sending $\mathbf{y}^\alpha \in \mathbb{K}[\mathbf{y}]_\delta$ either to $\mathbf{y}^{\alpha-\mathbf{e}_k}$ if $\alpha_k > 0$ or to 0. We recall also that the proper intersection of the linear spaces $\mathcal{L}$ and $\mathbf{v} = \mathbf{v}_1 \wedge \cdots \wedge \mathbf{v}_\sigma$, $\mathbf{v}_i \in \mathbb{K}[\mathbf{y}]_{\delta-1}$ respectively of dimension δ and $\sigma = s_{n,\delta-1} - \delta + 1$ is given by the intersection product (c.f. [**13**])

$$\mathcal{L} \cdot \mathbf{v} = \sum_{|\alpha|<t} I_\alpha(\Delta(\mathcal{L}), \mathbf{v}_1, \ldots, \mathbf{v}_\sigma)\, \mathbf{x}^\alpha \in \mathbb{K}[\mathbf{y}]_{\delta-1}$$

where $I_\alpha(\Delta(\mathcal{L}), \mathbf{v}_1, \ldots, \mathbf{v}_\sigma)$ is a multilinear function of $\Delta(\mathcal{L}), \mathbf{v}_1, \ldots, \mathbf{v}_\sigma$ for $|\alpha| < \delta$.

PROPOSITION 3.2. *The punctual Hilbert scheme* Hilb_δ *is the projective variety of linear spaces* $\mathcal{L} \in \mathscr{G}_\delta(\mathbb{K}[\mathbf{y}]_{\delta-1})$ *such that*

$$\partial_k(\mathcal{L} \cdot \mathbf{y}^{\alpha_1} \wedge \cdots \wedge \mathbf{y}^{\alpha_\sigma}) \wedge \mathcal{L} = 0 \textit{ for } |\alpha_i| < \delta, \quad 1 \le k \le n. \tag{9}$$

These are quadratic equations in the Plücker coordinates $\Delta(\mathcal{L})$ *of* $\mathcal{L}$.

PROOF. As the set of $\mathcal{L}\cdot\mathbf{v}_1\wedge\cdots\wedge\mathbf{v}_\sigma$ for $\mathbf{v}_i\in\mathbb{K}[\mathbf{y}]_{\delta-1}$ is the whole linear space $\mathcal{L}$, $\mathcal{L}$ is stable by ∂_k for $k=1,\ldots,b$ iff $\partial_k(\mathcal{L}\cdot\mathbf{v}_1\wedge\cdots\wedge\mathbf{v}_\sigma)\in\mathcal{L}$, that is,

$$\partial_k(\mathcal{L}\cdot\mathbf{v}_1\wedge\cdots\wedge\mathbf{v}_\sigma)\wedge\mathcal{L}=0$$

for all $\mathbf{v}_i\in\mathbb{K}[\mathbf{y}]_{\delta-1}, 1\leq k\leq n$. Since these are multilinear relations in $\mathbf{v}_1,\ldots,\mathbf{v}_\sigma$, they are satisfied iff they vanish when $\mathbf{v}_i$ is replaced by any element of the canonical basis $(\mathbf{y}^\alpha)_{|\alpha|<\delta}$. This yields the relations (9). □

We are going to analyse locally the punctual Hilbert scheme, fixing a basis B of the quotient $\mathbb{K}[\mathbf{x}]/Q$ where $Q=\mathcal{L}^\perp$:

DEFINITION 3.3. For B a set of monomials in $\mathbb{K}[\mathbf{x}]$ such that $|B|=\delta$, let $\mathrm{Hilb}_B=\{\mathcal{L}\in\mathrm{Hilb}_\delta\mid\Delta_B(\mathcal{L})\neq0\}$ be a (possibly empty) chart of Hilb_δ.

We verify that if $\Delta_B(\mathcal{L})\neq0$ where $B=\{\mathbf{x}^{\beta_j}\}_{j=1}^\delta$ then $\mathcal{L}$ admits a basis $\Lambda=\{\lambda_1,\ldots,\lambda_\delta\}$ such that $[\langle\lambda_i\,|\,\mathbf{x}^{\beta_j}\rangle]_{1\leq i,j\leq\delta}=\mathrm{Id}$, that is, (B,Λ) is a primal-dual basis pair. Moreover, $\Delta_B(\mathcal{L})\neq0$ iff B is a basis of $\mathbb{K}[\mathbf{x}]/Q$ where $Q=\mathcal{L}^\perp$.

It is clear that $\cup_{|B|=\delta}\mathrm{Hilb}_B$ is an open covering of Hilb_δ. According to Algorithm 2, every $\mathcal{L}\in\mathrm{Hilb}_\delta$ admits a graded primal-dual basis pair (B,Λ) such that $\Delta_B(\mathcal{L})\neq0$. Moreover, we can assume that B is closed by division.

Therefore we can restrict the covering to charts Hilb_B such that (B,Λ) is a graded primal-dual basis pair and B is closed by division.

3.2. The affine chart Hilb_B**.** Let $B=\{\mathbf{x}^{\beta_i}\}_{i=1}^\delta\subset(\mathbf{x}^\alpha)_{|\alpha|<\delta}$ closed by division with $|B|=\delta$, $\mathbf{x}^{\beta_1}=1$, $0=|\beta_1|\leq|\beta_2|\leq\cdots\leq|\beta_\delta|$.

We assume that there exists $\Lambda=\{\lambda_i\}_{i=1}^\delta\subset\mathbb{K}[\mathbf{y}]_{\delta-1}$ such that (B,Λ) is a graded primal-dual basis pair. As $[\langle\lambda_i\,|\,\mathbf{x}^{\beta_j}\rangle]_{1\leq i,j\leq\delta}=\mathrm{Id}$, the coefficient matrix of the dual basis Λ in the monomial basis of $\mathbb{K}[\mathbf{y}]_{\delta-1}$ is of the form

$$\begin{array}{c} \\ \lambda_1\\ \vdots\\ \lambda_\delta\end{array}\begin{array}{c}\begin{array}{ccccccc}\beta_1&\cdots&\beta_\delta&\gamma_1&\cdots&\gamma_l\end{array}\\ \left[\begin{array}{ccccccc}1&&\mathbf{0}&\mu_{1,\gamma_1}&\cdots&\mu_{1,\gamma_c}\\ &\ddots&&\vdots&&\vdots\\ \mathbf{0}&&1&\mu_{\delta,\gamma_1}&\cdots&\mu_{\delta,\gamma_c}\end{array}\right]\end{array}\tag{10}$$

where $C=\{\mathbf{y}^{\gamma_1},\ldots,\mathbf{y}^{\gamma_c}\}$ is the set of monomials of $\mathbb{K}[\mathbf{y}]_{\delta-1}$, which are not in $\mathbf{y}^{\beta_1},\ldots,\mathbf{y}^{\beta_\delta}$. For $t\in\mathbb{N}$, we denote $C_t=C\cap\mathbb{K}[\mathbf{y}]_t$. As $\mathbf{x}^{\beta_1}=1$, we have $\lambda_1=1$ and $\mu_{\beta_1,\gamma_j}=0$ for $j=1,\ldots,c$. Since (B,Λ) is graded, we have

$$\lambda_i=\mathbf{y}^{\beta_i}+\sum_{\gamma\in C_{|\beta_i|}}\mu_{\beta_i,\gamma}\,\mathbf{y}^\gamma$$

and $\mu_{\beta_i,\alpha}=0$ for $\alpha\in C\setminus C_{|\beta_i|}$. Let $\boldsymbol{\mu}=\{\mu_{\beta_i,\gamma},1\leq i\leq\delta,\gamma\in C_{|\beta_i|}\}$ be the set of non-constant parameters or variables, that we need to describe Λ. Let $r_B=|\boldsymbol{\mu}|$ its size, which is less than $r_\delta=(\delta-1)\times(s_{n,\delta-1}-\delta)$ where $s_{n,\delta-1}=\dim\mathbb{K}[\mathbf{x}]_{\delta-1}$. We define $\boldsymbol{\mu}'=\{\mu_{\beta_i,\gamma_j}\in\boldsymbol{\mu}\mid\mathbf{x}^{\gamma_j}\in\partial B\}$. Let $r'_B=|\boldsymbol{\mu}'|$ be the number of variables of $\boldsymbol{\mu}'$.

By properties of the determinant, the Plücker coordinates of $\mathcal{L}$ are such that

$$\mu_{\beta_i,\alpha}=\frac{\Delta_{\beta_1,\ldots,\beta_{i-1},\alpha,\beta_{i+1},\ldots,\beta_\delta}}{\Delta_{\beta_1,\ldots,\beta_\delta}}\qquad i=1,\ldots,\delta,\quad|\alpha|<\delta.\tag{11}$$

By Equation (11), we can replace the homogeneous Plücker coordinates

$$\Delta_{B^{i\leftarrow\gamma_j}}(\mathcal{L}):=\Delta_{\beta_1,\ldots,\beta_{i-1},\gamma_j,\beta_{i+1},\ldots,\beta_\delta}(\mathcal{L})$$

of $\mathcal{L} \in \mathrm{Hilb}_\delta$ by the affine coordinates $\boldsymbol{\mu} = (\mu_{\beta_i,\gamma_j})_{2\leq i\leq\delta, 1\leq\gamma_j\leq l}$. As the remaining Plücker coordinates $\Delta_{\alpha_1,\ldots,\alpha_\delta} = \det([\mu_{\beta_i,\alpha_j}])$ are polynomial functions of $\boldsymbol{\mu}$, we will associate the coordinate ring $\mathbb{K}[\boldsymbol{\mu}]$ to the chart Hilb_B.

By Equation (4), the matrices of the operators of multiplication by x_k or divided difference

$$\begin{aligned}\partial_k : \mathcal{L} &\to \mathcal{L}\\ \lambda &\mapsto x_k \star \lambda = \partial_k \lambda\end{aligned}$$

in the basis Λ of $\mathcal{L}$ are

$$(12)\qquad M_{\mu,k}^T = \begin{bmatrix} 0 & \mu_{\beta_2,\mathbf{e}_k} & \mu_{\beta_3,\mathbf{e}_k} & \cdots & \mu_{\beta_\delta,\mathbf{e}_k}\\ 0 & 0 & \mu_{\beta_3,\beta_2+\mathbf{e}_k} & \cdots & \mu_{\beta_\delta,\beta_2+\mathbf{e}_k}\\ \vdots & \vdots & & & \vdots\\ 0 & 0 & 0 & \cdots & \mu_{\beta_\delta,\beta_{\delta-1}+\mathbf{e}_k}\\ 0 & 0 & 0 & \cdots & 0 \end{bmatrix}$$

for $1 \leq k \leq n$. We refer to this operator

$$\begin{aligned}M_{\mu,k}^T : \mathrm{span}_{\mathbb{K}[\mu]}(\Lambda) &\to \mathrm{span}_{\mathbb{K}[\mu]}(\Lambda)\\ \lambda_i &\mapsto \sum_{j:|\beta_j|<|\beta_i|} \mu_{\beta_i,\beta_j+\mathbf{e}_k}\lambda_j\end{aligned}$$

as the *parametric operator of divided difference* ∂_k. The transpose operator

$$\begin{aligned}M_{\mu,k} : \mathrm{span}_{\mathbb{K}[\mu]}(B) &\to \mathrm{span}_{\mathbb{K}[\mu]}(B)\\ \mathbf{x}^{\beta_i} &\mapsto \sum_{j:|\beta_i|<|\beta_j|} \mu_{\beta_j,\beta_i+\mathbf{e}_k}\mathbf{x}^{\beta_j}\end{aligned}$$

is the *parametric operator of multiplication* by x_k in B. If $\mathcal{L} \in \mathrm{Hilb}_B$, then the commutation relations (6) is satisfied and the matrices $\mathbf{M}_{\boldsymbol{\mu}} = (M_{\mu,1}, \ldots, M_{\mu,n})$ commute.

The commutation relations (6) correspond to the coefficients

$$(13)\quad \langle (M_{\mu,l}^T M_{\mu,k}^T - M_{\mu,k}^T M_{\mu,l}^T)(\lambda_i) \mid \mathbf{x}^{\beta_j}\rangle = \langle \lambda_{\beta_j} \mid (M_{\mu,k}M_{\mu,l} - M_{\mu,l}M_{\mu,k})(\mathbf{x}^{\beta_h})\rangle.$$

that is, the entries of the matrix $M_{\mu,k}M_{\mu,l} - M_{\mu,l}M_{\mu,k}$. We are going to analyse in more details, which entries are not zero, by exploiting the structure of these multiplication operators.

Let $B_t = B \cap \mathbb{K}[\mathbf{x}]_t$, $B_{[t]} = B_t \setminus B_{t-1}$ the set of elements of B of degree t and $h_t = |B_{[t]}|$. We define $\omega_t = |(\partial B)_t|$.

DEFINITION 3.4. For $1 \leq k, l \leq n$ and $\mathbf{x}^\beta \in B$, we define the *commutator polynomial*

$$\mathcal{C}_\beta^{k,l}(\mathbf{x}, \boldsymbol{\mu}) := (M_{\mu,k}M_{\mu,l} - M_{\mu,l}M_{\mu,k})(\mathbf{x}^\beta) = \sum_{\varsigma\in B:|\varsigma|\geq|\beta|+2} \mathcal{C}_{\beta,\varsigma}^{k,l}(\boldsymbol{\mu})\,\mathbf{x}^\varsigma.$$

The *non-zero* coefficients $\mathcal{C}_{\beta,\varsigma}^{k,l}(\boldsymbol{\mu})$ are called *commutator relations*. We say that $\mathbf{x}^\beta$ is a *commutator monomial* with respect to k, l if $\mathcal{C}_{\beta,\varsigma}^{k,l}(\boldsymbol{\mu}) \neq 0$.

As the matrices $M_{\mu,k}$ are upper triangular and nilpotent, we have $\mathcal{C}_\beta^{k,l}(\mathbf{x}) = 0$ for $|\varsigma| \le |\beta| + 1$. We notice that $\mathcal{C}_{\beta,\varsigma}^{k,l} \in \mathbb{K}[\boldsymbol{\mu}']$. We verify from (13) that

$$\mathcal{C}_{\beta,\varsigma}^{k,l}(\boldsymbol{\mu}) = \sum_{\theta \in B : |\beta| < |\theta| < |\varsigma|} \mu_{\varsigma,\theta+\mathbf{e}_k}\mu_{\theta,\beta+\mathbf{e}_l} - \mu_{\varsigma,\theta+\mathbf{e}_l}\mu_{\theta,\beta+\mathbf{e}_k}.$$

LEMMA 3.5. *Assume that B is closed by division. If $|\varsigma| \le |\beta|+1$ or $|\{\mathbf{x}^\beta, \mathbf{x}^{\beta+\mathbf{e}_k}, \mathbf{x}^{\beta+\mathbf{e}_l}, \mathbf{x}^{\beta+\mathbf{e}_k+\mathbf{e}_l}\} \cap B| > 2$, then*

$$\mathcal{C}_{\beta,\varsigma}^{k,l}(\boldsymbol{\mu}) = 0.$$

PROOF. For $\mathbf{x}^\beta \in B$, we have

$$M_{\mu,i}(\mathbf{x}^\beta) = \sum_{i=1}^{\delta} \mu_{\beta_i,\beta+\mathbf{e}_k}\mathbf{x}^{\beta_i}.$$

By the orthogonality relations (7), $M_{\mu,i}(\mathbf{x}^\beta) = \mathbf{x}^{\beta+\mathbf{e}_k}$ if $\mathbf{x}^{\beta+\mathbf{e}_k} \in B$.

If $|\{\mathbf{x}^\beta, \mathbf{x}^{\beta+\mathbf{e}_k}, \mathbf{x}^{\beta+\mathbf{e}_l}, \mathbf{x}^{\beta+\mathbf{e}_k+\mathbf{e}_l}\} \cap \partial B| \ge 2$ then

- either $\mathbf{x}^{\beta+\mathbf{e}_k+\mathbf{e}_l}, \mathbf{x}^{\beta+\mathbf{e}_k}, \mathbf{x}^{\beta+\mathbf{e}_l} \in B$,
- or $\mathbf{x}^{\beta+\mathbf{e}_k+\mathbf{e}_l} \notin B$ and $\mathbf{x}^{\beta+\mathbf{e}_k}, \mathbf{x}^{\beta+\mathbf{e}_l} \in B$,

since B is closed by division. In the first case,

$$\mathcal{C}_\beta^{k,l}(\mathbf{x}) = M_{\mu,k}(\mathbf{x}^{\beta+\mathbf{e}_l}) - M_{\mu,l}(\mathbf{x}^{\beta+\mathbf{e}_k}) = \mathbf{x}^{\beta+\mathbf{e}_l+\mathbf{e}_k} - \mathbf{x}^{\beta+\mathbf{e}_k+\mathbf{e}_l} = 0.$$

In the second case,

$$\mathcal{C}_\beta^{k,l}(\mathbf{x}) = M_{\mu,k}(\mathbf{x}^{\beta+\mathbf{e}_l}) - M_{\mu,l}(\mathbf{x}^{\beta+\mathbf{e}_k}) = \sum_\theta \mu_{\theta,\beta+\mathbf{e}_l+\mathbf{e}_k}\mathbf{x}^\theta - \sum_\theta \mu_{\theta,\beta+\mathbf{e}_k+\mathbf{e}_l}\mathbf{x}^\theta = 0.$$

This proves that $\mathcal{C}_{\beta,\varsigma}^{k,l}(\boldsymbol{\mu}) = 0$ when $|\{\mathbf{x}^\beta, \mathbf{x}^{\beta+\mathbf{e}_k}, \mathbf{x}^{\beta+\mathbf{e}_l}, \mathbf{x}^{\beta+\mathbf{e}_k+\mathbf{e}_l}\} \cap B| > 2$.

When $|\varsigma| \le |\beta| + 1$, $\mathcal{C}_{\beta,\varsigma}^{k,l}(\boldsymbol{\mu}) = 0$ by construction, since the matrices $M_{\mu,k}$ are upper triangular and nilpotent. □

We can now provide a reduced set of equations in $\mathbb{K}[\boldsymbol{\mu}]$ defining Hilb_B.

THEOREM 3.6. *The affine chart* Hilb_B *is defined by the equations:*

$$(i)\quad \mu_{\beta,\alpha} - \sum_{k=1}^{n} \sum_{\theta \in B : |\theta| < |\beta|} \sum_{\gamma \notin B : |\gamma| \le |\theta|} \mu_{\beta,\theta+\mathbf{e}_k}\mu_{\theta,\gamma}\langle \oint_k \mathbf{y}^\gamma \,|\, \mathbf{x}^\alpha\rangle = 0 \qquad \textit{(integration)}$$

for $\beta \in B, |\alpha| \le |\beta|$ with $\alpha \notin B^+$.

$$(c)\quad \sum_{\theta \in B : |\beta| < |\theta| < |\varsigma|} \mu_{\varsigma,\theta+\mathbf{e}_k}\mu_{\theta,\beta+\mathbf{e}_l} - \mu_{\varsigma,\theta+\mathbf{e}_l}\mu_{\theta,\beta+\mathbf{e}_k} = 0 \qquad \textit{(commutation)}$$

for $\beta \in B$ commutator monomial with respect to $k \ne l \in \{1, \ldots, n\}$ and $|\beta| + 2 \le |\varsigma|$.

in the variables $\mu_{\beta_i,\gamma_j} = \frac{\Delta_{B^{i \leftarrow \gamma_j}}}{\Delta_B}$ where $(\Delta_A)_{|A|=\delta, A \subset \mathcal{M}_{\delta-1}}$ are the Plücker coordinates in $\mathscr{G}_\delta(\mathbb{K}[\mathbf{y}]_{\delta-1})$, with $\mu_{\beta_i,\beta_j} = \delta_{i,j}$ for $1 \le i, j \le \delta$ and $\mu_{\beta,\gamma} = 0$ if $|\beta| < |\gamma|$.

PROOF. By Theorem 2.7, $\mathcal{L} \in \mathrm{Hilb}_B$ iff we have, for $1 \le i \le \delta$, the integration relations

$$\lambda_i := \sum_{|\alpha|<\delta} \mu_{\beta_i,\alpha} \mathbf{y}^\alpha = \sum_{|\beta_j|<|\beta_i|} \sum_{k=1}^{n} \mu_{\beta_i,\beta_j+\mathbf{e}_k} \oint_k \lambda_{\beta_j}$$

and the commutation relations (6). By Lemma 3.5, it is enough to consider the relations (c), associated to commutator monomials. We deduce from the integration relations that

$$\mu_{\beta_i,\alpha} = \langle \lambda_{\beta_i} \,|\, \mathbf{x}^\alpha \rangle = \sum_{|\beta_j|<|\beta_i|} \sum_{k=1}^{n} \mu_{\beta_i,\beta_j+\mathbf{e}_k} \langle \oint_k \lambda_{\beta_j} \,|\, \mathbf{x}^\alpha \rangle,$$

which gives the relations (i) by linearity. To conclude the proof, let us show that if $\alpha \in B^+$, then the relation (i) is identically 0. If $\alpha = \mathbf{0}$, then $\mu_{\beta_i,\mathbf{0}} = 0$ and $\langle \oint_k \lambda_{\beta_j} \,|\, 1 \rangle = 0$, which shows that the relation (i) is identically 0.

Let $\mathbf{x}^\alpha = x_m \mathbf{x}^\beta \in B^+$ with $|\alpha| > 0$, $\mathbf{x}^\beta \in B$ and where m is the maximal index of a variable x_m such that $\mathbf{x}^\alpha = x_m \mathbf{x}^\beta$ with $\mathbf{x}^\beta \in B$. Then the integration relation (i) for $\alpha = \beta + \mathbf{e}_m$ writes as

$$\begin{aligned}
&\mu_{\beta_i,\alpha} - \sum_{|\beta_j|<|\beta_i|} \sum_{k=1}^{n} \mu_{\beta_i,\beta_j+\mathbf{e}_k} \langle \oint_k \lambda_{\beta_j} \,|\, \mathbf{x}^\alpha \rangle \\
&= \mu_{\beta_i,\alpha} - \sum_{|\beta_j|<|\beta_i|} \sum_{k=1}^{n} \mu_{\beta_i,\beta_j+\mathbf{e}_k} \langle \partial_m \oint_k \lambda_{\beta_j} \,|\, \mathbf{x}^\beta \rangle \\
&= \mu_{\beta_i,\alpha} - \sum_{|\beta_j|<|\beta_i|} \mu_{\beta_i,\beta_j+\mathbf{e}_m} \langle \lambda_{\beta_j} \,|\, \mathbf{x}^\beta \rangle \\
&= \mu_{\beta_i,\alpha} - \mu_{\beta_i,\beta+\mathbf{e}_m} = 0
\end{aligned}$$

by the orthogonality relations (7) and since $\alpha = \beta + \mathbf{e}_m$. This shows that the relations (i) are not needed if $\alpha \in B^+$, which concludes the proof. □

DEFINITION 3.7. Let $\mathcal{I}_B$ be the ideal of $\mathbb{K}[\boldsymbol{\mu}]$ generated by the equations (i), (c).

By Theorem 3.6, $\mathrm{Hilb}_B = \mathcal{V}(\mathcal{I}_B)$. We can remark that the equations (i) and (c) are filtered by degree. Recall that $B_t = B \cap \mathbb{K}[\mathbf{x}]_t$ is the set of elements in B of degree $\le t$ and let $\delta_t = |B_t|$. Let $\boldsymbol{\mu}_t = \{\mu_{\beta,\alpha} \in \boldsymbol{\mu} \text{ s.t. } |\beta| \le t\}$ and let $\mathcal{I}_{B,t}$ be the ideal of $\mathbb{K}[\boldsymbol{\mu}_t]$ generated by the relations (i) with $|\beta| \le t$ and the relations (c) with $|\varsigma| \le t$. Then by restriction, we deduce from Theorem 3.6 the following:

COROLLARY 3.8. For $t \le \delta$, the ideal $\mathcal{I}_{B,t} \subset \mathbb{K}[\boldsymbol{\mu}_t]$ defines $\mathrm{Hilb}_{B_t} \subset \mathrm{Hilb}_{\delta_t}$.

EXAMPLE 3.9. We consider $B = \{1, x_1, x_1^2\} \subset \mathbb{K}[x_1, x_2]$ with $\delta = 3$. We have $(\partial B)_2 = \{x_2, x_1x_2\}$.

$$\begin{array}{c} \\ \lambda_1 \\ \lambda_2 \\ \lambda_3 \end{array}
\begin{array}{cccccc} 1 & y_1 & y_1^2 & y_2 & y_1y_2 & y_2^2 \\ \end{array}
\begin{bmatrix} 1 & 0 & 0 & 0 & 0 & 0 \\ 0 & 1 & 0 & \mu_{1,0;0,1} & 0 & 0 \\ 0 & 0 & 1 & \mu_{2,0;0,1} & \mu_{2,0;1,1} & \mu_{2,0;0,2} \end{bmatrix}$$

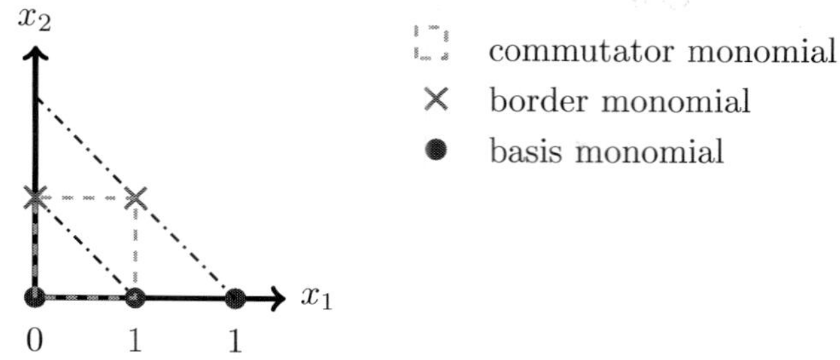

In this figure, the basis monomials in B are represented by the blue disks, the border monomials of degree ≤ 2 are the red crosses, the commutator monomial 1 is represented by the dashed square. The only monomial of degree $\leq 2 = \delta - 1$, which is not in $(\partial B)_2$ is x_2^2 and the corresponding integration relation (i) is

$$\mu_{2,0;0,2} - \mu_{2,0;1,1}\mu_{1,0;0,1} = 0.$$

The commutation relation (c) for the commutator monomial 1 is

$$\mu_{2,0;1,1} - \mu_{1,0;0,1} = 0.$$

These relations define Hilb_B as an irreducible algebraic variety parametrized by $\mu_{1,0;0,1}, \mu_{2,0;0,1}$ of dimension 2.

4. The geometry of Hilb_B

In this section, we discuss some properties of Hilb_B, exploiting the effective description of Section 3.2.

4.1. Transversality of Hilb_B. Having equations defining Hilb_B at hand, a natural question is whether $\mathcal{L} \in \mathrm{Hilb}_B$ is the transversal intersection of Hilb_B with the Grassmannian of δ-linear spaces, which are orthogonal to $Q = \mathcal{L}^\perp$. To analyse this property, we study the ideal defining this intersection in $\mathbb{K}[\boldsymbol{\mu}]$. For that purpose, we define the *parametric normal form*:

$$\begin{aligned} \mathcal{N} : \mathbb{K}[\mathbf{x}, \boldsymbol{\mu}] &\to \mathrm{span}_{\mathbb{K}[\mu]}(B) \\ p(\mathbf{x}, \boldsymbol{\mu}) &\mapsto \sum_{j=1}^{\delta} \langle \lambda_j \,|\, p \rangle \, \mathbf{x}^{\beta_j} \end{aligned} \tag{14}$$

We verify that for $|\alpha| < \delta$,

$$\mathcal{N}(\mathbf{x}^\alpha) = \sum_{j=1}^{\delta} \mu_{\beta_j,\alpha} \, \mathbf{x}^{\beta_j} = \sum_{j=1}^{\delta} \frac{\Delta_{B^{j\leftarrow\alpha}}}{\Delta_B} \mathbf{x}^{\beta_j}.$$

For $\alpha \in \mathbb{N}^n$, let $\mathbf{M}_{\boldsymbol{\mu}}^\alpha = M_{\mu,1}^{\alpha_1} \cdots M_{\mu,n}^{\alpha_n}$ and for $p = \sum_{\alpha\in\mathbb{N}^n} p_\alpha \, \mathbf{x}^\alpha \in \mathbb{K}[\mathbf{x}]$, let

$$p(\mathbf{M}_{\boldsymbol{\mu}}) = \sum_{\alpha\in\mathbb{N}^n} p_\alpha \, \mathbf{M}_{\boldsymbol{\mu}}^\alpha : \mathrm{span}_{\mathbb{K}[\mu]}(B) \to \mathrm{span}_{\mathbb{K}[\mu]}(B)$$

be the operator of multiplication by p. Then as $\beta_1 = \mathbf{0}$, $p(\mathbf{M}_{\boldsymbol{\mu}})(1)$ is a well defined element of $\mathrm{span}_{\mathbb{K}[\mu]}(B)$.

LEMMA 4.1. *For any $p \in \mathbb{K}[\mathbf{x}; \boldsymbol{\mu}]$, we have*

$$\mathcal{N}(p) - p(\mathbf{M}_{\boldsymbol{\mu}})(1) \equiv 0 \mod \mathcal{I}_B \tag{15}$$

where $\mathcal{N}$ is the parametric normal form defined in (14).

PROOF. We prove by induction on $|\alpha|$ that $\mathcal{N}(\mathbf{x}^\alpha) - \mathbf{M}_{\boldsymbol{\mu}}^\alpha(1) \mod \mathcal{I}_B$ for $\alpha \in \mathbb{N}^n$. The property is true for $|\alpha| = 0$ since $\mathcal{N}(1) = 1$. Assume it is true for $|\alpha'| = t-1$ and let $\alpha = \alpha' + \mathbf{e}_j$ for some $1 \leq j \leq n$. Then

$$\begin{aligned}
\mathcal{N}(\mathbf{x}^\alpha) &= \sum_{j=1}^{\delta} \langle \lambda_j \,|\, \mathbf{x}^\alpha \rangle\, \mathbf{x}^{\beta_j} = \sum_{j=1}^{\delta} \langle \partial_j \lambda_j \,|\, \mathbf{x}^{\alpha'} \rangle\, \mathbf{x}^{\beta_j} \\
&= M_{\mu,j} \left(\sum_{j=1}^{\delta} \langle \lambda_j \,|\, \mathbf{x}^{\alpha'} \rangle\, \mathbf{x}^{\beta_j} \right) = M_{\mu,j}\, \mathcal{N}(\mathbf{x}^{\alpha'}) \\
&= M_{\mu,j} M_{\boldsymbol{\mu}}^{\alpha'}(1) \mod \mathcal{I}_B \quad \text{(by induction hypothesis)} \\
&= M_{\boldsymbol{\mu}}^{\alpha}(1) \mod \mathcal{I}_B \quad \text{(by commutation of } M_{\mu,i} \text{ modulo } \mathcal{I}_B)
\end{aligned}$$

This proves by induction that the coefficient $\mu_{\beta_j,\alpha}$ of $\mathbf{x}^{\beta_j}$ is the coefficient $\langle \mathbf{y}^{\beta_j} \,|\, \mathbf{M}_{\boldsymbol{\mu}}^\alpha(1) \rangle$ of $\mathbf{x}^{\beta_j}$ in the the image of 1 by the multiplication operator $\mathbf{M}_{\boldsymbol{\mu}}^\alpha$, modulo $\mathcal{I}_B$. □

We now prove that the equations (i) and (c) combined with the relations (8) define a simple point, which is the inverse system associated to the isolated singular point $\mathbf{0}$. The proof follows the same lines as in [**21**].

For $p \in \mathbb{K}[\mathbf{x}]$, let $p^{(\perp)} = \{\mathcal{L} \in \mathscr{G}_\delta(\mathbb{K}[\mathbf{y}]_{\delta-1}) \mid \forall \lambda \in \mathcal{L}, \langle \lambda \,|\, p \rangle = 0\}$ be the Grasssmannian of δ-linear spaces, which elements are orthogonal to p. For $P \subset \mathbb{K}[\mathbf{x}]$, let $P^{(\perp)} = \cap_{p \in P}\, p^{(\perp)}$.

THEOREM 4.2. *Let $\mathcal{L} \in \mathrm{Hilb}_B$, $Q = \mathcal{L}^\perp$ and $\mathbf{f} = \{f_1, \ldots, f_s\} \subset \mathbb{K}[\mathbf{x}]$ such that Q is an isolated* **m***-primary component of the ideal $I = \langle \mathbf{f} \rangle$. Then, $\mathbf{f}^{(\perp)}$ intersects transversally* Hilb_B *at $\mathcal{L}$.*

PROOF. Let $\mathcal{L} \in \mathrm{Hilb}_B$ with Plücker coordinates $\boldsymbol{\nu} = (\nu_{\beta_i,\gamma}) \in \mathbb{K}^{r_B}$ satisfying equations (i) and (c), which generate the ideal $\mathcal{I}_B(\boldsymbol{\mu}) \subset \mathbb{K}[\boldsymbol{\mu}]$. Let $I_{\mathbf{f}} = \langle \langle \lambda_i \,|\, f_j \rangle \rangle$ be the ideal of $\mathbb{K}[\boldsymbol{\mu}]$ defining $\mathbf{f}^{(\perp)}$. Let $\mathcal{J} = \mathcal{I}_B + I_{\mathbf{f}} \subset \mathbb{K}[\boldsymbol{\mu}]$. By construction, we have

$$\mathcal{L} \in \mathcal{V}(\mathcal{J}) \text{ or } \mathcal{J} \subset \mathbf{m}_{\boldsymbol{\nu}},$$

where $\mathbf{m}_{\boldsymbol{\nu}} := (\mu_{\beta_i,\gamma} - \nu_{\beta_i,\gamma})$ is the maximal ideal defining $\boldsymbol{\nu}$.

Let us prove that $\mathbb{A} = \mathbb{K}[\boldsymbol{\mu}]/\mathcal{J}$ is of dimension 1, which implies that $\mathbf{f}^{(\perp)}$ intersects transversally Hilb_B and that $\mathcal{L}$ is the only intersection point. We denote consider the map

$$\begin{aligned}
\Phi : \mathbb{A}[\mathbf{x}] &\to \mathrm{span}_{\mathbb{A}}(B) \\
p &\mapsto p(M_{\boldsymbol{\mu}})(1)
\end{aligned}$$

where $p(\mathbf{M}_{\boldsymbol{\mu}})$ is the linear operator of $\mathrm{span}_{\mathbb{A}}(B)$ obtained by replacing x_i by $\mathbf{M}_{\boldsymbol{\mu},i}$.

Let K be its kernel. Since the matrices $M_{\mu,i}$ are commuting modulo $\mathcal{J}$, K is an ideal of $\mathbb{A}[\boldsymbol{\mu}]$. By Lemma 4.1, we have for all $p \in \mathbb{K}[\mathbf{x}]$,

$$\Phi(p) \equiv \sum_i \langle \lambda_i \,|\, p \rangle\, \mathbf{x}^{\beta_i} \mod \mathcal{J}. \tag{16}$$

As $i = 1, \ldots, \delta$ and for $i = 1, \ldots, s$, $\langle \lambda_i \,|\, f_j \rangle = 0 \mod \mathcal{J}$, we deduce that $\mathbf{f} \subset K$ and thus $I = \langle \mathbf{f} \rangle \subset K$.

Next we show that $Q \subset K$. Let $q \in Q$. As Q is the $(x_1, \ldots, x_n)$-primary component of I, there exists $p \in \mathbb{K}[\mathbf{x}]$ such that $p(\mathbf{0}) \neq 0$ and $p\,q \in I$. Since the matrices $M_{\mu,i}$ commute modulo $\mathcal{J}$, we have

$$\Phi(p\,q) = p(\mathbf{M}_{\boldsymbol{\mu}})\,\Phi(q) = 0.$$

Since $p(\mathbf{0}) \neq 0$ and $p(\mathbf{M}_{\boldsymbol{\mu}}) = p(\mathbf{0})\,\mathrm{Id} + \mathbf{L}$ with $\mathbf{L}$ lower triangular and nilpotent, $p(\mathbf{M}_{\boldsymbol{\mu}})$ is invertible. We deduce that $\Phi(q) = p(\mathbf{M}_{\boldsymbol{\mu}})^{-1}\Phi(pq) = 0$ and $q \in K$, which implies that $Q \subset K$.

The relation (16) and the orthogonality relations (7) implies that $\Phi(\mathbf{x}^{\beta_k}) = \mathbf{x}^{\beta_k}$ for $k = 1, \ldots, \delta$. This shows that the restriction of Φ on $\mathrm{span}_{\mathbb{A}}(B) \subset \mathbb{A}[\mathbf{x}]$ is the identity.

We can now prove that $\mathbf{m}_{\boldsymbol{\nu}} \subset \mathcal{J}$. Let substitute the parameters $\boldsymbol{\mu}$ by the values $\boldsymbol{\nu} \in \mathbb{K}^{r_B}$ in $M_{\nu,i}$. We get the multiplication by x_i in the basis B of $\mathbb{K}[\mathbf{x}]/Q$. Thus, for any $b \in B$ and $i = 1, \ldots, n$, we have $x_i\,b = M_{\nu,i}(b) + q$ with $q \in Q$. As $Q \subset K$ and $\Phi_{|\,\mathrm{span}_{\mathbb{A}}(B)} = \mathrm{Id}$, we deduce that for $\beta \in B$,

$$\Phi(x_i\,\mathbf{x}^{\beta}) = M_{\mu,i}\Phi(\mathbf{x}^{\beta}) = M_{\mu,i}(\mathbf{x}^{\beta}) = \Phi(M_{\nu,i}(\mathbf{x}^{\beta})) = M_{\nu,i}(\mathbf{x}^{\beta}).$$

This shows that for $\beta \in B$ and $|\alpha| < \delta$, $\mu_{\beta,\alpha} - \nu_{\beta,\alpha} \in \mathcal{J}$ and thus that $\mathbf{m}_{\boldsymbol{\nu}} = \mathcal{J}$, which concludes the proof. □

This property of transversal intersection is the basis for Newton-type methods developed in [**21**, **30**] to numerically compute the inverse system and the associated multiple point.

4.2. Irreducible components of Hilb_B **and commutation rank.** For $t \in \mathbb{N}$ and $\varsigma \in B_{[t]}$, let $\boldsymbol{\mu}_{\varsigma} = \{\mu_{\varsigma,\gamma} \in \boldsymbol{\mu} \text{ s.t. } |\gamma| \in (\partial B)_t\}$. Let $\omega_t = |\boldsymbol{\mu}_{\varsigma}| = |(\partial B)_t|$.

We verify that the commutation relations (c) are linear in the variables $\boldsymbol{\mu}_{\varsigma}$ (and in the variables $\boldsymbol{\mu}_{t-1} := \{\mu_{\beta,\alpha} \in \boldsymbol{\mu} \text{ s.t. } |\beta| \leq t-1, |\alpha| \leq |\beta|, \alpha \notin B\}$).

We exploit this property to describe the tangent space of Hilb_B at regular points. For $\varsigma \in B_{[t]}$, let $\mathcal{C}_{\varsigma} = \{C_{\beta,\varsigma}^{k,l} \neq 0\}$ be the set of non-zero commutator relations for ς. Let $H_{\varsigma} \in \mathbb{K}[\boldsymbol{\mu}_{t-1}]^{|\mathcal{C}_{\varsigma}| \times \omega_t}$ and $\overline{H}_{\varsigma} \in \mathbb{K}[\boldsymbol{\mu}_{t-1}]^{|\mathcal{C}_{\varsigma}|}$ be coefficient matrices of the commutator relations such that

$$H_{\varsigma} \cdot [\boldsymbol{\mu}_{\varsigma}] + \overline{H}_{\varsigma} = [C_{\beta,\varsigma}]_{\beta \in \mathcal{C}_{\varsigma}}$$

For $\mathcal{L} \in \mathrm{Hilb}_B$, the commutator relations are satisfied iff

$$\mathrm{rank}\, H_{\varsigma}(\mathcal{L}_{t-1}) = \mathrm{rank}[H_{\varsigma}(\mathcal{L}_{t-1}), \overline{H}_{\varsigma}(\mathcal{L}_{t-1})].$$

DEFINITION 4.3 (Commutation rank). The commutation rank of $\mathcal{L} \in \mathrm{Hilb}_B$ at $\varsigma \in B_{[t]}$ is $\mathrm{cr}_{\varsigma}(\mathcal{L}_{t-1}) := \mathrm{rank}\, H_{\varsigma}(\mathcal{L}_{t-1}) = \mathrm{rank}[H_{\varsigma}(\mathcal{L}_{t-1}), \overline{H}_{\varsigma}(\mathcal{L}_{t-1})]$.

In the case where the commutation relations have a solution, we have

$$\dim\{\boldsymbol{\mu}_{\varsigma} \in \mathbb{K}^{\omega_t} \text{ s.t. } H_{\varsigma}(\mathcal{L}_{t-1}) \cdot [\boldsymbol{\mu}_{\varsigma}] + \overline{H}_{\varsigma}(\mathcal{L}_{t-1}) = 0\} = \omega_t - \mathrm{cr}_{\varsigma}(\mathcal{L}_{t-1}). \tag{17}$$

For an irreducible component $\mathcal{V}$ of Hilb_B, the commutation rank $\mathrm{cr}_{\varsigma}(\mathcal{L}_{t-1})$ with $\mathcal{L} \in \mathcal{V}$ is generically maximal (by semi-continuity of the rank function). We say that $\mathcal{L} \in \mathcal{V}$ is *regular* if for all $1 \leq t < \deg(B)$ and all $\varsigma \in B_{[t]}$, $\mathrm{cr}_{\varsigma}(\mathcal{L}_{t-1})$ is the generic commutation rank at ς on $\mathcal{V}$.

THEOREM 4.4. *For an irreducible component $\mathcal{V}$ of* Hilb_B *and a regular point $\mathcal{L} \in \mathcal{V}$, we have*

$$\dim \mathcal{V} = \sum_{t \le \deg(B)} \Big(|B_{[t]}|\,|(\partial B)_t| - \sum_{\varsigma \in B_{[t]}} \mathrm{cr}_\varsigma(\mathcal{L}_{t-1}) \Big)$$

PROOF. For $t \in \mathbb{N}$, let π_t be the projection from $\mathbb{K}[\mathbf{y}]_\delta$ to $\mathbb{K}[\mathbf{y}]_t$ obtained by truncation in degree $\le t$. We denote by $\mathcal{V}_t := \pi_t(\mathcal{V})$ the induced projection of $\mathcal{V}$. Then $\mathcal{V}_t$ is irreducible and projects on $\mathcal{V}_{t-1}$. Let $\boldsymbol{\mu}' = \cup_{\varsigma \in B} \boldsymbol{\mu}_\varsigma$ be the set of border variables. By the integration relations (i) we express the coefficients which are not in $\boldsymbol{\mu}'$ as polynomials in the variables $\boldsymbol{\mu}'$. By (17), we deduce that the fiber of π_{t-1} is generically irreducible and of dimension $\sum_{\varsigma \in} \omega_t - \mathrm{cr}_\varsigma(\mathcal{L}_{t-1})$ for a regular point $\mathcal{L} \in \mathcal{V}$. Summing up the dimensions of the fibers for $t = 1, \dots, \deg(B)$, we obtain the dimension of the irreducible component $\mathcal{V}$. □

THEOREM 4.5. *The irreducible components of* Hilb_B *are birational varieties.*

PROOF. Let $\mathcal{V}$ be an irreducible component of Hilb_B. We prove by induction on t, that $\mathcal{V}_t = \pi_t(\mathcal{V})$ is birationaly parametrised. The property is clearly true for $\mathcal{V}_0 = \{1\}$. If $\mathcal{L}_{t-1} \in \mathcal{V}_{t-1}$ is birationaly parametrised then the affine solution set of the system in (17) is also birationaly parametrised. Thus $\mathcal{V}_t$ is birational. This proves by induction that $\mathcal{V}$ is birational. □

4.3. Examples. Let us illustrate the previous results on few simple examples.

EXAMPLE 4.6. We consider first $B = \{1, x_1, x_2, x_1x_2\}$ with $\delta = 4$.

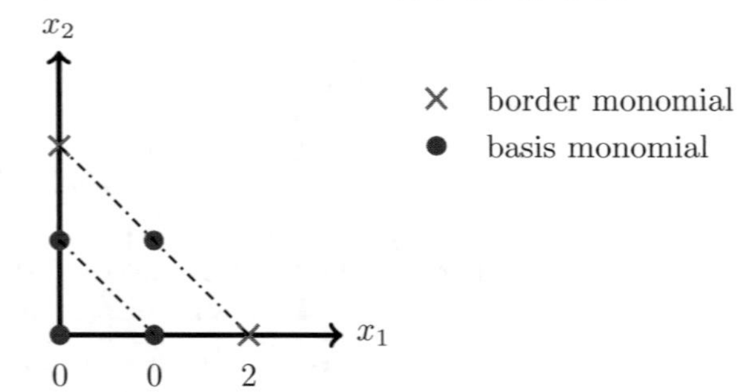

We have $\deg(B) = 2$. There is no monomial of degree ≤ 2, which is not in $(\partial B)_2$. Thus we have no integration relation (i). There is no commutator relation (c). Thus $\mathcal{I}_B = \langle 0 \rangle \subset \mathbb{K}[\mu_{1,1;2,0}, \mu_{1,1;0,2}]$ and Hilb_B is irreducible of dimension 2.

EXAMPLE 4.7. We consider now $B = \{1, x_1, x_2, x_1^2, x_1x_2, x_1^3\}$ with $\delta = 6$.

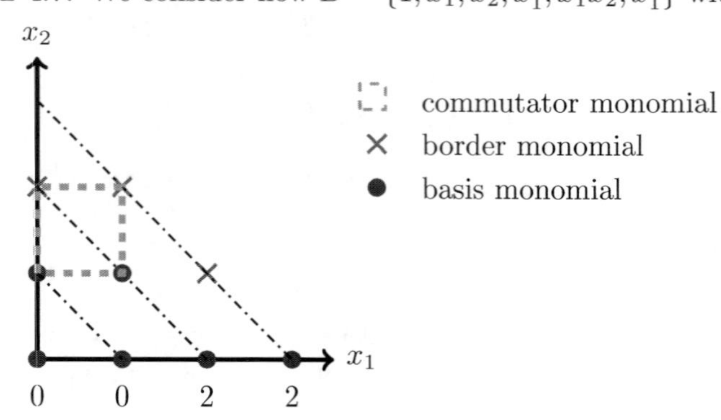

We have $\deg(B) = 3$. There is one integration relation (i) for the monomial x_2^3, which is not in $(\partial B)_3$. It expresses $\mu_{3,0;0,3}$ as a polynomial function of the variables $\boldsymbol{\mu}'$. There is one non-zero commutation relation (c):

$$\mu_{3,0;1,2} - \mu_{3,0;2,1}\mu_{1,1;0,2} - \mu_{2,0;0,2} = 0$$

which is irreducible and expresses $\mu_{3,0;1,2}$ in terms of the other variables of $\boldsymbol{\mu}'$. Thus Hilb_B is irreducible and birational. By Theorem 4.4, we have

$$\dim \mathrm{Hilb}_B = 2 \times 2 + (1 \times 3 - 1) = 4.$$

EXAMPLE 4.8. We take $B = \{1, x_1, x_2, x_3, x_1x_2, x_1x_3, x_2x_3, x_1^3, x_1x_2x_3\}$ with $\delta = 10$.

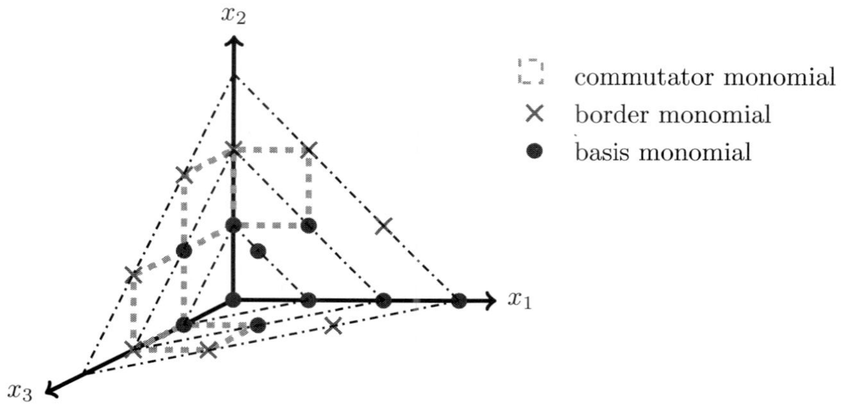

We have $\deg(B) = 3$. There are 2 integration relations (i) for the monomials x_2^3, x_3^2. We have 4 commutation relations (c) corresponding to the commutator monomial x_2 with respect to (x_1, x_2) and (x_2, x_3), x_3 with respect to (x_1, x_3) and (x_2, x_3). An explicit computation shows that they yield polynomial expressions of $\mu_{\varsigma;1,2;0}, \mu_{\varsigma;0,2,1}, \mu_{\varsigma;0,1,2}, \mu_{\varsigma;1,0,2}$ for $\varsigma \in \{(3,0,0), (1,1,1)\} = B_{[3]}$ as polynomial functions of the other variables. Thus Hilb_B is irreducible and birational and the generic commutation rank cr_ς is 4 for $\varsigma \in B_{[3]}$. By Theorem 4.4, we have

$$\dim \mathrm{Hilb}_B = 3 \times 4 + (2 \times 8 - 4 - 4) = 20.$$

References

[1] Mariemi Alonso, Jérome Brachat, and Bernard Mourrain. The Hilbert scheme of points and its link with border basis, 2010. arXiv:0911.3503.

[2] D. Bejleri and D. Stapleton, *The tangent space of the punctual Hilbert scheme*, Michigan Math. J. **66** (2017), no. 3, 595–610, DOI 10.1307/mmj/1501120900. MR3695354

[3] J. Bertin, *The punctual Hilbert scheme: an introduction* (English, with English and French summaries), Geometric methods in representation theory. I, Sémin. Congr., vol. 24, Soc. Math. France, Paris, 2012, pp. 1–102. MR3202701

[4] C. Bertone, F. Cioffi, P. Lella, and M. Roggero, *Upgraded methods for the effective computation of marked schemes on a strongly stable ideal*, J. Symbolic Comput. **50** (2013), 263–290, DOI 10.1016/j.jsc.2012.07.006. MR2996879

[5] C. Bertone, P. Lella, and M. Roggero, *A Borel open cover of the Hilbert scheme*, J. Symbolic Comput. **53** (2013), 119–135, DOI 10.1016/j.jsc.2013.01.001. MR3027986

[6] J. Brachat, P. Lella, B. Mourrain, and M. Roggero, *Extensors and the Hilbert scheme*, Ann. Sc. Norm. Super. Pisa Cl. Sci. (5) **16** (2016), no. 1, 65–96. MR3524665

[7] J. Briançon, Description de *Hilb*^{n}*C*$\{x, y\}$, *Inventiones mathematicae* **41** (1977), 45–90.

[8] J. Briançon and A. Iarrobino, *Dimension of the punctual Hilbert scheme*, J. Algebra **55** (1978), no. 2, 536–544, DOI 10.1016/0021-8693(78)90236-3. MR523473

[9] D. A. Cartwright, D. Erman, M. Velasco, and B. Viray, *Hilbert schemes of 8 points*, Algebra Number Theory **3** (2009), no. 7, 763–795, DOI 10.2140/ant.2009.3.763. MR2579394

[10] F. Cioffi, P. Lella, M. G. Marinari, and M. Roggero, *Segments and Hilbert schemes of points*, Discrete Math. **311** (2011), no. 20, 2238–2252, DOI 10.1016/j.disc.2011.07.011. MR2825669

[11] B. H. Dayton, *Numerical local rings and local solution of nonlinear systems*, SNC'07, ACM, New York, 2007, pp. 79–86. MR2404917

[12] B. H. Dayton and Z. Zeng, *Computing the multiplicity structure in solving polynomial systems*, ISSAC'05, ACM, New York, 2005, pp. 116–123, DOI 10.1145/1073884.1073902. MR2280537

[13] P. Doubilet, G.-C. Rota, and J. Stein, *On the foundations of combinatorial theory. IX. Combinatorial methods in invariant theory*, Studies in Appl. Math. **53** (1974), 185–216, DOI 10.1002/sapm1974533185. MR498650

[14] M. Elkadi and B. Mourrain, *Introduction à la résolution des systèmes polynomiaux* (French), Mathématiques & Applications (Berlin) [Mathematics & Applications], vol. 59, Springer, Berlin, 2007, DOI 10.1007/978-3-540-71647-1. MR2322065

[15] J. Emsalem, *Géométrie des points épais* (French, with English summary), Bull. Soc. Math. France **106** (1978), no. 4, 399–416. MR518046

[16] M. Granger, *Géométrie des schémas de Hilbert ponctuels* (French, with English summary), Mém. Soc. Math. France (N.S.) **8** (1983), 84. MR716783

[17] A. Grothendieck, *Techniques de construction et théorèmes d'existence en géométrie algébrique. IV. Les schémas de Hilbert* (French), Séminaire Bourbaki, Vol. 6, Soc. Math. France, Paris, 1995, pp. Exp. No. 221, 249–276. MR1611822

[18] T. S. Gustavsen, D. Laksov, and R. M. Skjelnes, *An elementary, explicit, proof of the existence of Hilbert schemes of points*, J. Pure Appl. Algebra **210** (2007), no. 3, 705–720, DOI 10.1016/j.jpaa.2006.11.003. MR2324602

[19] M. Haiman, *t, q-Catalan numbers and the Hilbert scheme*, Discrete Math. **193** (1998), no. 1-3, 201–224, DOI 10.1016/S0012-365X(98)00141-1. Selected papers in honor of Adriano Garsia (Taormina, 1994). MR1661369

[20] M. Haiman and B. Sturmfels, *Multigraded Hilbert schemes*, J. Algebraic Geom. **13** (2004), no. 4, 725–769, DOI 10.1090/S1056-3911-04-00373-X. MR2073194

[21] J. D. Hauenstein, B. Mourrain, and A. Szanto, *On deflation and multiplicity structure*, J. Symbolic Comput. **83** (2017), 228–253, DOI 10.1016/j.jsc.2016.11.013. MR3645652

[22] M. E. Huibregtse, *A description of certain affine open subschemes that form an open covering of* $\mathrm{Hilb}^n_{\mathbf{A}^2_k}$, Pacific J. Math. **204** (2002), no. 1, 97–143, DOI 10.2140/pjm.2002.204.97. MR1905194

[23] A. A. Iarrobino, *Punctual Hilbert schemes*, Mem. Amer. Math. Soc. **10** (1977), no. 188, viii+112, DOI 10.1090/memo/0188. MR485867

[24] A. Iarrobino and V. Kanev, *Power sums, Gorenstein algebras, and determinantal loci*, Lecture Notes in Mathematics, vol. 1721, Springer-Verlag, Berlin, 1999. Appendix C by Iarrobino and Steven L. Kleiman, DOI 10.1007/BFb0093426. MR1735271

[25] M. Kreuzer and L. Robbiano, *Deformations of border bases*, Collect. Math. **59** (2008), no. 3, 275–297, DOI 10.1007/BF03191188. MR2452308

[26] M. Kreuzer and L. Robbiano, *The geometry of border bases*, J. Pure Appl. Algebra **215** (2011), no. 8, 2005–2018, DOI 10.1016/j.jpaa.2010.11.011. MR2776440

[27] M. Lederer, *Gröbner strata in the Hilbert scheme of points*, J. Commut. Algebra **3** (2011), no. 3, 349–404, DOI 10.1216/JCA-2011-3-3-349. MR2826478

[28] F. S. Macaulay, *The algebraic theory of modular systems*, Cambridge Mathematical Library, Cambridge University Press, Cambridge, 1994. Revised reprint of the 1916 original; With an introduction by Paul Roberts. MR1281612

[29] A. Mantzaflaris and B. Mourrain, *Deflation and certified isolation of singular zeros of polynomial systems*, ISSAC 2011—Proceedings of the 36th International Symposium on Symbolic and Algebraic Computation, ACM, New York, 2011, pp. 249–256, DOI 10.1145/1993886.1993925. MR2895219

[30] A. Mantzaflaris, B. Mourrain, and A. Szanto, *A certified iterative method for isolated singular roots*, J. Symbolic Comput. **115** (2023), 223–247, DOI 10.1016/j.jsc.2022.08.006. MR4469152

[31] B. Mourrain, *Isolated points, duality and residues*, J. Pure Appl. Algebra **117/118** (1997), 469–493, DOI 10.1016/S0022-4049(97)00023-6. Algorithms for algebra (Eindhoven, 1996). MR1457851

[32] B. Mourrain, *A new criterion for normal form algorithms*, Applied algebra, algebraic algorithms and error-correcting codes (Honolulu, HI, 1999), Lecture Notes in Comput. Sci., vol. 1719, Springer, Berlin, 1999, pp. 430–443, DOI 10.1007/3-540-46796-3_41. MR1846517

[33] B. Mourrain, *Polynomial-exponential decomposition from moments*, Found. Comput. Math. **18** (2018), no. 6, 1435–1492, DOI 10.1007/s10208-017-9372-x. MR3875844

[34] R. Notari and M. L. Spreafico, *A stratification of Hilbert schemes by initial ideals and applications*, Manuscripta Math. **101** (2000), no. 4, 429–448, DOI 10.1007/s002290050225. MR1759253

[35] A. A. Reeves, *The radius of the Hilbert scheme*, J. Algebraic Geom. **4** (1995), no. 4, 639–657. MR1339842

[36] L. Robbiano, *On border basis and Gröbner basis schemes*, Collect. Math. **60** (2009), no. 1, 11–25, DOI 10.1007/BF03191213. MR2490747

Inria at Université Côte d'Azur, 2004 route des Lucioles, BP 93, 06902 Sophia Antipolis, France

Email address: bernard.mourrain@inria.fr

Contemporary Mathematics
Volume **805**, 2024
https://doi.org/10.1090/conm/805/16136

Dg module structures and minimal free resolutions modulo an exact zero divisor

Liana M. Şega and Deepak Sireeshan

Abstract. Let Q be a local ring with maximal ideal $\mathfrak{n}$ and let $f, g \in \mathfrak{n} \smallsetminus \mathfrak{n}^2$ with $fg = 0$. When M is a finite Q-module with $fM = 0$, we show that a minimal free resolution of M over Q has a differential graded module structure over the differential graded algebra $Q\langle y, t \mid \partial(y) = f, \partial(t) = gy\rangle$. When (f, g) is an exact pair of zero divisors, we use this structure to describe a minimal free resolution of M over $Q/(f)$.

Introduction

Differential graded (dg) structures provide an effective and elegant tool in commutative algebra, with particularly strong applications in the study of homological behavior under a change of ring, see Avramov [**1**] for the basics of the theory.

Let $\varphi\colon Q \to R$ be a ring homomorphism and M an R-module. One would like to be able to describe a projective resolution of M over R, given a projective resolution A of R over Q and a projective resolution U of M over Q. Iyengar [**5**] provides such a construction when the resolutions A and U have appropriate differential graded (dg) structures and discusses the existence of such structures. When Q is local (meaning also commutative noetherian) with maximal ideal $\mathfrak{n}$ and φ is surjective, we are particularly interested in minimal free resolutions. In this case, if A is a resolution of R over Q admitting a dg algebra structure and U is a minimal free resolution of M over Q admitting a semi-free dg A-module structure, it is known that $U \otimes_A R$ is a minimal free resolution of M over R, see Theorem 1.2.

When $f \in \mathfrak{n} \smallsetminus \mathfrak{n}^2$ and $R = Q/(f)$, a construction of Shamash can be used to build a semi-free dg module structure of U over the Koszul complex on f, see [**1**, Proposition 2.2.2]. Consequently, this leads to a description of a minimal free resolution of M over R when f is, in addition, a non-zero divisor, cf. [**1**, Theorem 2.2.3]. In this paper, we extend the Shamash construction when f is a zero divisor.

2020 *Mathematics Subject Classification.* Primary 13D02.
Key words and phrases. Free resolution, differential graded module, exact zero divisor.
This work was supported in part by a Simons Foundation grant (#354594, Liana Şega).

THEOREM. *Let* $(Q, \mathfrak{n})$ *be a local ring and let* $f, g \in \mathfrak{n} \setminus \mathfrak{n}^2$ *such that* $fg = 0$. *Set* $R = Q/(f)$ *and* $A = Q\langle y, t \mid \partial(y) = f, \partial(t) = gy\rangle$. *If* M *is a finitely generated* R*-module and* U *is a minimal free resolution of* M *over* Q, *then:*

(a) U *has a structure of semi-free dg module over* A.
(b) *Furthermore, if* $\operatorname{ann}(f) = (g)$ *and* $\operatorname{ann}(g) = (f)$, *then* $U \otimes_A R$ *is a minimal free resolution of* M *over* R.

The dg algebra A in the theorem is obtained from Q by adjoining an exterior algebra variable y in degree 1 and a divided powers variable t in degree 2, see Section 1.2 for details on the notation, which is based on a construction usually referred to as a *Tate construction*. When the hypothesis of part (b) holds, we say that (f, g) is an *exact pair of zero divisors* and that f, respectively g, is an *exact zero divisor*. In this case, A is a minimal free resolution of R over Q. Although not all rings admit exact zero divisors, there exist significant classes of rings that do. See, for example, Henriques and Şega [**4**] for a treatment of a class of rings that are known to admit exact zero divisors, due to a result of Conca, Rossi, and Valla. Such elements can play an important role in understanding the structure and homological properties of artinian algebras.

Let $Q \to R$ be a surjective homomorphism of local rings such that a minimal free resolution A of R over Q admits a dg algebra structure and M is a finite R-module. The question of whether a minimal free resolution U of M over Q admits a dg A-module structure was originally asked by Buchsbaum and Eisenbud [**3**], in the case when A is a Koszul complex. Subsequently, this question was studied in the work of several authors, including Avramov, Kustin, Iyengar, Miller, and Srinivasan, see [**2**], [**5**], [**6**], [**7**], [**8**], [**9**], with both positive and negative answers. In all previously known results in which existence is established, either A is a Koszul complex, or the resolution U has a finite length. Our result provides a first positive answer (to our knowledge) in a case when both A and U are infinite.

The minimal free resolution of M over R constructed in the theorem can be used to recover a known relationship [**4**, Theorem 1.7] between the betti numbers of M over Q and the betti numbers of M over R when (f, g) is an exact pair of zero divisors, see Theorem 2.3. We note, however, that part (b) of the theorem, together with the construction of the dg module structure in the proof of (a) and the more detailed description of the complex $U \otimes_A R$ in Construction 1.1, give a full description of the differential, and not just the betti numbers.

1. The Tate construction

In this section, we describe in detail the construction of the dg algebra A in the main theorem, and we set up notation that will be used in the proof of the theorem.

1.1. Dg algebras and dg modules. Let Q denote a commutative ring. If W is a complex of Q-modules, we denote by $W^{\#}$ the underlying graded Q-module and we use $|w|$ for the degree of a homogeneous element w of W.

A *dg algebra* over Q is a complex $A = (A, \partial)$, concentrated in non-negative homological degrees, where $A^{\#}$ is a Q-algebra that is (graded) commutative, meaning that

$$ab = (-1)^{|a||b|}ba \quad \text{for } a, b \in A \quad \text{and} \quad a^2 = 0 \text{ when } |a| \text{ is odd},$$

and such that the Leibniz formula holds:

$$\partial(ab) = (\partial a)b + (-1)^{|a|}a(\partial b) \quad \text{for } a, b \in A.$$

If A and B are dg Q-algebras, then $A \otimes_Q B$ is also a dg-algebra, with multiplication given by

$$(a \otimes b)(a' \otimes b') = (-1)^{|a'||b|} aa' \otimes bb' \tag{1.1}$$

and the usual differential defined on a tensor product of complexes. We usually identify A and B with their images in $A \otimes_Q B$ and we write ab instead of $a \otimes b$.

A *dg module* U over A is a complex (U, ∂) with $U^{\#}$ an $A^{\#}$-module such that

$$\partial(au) = (\partial a)u + (-1)^{|a|}a(\partial u) \quad \text{for } a \in A \text{ and } u \in U.$$

A bounded below dg A-module U is said to be *semi-free* if $U^{\#}$ is free over $A^{\#}$.

Existence of semi-free dg module structures is particularly useful, as they give an easy construction of free resolutions over R from those over Q. For a more general construction, that does not require the semi-free assumption, see [**5**]. The benefit of the setting in Lemma 1.2 below is that, in the local case, one obtains a minimal resolution when starting with a minimal one.

Construction 1.1. Let $Q \to R$ be a surjective homomorphism of commutative rings. Let A be a dg Q-algebra algebra with $H_0(A) = R$ and let $U = (U, \partial)$ be a dg semi-free A-module. Let $\{e_\lambda\}_{\lambda \in \Lambda}$ denote a basis of $U^{\#}$ over $A^{\#}$. For each integer n, let V_n denote the free Q-summand of U_n with basis $\{e_\lambda \mid \lambda \in \Lambda, |e_\lambda| = n\}$ and let $\pi_{n-1}\colon U_{n-1} \to V_{n-1}$ denote the projection map. Then $U \otimes_A R$ can be identified with the complex $U' = (U', \partial')$, with

$$U'_n = V_n \otimes_Q R \quad \text{and} \quad \partial'_n = \pi_{n-1}\partial_n\big|_{V_n} \otimes_Q 1_R.$$

Indeed, assume $R = Q/I$ for an ideal I, and $Q \to R$ is the natural projection. Let $\varepsilon\colon A \to R$ denote the augmentation map and set $J = \operatorname{Ker}(\varepsilon)$. We have then $R = A/J$, and thus $U \otimes_A R \cong U/JU$. Let $n \geq 0$. The definition of V_n gives $U_n = V_n \oplus W_n$, where $W_n = (A_{\geq 1}U)_n$. We have thus:

$$(U \otimes_A R)_n \cong \left(\frac{U}{JU}\right)_n = \frac{U_n}{IU_n + W_n} \cong \frac{U_n}{W_n} \otimes_Q R \cong V_n \otimes_Q R.$$

The differential of U' is induced by the one on $U \otimes_A R$, which is in turn induced by the differential of U, and this yields directly the claimed expression for ∂'.

The following is a standard argument that appears, for instance, in the proof of [**1**, Theorem 2.2.3].

Lemma 1.2. *Let $Q \to R$ be a surjective homomorphism of commutative rings. Let A be a free resolution of R over Q that has a structure of dg algebra, and let $U = (U, \partial)$ be a free resolution of M over Q, that has a structure of semi-free dg-module over A. Then $U \otimes_A R$ is a free resolution of M over R. In particular, if Q is local and U is minimal, then $U \otimes_A R$ is a minimal free resolution of M over R.*

Proof. Since A is a free resolution of R over Q, the augmentation map ε : $A \to R$ is a quasi-isomorphism (that is, it induces an isomorphism in homology). By [**1**, Proposition 1.3.2], $\varepsilon\colon A \to R$ induces a quasi-isomorphism $U \to U \otimes_A R$. This shows that $U \otimes_A R$ is a free resolution of M over R. The last statement, regarding minimality, follows directly from Construction 1.1. □

1.2. The Tate construction. We describe below a basic procedure of adjoining variables to create new dg algebras, originally used by Cartan and further adapted by Tate. We refer to [**1**] for the general construction, and we adapt it here to the case of interest.

We denote by $Q\langle y\rangle$ the graded free Q-module with basis 1, y, where $|y| = 1$, and define a Q-linear multiplication by setting

$$y^2 = 0.$$

Let $f \in Q$. We write $K = Q\langle y \mid \partial(y) = f\rangle$ for the dg Q-algebra with underlying graded algebra $Q\langle y\rangle$ and differential satisfying

$$\partial(y) = f.$$

This is precisely a Koszul complex on f.

We denote by $Q\langle t\rangle$ the graded free module with basis $\{t^{(i)}\}$, where $|t^{(i)}| = 2i$, and define a Q-linear multiplication by

$$t^{(0)} = 1 \quad \text{and} \quad t^{(i)}t^{(j)} = \frac{(i+j)!}{i!j!}t^{(i+j)}.$$

We set $t = t^{(1)}$; this element is called a *divided powers variable* and the basis elements $\{t^{(i)}\}$ are called a *system of divided powers* of t.

Let $g \in Q$ with $fg = 0$, so that gy is a cycle in K. We write

$$A = Q\langle y, t \mid \partial(y) = f, \partial(t) = gy\rangle$$

for the dg Q-algebra with underlying graded algebra $K \otimes_Q Q\langle t\rangle$, with differential extending the one on K and satisfying

$$\partial(t^{(i)}) = gyt^{(i-1)}.$$

Note that we are using product notation for tensor products of elements, e.g. we are writing $yt^{(n)}$ instead of $y \otimes t^{(n)}$. The multiplication on A comes from the multiplication on K and that on $Q\langle t\rangle$ using formula (1.1). In particular, we have

$$yt^{(n)} = t^{(n)}y \quad \text{for all } n \geq 0.$$

It is known and easy to check that, with these definitions, A is indeed a dg algebra, that is referred to as the *Tate construction* on f, g.

1.3. Notation. We set

$$f_i = \begin{cases} f & \text{if } i \text{ is odd} \\ g & \text{if } i \text{ is even.} \end{cases}$$

For all $i \geq 0$ we set

$$y_{2i} = t^{(i)} \quad \text{and} \quad y_{2i+1} = t^{(i)}y.$$

We set $y_i = 0$ if $i < 0$. With this notation, note that A is a free Q-module with basis $\{y_n\}_{n\geq 0}$ such that $|y_n| = n$ for all n, and

$$\partial_i(y_i) = f_i y_{i-1} \quad \text{for all } i.$$

The multiplicative structure of A is given by the unit y_0 and the relations

$$
\begin{aligned}
y_{2i+1}y_{2j+1} &= 0 \\
y_{2i}y_{2j} &= \binom{i+j}{i} y_{2(i+j)} \\
y_{2i}y_{2j+1} &= \binom{i+j}{i} y_{2(i+j)+1} = y_{2j+1}y_{2i}
\end{aligned} \tag{1.2}
$$

for all $i, j \geq 0$. Using these relations, define $\alpha_{i,j} \in Q$ such that

$$y_i y_j = \alpha_{i,j} y_{i+j} \quad \text{for all } i \geq 0,\ j \geq 0.$$

More precisely, we have $\alpha_{i,j} = 0$ when both i and j are odd, $\alpha_{i,j} = \binom{\frac{1}{2}(i+j)}{\frac{i}{2}}$ if both i, j are even, $\alpha_{i,j} = \binom{\frac{1}{2}(i+j-1)}{\frac{i}{2}}$ if i is even and j is odd, and $\alpha_{i,j} = \binom{\frac{1}{2}(i+j-1)}{\frac{i-1}{2}}$ if i is odd and j is even.

For all $i, j \geq 0$, note that

$$\alpha_{i,j} = \alpha_{j,i}.$$

Further, one can easily verify that

$$\alpha_{i,j}\alpha_{i+j,k} = \alpha_{j,k}\alpha_{i,j+k} \qquad \text{for all } i, j, k \geq 0 \tag{1.3}$$

$$\alpha_{i-1,j} f_i + (-1)^i \alpha_{i,j-1} f_j = \alpha_{i,j} f_{i+j} \qquad \text{for all } i, j \geq 1. \tag{1.4}$$

In fact, (1.3) is equivalent to the associativity of A, namely $(y_i y_j)y_k = y_i(y_j y_k)$, and (1.4) is equivalent to the Leibniz rule for A, namely:

$$\partial(y_i y_j) = \partial(y_i)y_j + (-1)^i y_i \partial(y_j).$$

Finally, observe that the Tate construction A is a periodic complex of rank 1 free Q-modules:

$$\cdots \to Qy_{2n+1} \xrightarrow{f} Qy_{2n} \xrightarrow{g} Qy_{2n-1} \xrightarrow{f} Qy_{2n-2} \to \cdots \to Qy_2 \xrightarrow{g} Qy_1 \xrightarrow{f} Qy_0 \to 0.$$

REMARK 1.3. Let A be as in 1.2, and set $R = Q/(f)$. Then, with the assumptions and notation introduced in Construction 1.1, we have an isomorphism of free Q-modules

$$U_n = V_n \oplus y_1 V_{n-1} \oplus y_2 V_{n-2} \oplus \cdots \oplus y_n V_0 \cong \bigoplus_{i=0}^{n} V_n. \tag{1.5}$$

DEFINITION 1.4 (**Exact zero divisors**). Let $f, g \in Q$ nonzero. We say that (f, g) is an *exact pair of zero divisors* if $\operatorname{ann}(f) = (g)$ and $\operatorname{ann}(g) = (f)$; we also refer to f or g as an *exact zero divisor*. In this case, the Tate construction on f, g is a minimal free resolution of $Q/(f)$ over Q.

2. Proof of the main theorem

Our main theorem is an extension of a construction of Shamash [**10**], along the lines of the proof in [**1**, Proposition 2.2.2, Theorem 2.2.3]. We recall below the statement of the main theorem from the introduction and then proceed to prove it. A corollary about betti numbers is given at the end of the section.

THEOREM 2.1. *Let $(Q, \mathfrak{n})$ be a local ring and let $f, g \in \mathfrak{n} \smallsetminus \mathfrak{n}^2$ such that $fg = 0$. Set $R = Q/(f)$ and $A = Q\langle y, t \mid \partial(y) = f, \partial(t) = gy\rangle$. If M is a finitely generated R-module and U is a minimal free resolution of M over Q, then:*

(a) *U has a structure of semi-free dg module over A.*
(b) *Furthermore, if $\operatorname{ann}(f) = (g)$ and $\operatorname{ann}(g) = (f)$, then $U \otimes_A R$ is a minimal free resolution of M over R.*

PROOF. (a) For all integers $i \geq 0$ and n we will construct homomorphisms

$$\sigma_{i,n} \colon U_n \to U_{i+n}$$

and free Q-modules V_n with $V_n \subseteq U_n$ such that:

(1) $\sigma_{0,n} = \mathrm{id}_{U_n}$ for all n.
(2) For all n and $i \geq 1$ the following condition holds:

$$\mathcal{A}(i,n): \qquad \partial_{i+n}\sigma_{i,n} + (-1)^{i+1}\sigma_{i,n-1}\partial_n = f_i\sigma_{i-1,n}. \tag{2.1}$$

(3) For all n and all i, j with $i \geq 0$ and $j \geq 0$ the following condition holds:

$$\mathcal{B}(i,j,n): \qquad \sigma_{i,n}\sigma_{j,n-j} = \alpha_{i,j}\sigma_{i+j,n-j}.$$

(4) For all n the following condition holds:

$\mathcal{C}(n)$: The map $\Phi_n \colon V_0 \oplus V_1 \oplus \cdots \oplus V_n \to U_{n+1}$ given by

$$\Phi_n(x_0, x_1, \ldots, x_n) = \sum_{i=1}^{n+1} \sigma_{i,n+1-i}(x_{n+1-i})$$

is a split injection.

(5) For all n the following holds:

$$\mathcal{D}(n): \quad U_n = \sum_{k=0}^{n} \sigma_{k,n-k}(U_{n-k}) = \bigoplus_{k=0}^{n} \sigma_{k,n-k}(V_{n-k}),$$

where the direct sum indicates an *internal* direct sum, that is, every element of U_n can be written uniquely as a sum of elements in $\sigma_{k,n-k}(V_{n-k})$.

Once these conditions are established, we define

$$y_i u = \sigma_{i,n}(u) \quad \text{for all } n,\ i \geq 0 \text{ and } u \in U_n$$

and then extend linearly to a multiplication $A \times U \to U$. Conditions (1)-(3) imply that this rule defines a dg A-module structure on U. Item (5) shows that U is a semi-free dg A-module. Indeed, condition $\mathcal{D}(n)$ can be re-written as $U_n = \bigoplus_{k=0}^{n} y_k V_{n-k}$, implying that a basis of $U^{\#}$ over $A^{\#}$ is $\cup_{i \geq 0} B_i$, where B_i denotes a Q-basis of V_i.

Before proceeding with the bulk of the proof, we show:

CLAIM. Let n be an integer. Assume $\mathcal{C}(m)$, $\mathcal{D}(m)$ and $\mathcal{B}(i,j,m)$ hold for all m with $0 \leq m \leq n-1$ and all $i, j \geq 0$. Then

$$\sum_{k=j}^{n} \sigma_{k,n-k}(U_{n-k}) = \bigoplus_{k=j}^{n} \sigma_{k,n-k}(V_{n-k}) \subseteq U_n \quad \text{for all } j \geq 1. \tag{2.2}$$

Proof of Claim. Let $j \geq 1$. We have:

$$\begin{aligned}\sum_{k=j}^{n} \sigma_{k,n-k}(U_{n-k}) &= \sum_{k=j}^{n}\sum_{k'=0}^{n-k} \sigma_{k,n-k}\sigma_{k',n-k-k'}(V_{n-k-k'}) \\ &\subseteq \sum_{k=j}^{n}\sum_{k'=0}^{n-k} \sigma_{k+k',n-(k+k')}(V_{n-(k+k')}) \\ &\subseteq \sum_{l=j}^{n} \sigma_{l,n-l}(V_{n-l}) \\ &\subseteq \sum_{k=j}^{n} \sigma_{k,n-k}(U_{n-k}).\end{aligned}$$

We used $\mathcal{D}(n-k)$ in the first line, $\mathcal{B}(k,k',n-k)$ in the second, and $V_i \subseteq U_i$ in the fourth. (Note that $k \geq j \geq 1$ and thus $n-k \leq n-1$.) It follows that all inclusions above are in fact equalities. Further, the sum $\sum_{k=j}^{n} \sigma_{k,n-k}(V_{n-k})$ is an internal direct sum, as can be seen from $\mathcal{C}(n-1)$. This finishes the proof of the claim.

We now proceed to construct the maps $\sigma_{i,n}$ and the modules V_n. We start by setting $\sigma_{*,n} = 0$ and $V_n = 0$ whenever $n < 0$. With this definition, note that (1)-(5) hold when $n < 0$. We proceed by induction.

Consider the following statement, depending on an integer k.

INDUCTION STATEMENT I. The maps $\sigma_{i,k}\colon U_k \to U_{i+k}$ are defined for all $i \geq 0$ and the free Q-module V_k with $V_k \subseteq U_k$ are defined, and satisfy the properties:

- $\sigma_{0,k} = \mathrm{id}_{U_k}$
- $\mathcal{A}(i,k)$ holds for all $i \geq 1$
- $\mathcal{B}(i,j,k)$ holds for all $i \geq 0$, $j \geq 0$
- $\mathcal{C}(k)$ and $\mathcal{D}(k)$ hold.

As noted above, this statement holds when $k < 0$. Let $n \geq 0$. Assume that Induction statement I holds for all $k \leq n-1$. To complete the induction, we define next $\sigma_{*,n}$ and V_n, and we show that all four items in Induction statement I hold when $k = n$.

We start by defining $\sigma_{0,n}$ to be the identity map on U_n. Then, we define the free module V_n as follows: Condition $\mathcal{C}(n-1)$ gives that $\mathrm{Im}\Phi_{n-1}$ is a direct summand of U_n. We define V_n to be the complementary direct summand, so that

$$U_n = \mathrm{Im}\Phi_{n-1} \oplus V_n = \mathrm{Im}\Phi_{n-1} \oplus \sigma_{0,n}(V_n).$$

Observe that $\mathcal{D}(n)$ must then hold, as it is a direct consequence of this definition and of $\mathcal{C}(n-1)$. We need to define now $\sigma_{i,n}$ for $i \geq 1$. We proceed by induction.

With n fixed as above, consider the following statement, depending on an integer $l \geq 0$:

INDUCTION STATEMENT II. The map $\sigma_{l,n}\colon U_n \to U_{l+n}$ is defined such that

- $\mathcal{A}(l,n)$ holds
- $\mathcal{B}(l,j,n)$ holds for all $j \geq 0$.

Note that these two conditions hold trivially when $l = 0$. Let $i \geq 1$ and assume that Induction statement II holds for all l with $l \leq i-1$. We now define $\sigma_{i,n}$ and we show that the two items in Induction Statement II also hold with $l = i$.

In view of the direct sum decomposition of U_n given by condition $\mathcal{D}(n)$ (which we know holds, see above), in order to define $\sigma_{i,n}$ it suffices to define the restriction functions $\sigma_{i,n}\big|_{\sigma_{k,n-k}(V_{n-k})}$ for all k with $0 \le k \le n$.

Assume first $k > 0$. Note that $\mathcal{C}(n-k)$ implies that the restricted function

$$\sigma_{k,n-k}\big|_{V_{n-k}} : V_{n-k} \to \sigma_{k,n-k}(V_{n-k})$$

is bijective. We define then

$$\sigma_{i,n}\big|_{\sigma_{k,n-k}(V_{n-k})} := \alpha_{i,k}\sigma_{i+k,n-k}\big|_{V_{n-k}}(\sigma_{k,n-k}\big|_{V_{n-k}})^{-1}.$$

In other words, this definition ensures that

$$\sigma_{i,n}\sigma_{k,n-k}\big|_{V_{n-k}} = \alpha_{i,k}\sigma_{i+k,n-k}\big|_{V_{n-k}} \quad \text{when } k > 0. \tag{2.3}$$

We need to define now $\sigma_{i,n}\big|_{V_n}$. To do so, we first claim that

$$\left(f_i\sigma_{i-1,n} - (-1)^{i+1}\sigma_{i,n-1}\partial_n\right)(U_n) \subseteq \partial_{i+n}(U_{i+n}). \tag{2.4}$$

Assuming (2.4) holds and recalling that V_n is a free Q-module, we can define $\sigma_{i,n}\big|_{V_n} : V_n \to U_{i+n}$ so that

$$\partial_{i+n}\sigma_{i,n}\big|_{V_n} = f_i\sigma_{i-1,n} - (-1)^{i+1}\sigma_{i,n-1}\partial_n\big|_{V_n}.$$

Indeed, fix a basis of V_n and define $\sigma_{i,n}\big|_{V_n}(e)$ for a basis element e to be the preimage of $\left(f_i\sigma_{i-1,n} - (-1)^{i+1}\sigma_{i,n-1}\partial_n\right)(e)$ under ∂_{i+n}, and then extend by linearity. Note that this definition depends on the choice of preimage, and thus is not unique.

To prove (2.4), we compute:

$$\begin{aligned}
\partial_{i+n-1}&(f_i\sigma_{i-1,n} - (-1)^{i+1}\sigma_{i,n-1}\partial_n)\\
&= f_i\partial_{i+n-1}\sigma_{i-1,n} + (-1)^i\partial_{i+n-1}\sigma_{i,n-1}\partial_n\\
&= f_i\left(f_{i-1}\sigma_{i-2,n} - (-1)^i\sigma_{i-1,n-1}\partial_n\right) +\\
&\qquad\qquad + (-1)^i\left(f_i\sigma_{i-1,n-1}\partial_n - (-1)^{i+1}\sigma_{i-1,n-2}\partial_{n-1}\partial_n\right)\\
&= 0,
\end{aligned}$$

where for the second equality we used the induction hypothesis that $\mathcal{A}(i-1,n)$ and $\mathcal{A}(i,n-1)$ hold, and for the third equality we used $f_if_{i-1} = fg = 0$ and $\partial^2 = 0$. The computation above shows

$$\left(f_i\sigma_{i-1,n} - (-1)^{i+1}\sigma_{i,n-1}\partial_n\right)(U_n) \subseteq \operatorname{Ker}(\partial_{i+n-1}).$$

Since $i \ge 1$ and $n \ge 0$, we have $i+n-1 \ge 0$. When $i+n-1 > 0$, we use the fact that U is acyclic, and hence $\operatorname{Ker}(\partial_{i+n-1}) = \operatorname{Im}(\partial_{i+n})$ and thus (2.4) holds. When $i = 1$ and $n = 0$ we have

$$f_1\sigma_{0,0} - (-1)^2\sigma_{1,-1}\partial_0 = f_1\mathrm{id}_{U_0}.$$

Since U is a minimal free resolution of M, we have $M = U_0/\partial_1(U_1)$. Since $fM = 0$, we have $f_1U_0 \subseteq \partial_1(U_1)$. This finishes the proof of (2.4) and the definition $\sigma_{i,n}$.

Let $j \ge 0$. We want to prove that $\mathcal{B}(i,j,n)$ holds. Since $\sigma_{0,n}$ is the identity map, $\mathcal{B}(i,0,n)$ holds.

Assume now $j > 0$. In view of the direct sum decomposition provided by $\mathcal{D}(n-j)$, in order to prove $\mathcal{B}(i,j,n)$, it suffices to show

$$\sigma_{i,n}\sigma_{j,n-j}\big|_{\sigma_{k,n-j-k}(V_{n-j-k})} = \alpha_{i,j}\sigma_{i+j,n-j}\big|_{\sigma_{k,n-j-k}(V_{n-j-k})}$$

for all k with $0 \leq k \leq n-j$. We need to check that

$$\sigma_{i,n}\sigma_{j,n-j}\sigma_{k,n-j-k}\big|_{V_{n-j-k}} = \alpha_{i,j}\sigma_{i+j,n-j}\sigma_{k,n-j-k}\big|_{V_{n-j-k}}.$$

Indeed, we have:

$$\begin{aligned}
\sigma_{i,n}\sigma_{j,n-j}\sigma_{k,n-j-k}\big|_{V_{n-j-k}} &= \alpha_{j,k}\sigma_{i,n}\sigma_{j+k,n-(j+k)}\big|_{V_{n-(j+k)}} \\
&= \alpha_{j,k}\alpha_{i,j+k}\sigma_{i+j+k,n-(j+k)}\big|_{V_{n-(j+k)}} \\
&= \alpha_{i,j}\alpha_{i+j,k}\sigma_{i+j+k,n-(j+k)}\big|_{V_{n-(j+k)}} \\
&= \alpha_{i,j}\sigma_{i+j,n-j}\sigma_{k,n-j-k}\big|_{V_{n-j-k}}.
\end{aligned}$$

Here, we used $\mathcal{B}(j,k,n-j)$ in the first equality, (2.3) in the second, (1.3) in the third, and $\mathcal{B}(i+j,k,n-j)$ in the last. This finishes the proof of $\mathcal{B}(i,j,n)$.

Next, we want to prove that $\mathcal{A}(i,n)$ holds. Given the definition of $\sigma_{i,n}\big|_{V_n}$, we know that the relation holds when we restrict the functions to V_n. We need to also check this relation when the functions are restricted to $\sigma_{k,n-k}(V_{n-k})$ with $k > 0$. To this extent, it suffices to show

$$(f_i\sigma_{i-1,n} - (-1)^{i+1}\sigma_{i,n-1}\partial_n)\sigma_{k,n-k} = \partial_{i+n}\sigma_{i,n}\sigma_{k,n-k} \quad \text{for all } k > 0.$$

Let $k > 0$. The computation below achieves the desired conclusion:

$$\begin{aligned}
&(f_i\sigma_{i-1,n} - (-1)^{i+1}\sigma_{i,n-1}\partial_n)\sigma_{k,n-k} \\
&\quad = f_i\sigma_{i-1,n}\sigma_{k,n-k} - (-1)^{i+1}\sigma_{i,n-1}\partial_n\sigma_{k,n-k} \\
&\quad = \alpha_{i-1,k}f_i\sigma_{i-1+k,n-k} - (-1)^{i+1}\sigma_{i,n-1}(f_k\sigma_{k-1,n-k} - (-1)^{k+1}\sigma_{k,n-k-1}\partial_{n-k}) \\
&\quad = \alpha_{i-1,k}f_i\sigma_{i-1+k,n-k} + (-1)^{i}\alpha_{i,k-1}f_k\sigma_{i+k-1,n-k} + (-1)^{i+k}\alpha_{i,k}\sigma_{i+k,n-k-1}\partial_{n-k} \\
&\quad = \alpha_{i,k}f_{i+k}\sigma_{i+k-1,n-k} - \alpha_{i,k}(-1)^{i+k+1}\sigma_{i+k,n-k-1}\partial_{n-k} \\
&\quad = \alpha_{i,k}\partial_{i+n}\sigma_{i+k,n-k} \\
&\quad = \partial_{i+n}\sigma_{i,n}\sigma_{k,n-k}.
\end{aligned}$$

We used $\mathcal{B}(i-1,k,n)$ and $\mathcal{A}(k,n-k)$ for the second equality, $\mathcal{B}(i,k-1,n-1)$ and $\mathcal{B}(i,k,n-1)$ for the third, (1.4) for the fourth, $\mathcal{A}(i+k,n-k)$ for the fifth and $\mathcal{B}(i,k,n)$ (which was proved earlier) for the last equality.

At this point, we verified Induction statement II for $l = i$, hence we have a definition of $\sigma_{i,n}$ for all i, such that $\mathcal{A}(i,n)$ holds and $\mathcal{B}(i,j,n)$ holds for all $i,j \geq 0$. We need to finalize the proof of the Induction statement I for $k = n$. The only remaining item is to show $\mathcal{C}(n)$, namely that the map Φ_n is a split injective. We will show that if $\Phi_n(x) \in \mathfrak{n}U_{n+1}$ for

$$x = (x_0, x_1, \ldots, x_n) \in V_0 \oplus V_1 \oplus \cdots \oplus V_n,$$

then $x_i \in \mathfrak{n}V_i$ for all i.

Assume $\Phi_n(x) \in \mathfrak{n}U_{n+1}$. Then $\partial_{n+1}\Phi_n(x) \in \mathfrak{n}^2U_n$. For the purposes of the next computation, we set $x_{n+1} = 0$. We have then:

$$
\begin{aligned}
\partial_{n+1}\Phi_n(x) &= \sum_{i=1}^{n+1} \partial_{n+1}\sigma_{i,n+1-i}(x_{n+1-i}) \\
&= \sum_{i=1}^{n+1} f_i\sigma_{i-1,n+1-i}(x_{n+1-i}) - (-1)^{i+1}\sigma_{i,n-i}\partial_{n+1-i}(x_{n+1-i}) \\
&= \sum_{j=0}^{n} f_{j+1}\sigma_{j,n-j}(x_{n-j}) + \sum_{j=0}^{n}(-1)^j\sigma_{j,n-j}\partial_{n+1-j}(x_{n+1-j}) \\
&= \sum_{j=0}^{n} \sigma_{j,n-j}\left(f_{j+1}x_{n-j} - \partial_{n-j+1}(x_{n-j+1})\right).
\end{aligned} \tag{2.5}
$$

We used $\mathcal{A}(i, n+1-i)$ for the second equality. Next, we set

$$
w_j = \sigma_{j,n-j}\left(f_{j+1}x_{n-j} - \partial_{n-j+1}(x_{n-j+1})\right),
$$

so that (2.5) becomes

$$
\partial_{n+1}\Phi_n(x) = \sum_{j=0}^{n} w_j.
$$

We prove by induction on i with $0 \le i \le n+1$ that $x_{n+1-i} \in \mathfrak{n}V_{n+1-i}$. This is true when $i = 0$, as we defined $x_{n+1} = 0$. Let k be so that $0 \le k \le n$ and assume that $x_{n+1-i} \in \mathfrak{n}V_{n+1-i}$ for all i with $0 \le i \le k$. The fact that that $f_i \in \mathfrak{n}$ for all i and the minimality of U imply

$$
w_j \in \mathfrak{n}^2V_{n-j} \quad \text{for all} \quad j \le k-1.
$$

To complete the induction argument, we show that $x_{n-k} \in \mathfrak{n}V_{n-k}$. We have

$$
\begin{aligned}
f_{k+1}\sigma_{k,n-k}(x_{n-k}) + \sum_{j=k+1}^{n} w_j &= w_k + \sigma_{k,n-k}\partial_{n-k+1}(x_{n-k+1}) + \sum_{j=k+1}^{n} w_j \\
&= \sigma_{k,n-k}\partial_{n-k+1}(x_{n-k+1}) + \partial_{n+1}\Phi_n(x) - \sum_{j=0}^{k-1} w_j \\
&\in \mathfrak{n}^2U_n.
\end{aligned}
$$

Recall the direct sum decomposition of U_n given by property $\mathcal{D}(n)$, and notice that $f_{k+1}\sigma_{k,n-k}(x_{n-k})$ and $\sum_{j=k+1}^{n} w_j$ belong to distinct summands in this decomposition. Indeed, we have $f_{k+1}\sigma_{k,n-k}(x_{n-k}) \in \sigma_{k,n-k}(V_{n-k})$, while (2.2) implies

$$
\sum_{j=k+1}^{n} w_j \in \sum_{j=k+1}^{n} \sigma_{j,n-j}(U_{n-j}) = \bigoplus_{j=k+1}^{n} \sigma_{j,n-j}(V_{n-j}).
$$

We conclude $f_{k+1}\sigma_{k,n-k}(x_{n-k}) \in \mathfrak{n}^2U_n$, and hence $\sigma_{k,n-k}(x_{n-k}) \in \mathfrak{n}U_n$, since $f_{k+1} \notin \mathfrak{n}^2$. Since Φ_{n-1} is a split injection by $\mathcal{C}(n-1)$, it follows $x_{n-k} \in \mathfrak{n}V_{n-k}$.

We showed thus that $\mathcal{C}(n)$ holds. This finishes the proof of Induction Statement I and concludes the proof of (a).

(b) The fact that (f, g) is an exact pair of zero divisors implies that A is a minimal free resolution of R over Q, as noted in Theorem 1.4. The conclusion follows then from Theorem 1.2, making use of (a). □

REMARK 2.2. The proof of the Main Theorem is inspired by the construction of Shamash described in [**1**, Proposition 2.2]. It is proved there that if $f \in \mathfrak{n} \smallsetminus \mathfrak{n}^2$, then, with notation as in our theorem, U has a structure of semi-free dg module over the Koszul complex $K = Q\langle y \mid \partial(y) = f\rangle$. As noted in [**1**, Remark 2.2.1], since $f\mathrm{id}^U$ and 0^U both induce the zero map on M, they are homotopic, say $f\mathrm{id}^U = \partial\sigma + \sigma\partial$, and the existence of the desired dg module structure is equivalent with the existence of a homotopy σ with $\sigma^2 = 0$. The map $\sigma_{1,*}$ constructed in our proof above is exactly the homotopy σ constructed in the proof of [**1**, Proposition 2.2]. In fact, all maps $\sigma_{i,*}$ with $i \geq 1$ are homotopies. Indeed, since $f_{i-1}f_i = 0$, we see from condition $\mathcal{A}(i,n)$ in (2.1) that $f_{i-1}\sigma_{i,*}$ is a chain map of degree i on U, for all $i \geq 1$. Consequently, $\mathcal{A}(i,n)$ shows that $\sigma_{i,*}$ is a homotopy between $f_i\sigma_{i-1,*}$ and the zero map on U.

REMARK 2.3. Assume that the hypothesis of the Main Theorem holds. In the proof of the theorem, we saw that a basis for $U^\#$ over $A^\#$ can be taken to consist of the union of the bases of the free Q-modules $V_0, V_1, V_2, \ldots$. With this choice of basis, the modules V_n defined in the proof of the theorem coincide with the modules V_n introduced in Construction 1.1, and thus the differential of the complex $U \otimes_A R$ can be understood as described there. Further, Construction 1.1 and (1.5) give

$$U_n \cong \bigoplus_{i=0}^{n} V_i \qquad \text{and} \qquad (U \otimes_A R)_n \cong V_n \otimes_Q R. \tag{2.6}$$

When R is a local ring and M is a finitely generated R-module, we let $\beta_n^R(M)$ denote the nth *Betti number* of M over R, that is, the rank of the nth free R-module in a minimal free resolution of M, and we let

$$P_M^R(t) = \sum_{i=0}^{\infty} \beta_i^R(M)t^i$$

denote the *Poincaré series* of M over R.

Corollary 2.4 below recovers the Poincaré series formula in [**4**, Theorem 1.7], and shows that the hypothesis on f, g in part (b) of Theorem 2.1 is necessary.

COROLLARY 2.4. *Let $(Q, \mathfrak{n})$ be a local ring and $f, g \in \mathfrak{n} \smallsetminus \mathfrak{n}^2$ with $fg = 0$. Set $R = Q/(f)$ and $A = Q\langle y, t \mid \partial(y) = f, \partial(t) = gy\rangle$. The following are equivalent:*

(1) *(f, g) is an exact pair of zero divisors;*
(2) *For all finitely generated R-modules M, if U is a minimal free resolution of M over Q, then $U \otimes_A R$ is a minimal free resolution of M over R;*
(3) *For all finitely generated R-modules M, $P_M^R(t) = (1-t)P_M^Q(t)$.*

PROOF. The implication (1) $\Longrightarrow$ (2) is established in Theorem 2.1(b).

(2) $\Longrightarrow$ (3): Let $n \geq 0$. Under the hypothesis of (2), we have $\beta_n^R(M) = \operatorname{rank}_R(U \otimes_A R)_n$ and $\beta_n^Q(M) = \operatorname{rank}_Q(U_n)$. Then (2.6) gives

$$\beta_n^Q(M) = \sum_{i=0}^{n} \beta_i^R(M)$$

and the Poincaré series formula follows from here.

(3) $\Longrightarrow$ (1): If (3) holds, take $M = R$ in the Poincaré series formula to conclude $P_R^Q(t) = (1-t)^{-1}$, and hence $\beta_n^Q(R) = 1$ for all $n \geq 0$. Since $\operatorname{ann}(f)$ is a second syzygy in a minimal free resolution of $R = Q/(f)$ over Q, we conclude that $\operatorname{ann}(f)$

is principal. If $\operatorname{ann}(f) = (h)$ for some $h \in \mathfrak{n}$, then, since $g \in (h)$ and $g \notin \mathfrak{n}^2$, we must have $g = uh$ for some unit u, and hence $\operatorname{ann}(f) = (h) = (g)$. Then $\operatorname{ann}(g)$ is a third syzygy in a minimal free resolution of R over Q, and we conclude it must be principal as well. Similarly, we obtain $\operatorname{ann}(g) = (f)$. □

Acknowledgments

We are thankful to the referee for a careful reading, and useful suggestions.

References

[1] L. L. Avramov, *Infinite free resolutions*, Six lectures on commutative algebra (Bellaterra, 1996), Progr. Math., vol. 166, Birkhäuser, Basel, 1998, pp. 1–118. MR1648664

[2] L. L. Avramov, *Obstructions to the existence of multiplicative structures on minimal free resolutions*, Amer. J. Math. **103** (1981), no. 1, 1–31, DOI 10.2307/2374187. MR601460

[3] D. A. Buchsbaum and D. Eisenbud, *Algebra structures for finite free resolutions, and some structure theorems for ideals of codimension* 3, Amer. J. Math. **99** (1977), no. 3, 447–485, DOI 10.2307/2373926. MR453723

[4] I. Bonacho Dos Anjos Henriques and L. M. Şega, *Free resolutions over short Gorenstein local rings*, Math. Z. **267** (2011), no. 3-4, 645–663, DOI 10.1007/s00209-009-0639-z. MR2776052

[5] S. Iyengar, *Free resolutions and change of rings*, J. Algebra **190** (1997), no. 1, 195–213, DOI 10.1006/jabr.1996.6901. MR1442152

[6] A. R. Kustin, *Gorenstein algebras of codimension four and characteristic two*, Comm. Algebra **15** (1987), no. 11, 2417–2429, DOI 10.1080/00927878708823544. MR912779

[7] A. R. Kustin and M. Miller, *Algebra structures on minimal resolutions of Gorenstein rings of embedding codimension four*, Math. Z. **173** (1980), no. 2, 171–184, DOI 10.1007/BF01159957. MR583384

[8] H. Srinivasan, *The nonexistence of a minimal algebra resolution despite the vanishing of Avramov obstructions*, J. Algebra **146** (1992), no. 2, 251–266, DOI 10.1016/0021-8693(92)90067-V. MR1152904

[9] H. Srinivasan, *A grade five Gorenstein algebra with no minimal algebra resolutions*, J. Algebra **179** (1996), no. 2, 362–379, DOI 10.1006/jabr.1996.0016. MR1367854

[10] J. Shamash, *The Poincaré series of a local ring*, J. Algebra **12** (1969), 453–470, DOI 10.1016/0021-8693(69)90023-4. MR241411

Liana M. Şega, School of Science and Engineering, University of Missouri - Kansas City, Kansas City, Missouri 64110
Email address: segal@umkc.edu

Deepak Sireeshan, School of Science and Engineering, University of Missouri - Kansas City, Kansas City, Missouri 64110
Email address: dsbx7@mail.umkc.edu

Selected Published Titles in This Series

805 **Anthony Iarrobino, Pedro Macias Marques, Maria Evelina Rossi, and Jean Vallès, Editors,** Deformation of Artinian Algebras and Jordan Type, 2024

798 **Marat V. Markin, Igor V. Nikolaev, and Carsten Trunk, Editors,** Advances in Functional Analysis and Operator Theory, 2024

796 **John Cremona, John Jones, Jennifer Paulhus, Andrew V. Sutherland, and John Voight, Editors,** LuCaNT: LMFDB, Computation, and Number Theory, 2024

795 **Michael A. Jones, David McCune, and Jennifer M. Wilson, Editors,** Mathematical Analyses of Decisions, Voting and Games, 2024

794 **Jean-Pierre Magnot, Editor,** Recent Advances in Diffeologies and Their Applications, 2024

793 **Abba Gumel, Editor,** Mathematical and Computational Modeling of Phenomena Arising in Population Biology and Nonlinear Oscillations, 2024

792 **Shaoming Guo, Zane Kun Li, and Brian Street, Editors,** Recent Developments in Harmonic Analysis and its Applications, 2024

791 **Mee Seong Im, Bach Nguyen, and Arik Wilbert, Editors,** Algebraic and Topological Aspects of Representation Theory, 2024

790 **Peter Crooks and Alexandru I. Suciu, Editors,** Compactifications, Configurations, and Cohomology, 2023

789 **I. S. Krasil′shchik, A. B. Sossinsky, and A. M. Verbovetsky, Editors,** The Diverse World of PDEs, 2023

788 **I. S. Krasil′shchik, A. B. Sossinsky, and A. M. Verbovetsky, Editors,** The Diverse World of PDEs, 2023

787 **Hemen Dutta, Editor,** Mathematical Modelling, 2023

786 **Hemen Dutta, Editor,** Mathematical Modelling, 2023

785 **A. Leroy and S. K. Jain, Editors,** Algebra and Coding Theory, 2023

784 **Dinh-Liem Nguyen, Loc Hoang Nguyen, and Thi-Phong Nguyen, Editors,** Advances in Inverse Problems for Partial Differential Equations, 2023

783 **Alla Detinko, Michael Kapovich, Alex Kontorovich, Peter Sarnak, and Richard Schwartz, Editors,** Computational Aspects of Discrete Subgroups of Lie Groups, 2023

782 **Galina Filipuk, Alberto Lastra, and Sławomir Michalik, Editors,** Recent Trends in Formal and Analytic Solutions of Diff. Equations, 2023

781 **Emanuel Indrei, Diego Marcon, and Levon Nurbekyan, Editors,** Geometric and Functional Inequalities and Recent Topics in Nonlinear PDEs, 2023

780 **Erik Koelink, Stefan Kolb, Nicolai Reshetikhin, and Bart Vlaar, Editors,** Hypergeometry, Integrability and Lie Theory, 2022

779 **Samuele Anni, Valentijn Karemaker, and Elisa Lorenzo García, Editors,** Arithmetic, Geometry, Cryptography, and Coding Theory 2021, 2022

778 **Carlos Galindo, Alejandro Melle Hernández, Julio José Moyano-Fernández, and Wilson A. Zúñiga-Galindo, Editors,** p-Adic Analysis, Arithmetic and Singularities, 2022

777 **Bang-Yen Chen, Nicholas D. Brubaker, Takashi Sakai, Bogdan D. Suceavă, Makiko Sumi Tanaka, Hiroshi Tamaru, and Mihaela B. Vajiac, Editors,** Differential Geometry and Global Analysis, 2022

776 **Aaron Wootton, S. Allen Broughton, and Jennifer Paulhus, Editors,** Automorphisms of Riemann Surfaces, Subgroups of Mapping Class Groups and Related Topics, 2022

775 **Fernando Galaz-García, Cecilia González-Tokman, and Juan Carlos Pardo Millán, Editors,** Mexican Mathematicians in the World, 2021

For a complete list of titles in this series, visit the AMS Bookstore at **www.ams.org/bookstore/conmseries/**.